Access Your Online Resources

Don't miss out on the Online Resources included with your purchase!

Your purchase of this product unlocks access to our Online Resources page. Elevate your study experience with our **interactive practice test interface**, along with all of the additional resources that we couldn't include in this book.

Flip to the Online Resources section at the end of this book to find the link and a QR code to get started!

Mometrix
TEST PREPARATION

CHST®

Exam Secrets

Study Guide

Your Key to Exam Success

Mometrix
TEST PREPARATION

Written and edited by the Mometrix Safety Certification Test Team

Mometrix offers volume discount pricing to institutions. For more information or a price quote, please contact our sales department at sales@mometrix.com or 888-248-1219.

Paperback
ISBN 13: 978-1-60971-350-8
ISBN 10: 1-60971-350-8

Ebook
ISBN 13: 978-1-62120-420-6
ISBN 10: 1-62120-420-0

Hardback
ISBN 13: 978-1-5167-1356-1
ISBN 10: 1-5167-1356-7

DEAR FUTURE EXAM SUCCESS STORY

First of all, **THANK YOU** for purchasing Mometrix study materials!

Second, congratulations! You are one of the few determined test-takers who are committed to doing whatever it takes to excel on your exam. **You have come to the right place.** We developed these study materials with one goal in mind: to deliver you the information you need in a format that's concise and easy to use.

In addition to optimizing your guide for the content of the test, we've outlined our recommended steps for breaking down the preparation process into small, attainable goals so you can make sure you stay on track.

We've also analyzed the entire test-taking process, identifying the most common pitfalls and showing how you can overcome them and be ready for any curveball the test throws you.

Standardized testing is one of the biggest obstacles on your road to success, which only increases the importance of doing well in the high-pressure, high-stakes environment of test day. Your results on this test could have a significant impact on your future, and this guide provides the information and practical advice to help you achieve your full potential on test day.

Your success is our success

We would love to hear from you! If you would like to share the story of your exam success or if you have any questions or comments in regard to our products, please contact us at **800-673-8175** or **support@mometrix.com**.

Thanks again for your business and we wish you continued success!

Sincerely,
The Mometrix Test Preparation Team

Need more help? Check out our flashcards at:
http://MometrixFlashcards.com/CHST

TABLE OF CONTENTS

Introduction

Thank you for purchasing this resource! You have made the choice to prepare yourself for a test that could have a huge impact on your future, and this guide is designed to help you be fully ready for test day. Obviously, it's important to have a solid understanding of the test material, but you also need to be prepared for the unique environment and stressors of the test, so that you can perform to the best of your abilities.

For this purpose, the first section that appears in this guide is the **Secret Keys**. We've devoted countless hours to meticulously researching what works and what doesn't, and we've boiled down our findings to the five most impactful steps you can take to improve your performance on the test. We start at the beginning with study planning and move through the preparation process, all the way to the testing strategies that will help you get the most out of what you know when you're finally sitting in front of the test.

We recommend that you start preparing for your test as far in advance as possible. However, if you've bought this guide as a last-minute study resource and only have a few days before your test, we recommend that you skip over the first two Secret Keys since they address a long-term study plan.

If you struggle with **test anxiety**, we strongly encourage you to check out our recommendations for how you can overcome it. Test anxiety is a formidable foe, but it can be beaten, and we want to make sure you have the tools you need to defeat it.

Secret Key #1 – Plan Big, Study Small

There's a lot riding on your performance. If you want to ace this test, you're going to need to keep your skills sharp and the material fresh in your mind. You need a plan that lets you review everything you need to know while still fitting in your schedule. We'll break this strategy down into three categories.

Information Organization

Start with the information you already have: the official test outline. From this, you can make a complete list of all the concepts you need to cover before the test. Organize these concepts into groups that can be studied together, and create a list of any related vocabulary you need to learn so you can brush up on any difficult terms. You'll want to keep this vocabulary list handy once you actually start studying since you may need to add to it along the way.

Time Management

Once you have your set of study concepts, decide how to spread them out over the time you have left before the test. Break your study plan into small, clear goals so you have a manageable task for each day and know exactly what you're doing. Then just focus on one small step at a time. When you manage your time this way, you don't need to spend hours at a time studying. Studying a small block of content for a short period each day helps you retain information better and avoid stressing over how much you have left to do. You can relax knowing that you have a plan to cover everything in time. In order for this strategy to be effective though, you have to start studying early and stick to your schedule. Avoid the exhaustion and futility that comes from last-minute cramming!

Study Environment

The environment you study in has a big impact on your learning. Studying in a coffee shop, while probably more enjoyable, is not likely to be as fruitful as studying in a quiet room. It's important to keep distractions to a minimum. You're only planning to study for a short block of time, so make the most of it. Don't pause to check your phone or get up to find a snack. It's also important to **avoid multitasking**. Research has consistently shown that multitasking will make your studying dramatically less effective. Your study area should also be comfortable and well-lit so you don't have the distraction of straining your eyes or sitting on an uncomfortable chair.

The time of day you study is also important. You want to be rested and alert. Don't wait until just before bedtime. Study when you'll be most likely to comprehend and remember. Even better, if you know what time of day your test will be, set that time aside for study. That way your brain will be used to working on that subject at that specific time and you'll have a better chance of recalling information.

Finally, it can be helpful to team up with others who are studying for the same test. Your actual studying should be done in as isolated an environment as possible, but the work of organizing the information and setting up the study plan can be divided up. In between study sessions, you can discuss with your teammates the concepts that you're all studying and quiz each other on the details. Just be sure that your teammates are as serious about the test as you are. If you find that your study time is being replaced with social time, you might need to find a new team.

Secret Key #2 – Make Your Studying Count

You're devoting a lot of time and effort to preparing for this test, so you want to be absolutely certain it will pay off. This means doing more than just reading the content and hoping you can remember it on test day. It's important to make every minute of study count. There are two main areas you can focus on to make your studying count.

Retention

It doesn't matter how much time you study if you can't remember the material. You need to make sure you are retaining the concepts. To check your retention of the information you're learning, try recalling it at later times with minimal prompting. Try carrying around flashcards and glance at one or two from time to time or ask a friend who's also studying for the test to quiz you.

To enhance your retention, look for ways to put the information into practice so that you can apply it rather than simply recalling it. If you're using the information in practical ways, it will be much easier to remember. Similarly, it helps to solidify a concept in your mind if you're not only reading it to yourself but also explaining it to someone else. Ask a friend to let you teach them about a concept you're a little shaky on (or speak aloud to an imaginary audience if necessary). As you try to summarize, define, give examples, and answer your friend's questions, you'll understand the concepts better and they will stay with you longer. Finally, step back for a big picture view and ask yourself how each piece of information fits with the whole subject. When you link the different concepts together and see them working together as a whole, it's easier to remember the individual components.

Finally, practice showing your work on any multi-step problems, even if you're just studying. Writing out each step you take to solve a problem will help solidify the process in your mind, and you'll be more likely to remember it during the test.

Modality

Modality simply refers to the means or method by which you study. Choosing a study modality that fits your own individual learning style is crucial. No two people learn best in exactly the same way, so it's important to know your strengths and use them to your advantage.

For example, if you learn best by visualization, focus on visualizing a concept in your mind and draw an image or a diagram. Try color-coding your notes, illustrating them, or creating symbols that will trigger your mind to recall a learned concept. If you learn best by hearing or discussing information, find a study partner who learns the same way or read aloud to yourself. Think about how to put the information in your own words. Imagine that you are giving a lecture on the topic and record yourself so you can listen to it later.

For any learning style, flashcards can be helpful. Organize the information so you can take advantage of spare moments to review. Underline key words or phrases. Use different colors for different categories. Mnemonic devices (such as creating a short list in which every item starts with the same letter) can also help with retention. Find what works best for you and use it to store the information in your mind most effectively and easily.

Secret Key #3 – Practice the Right Way

Your success on test day depends not only on how many hours you put into preparing, but also on whether you prepared the right way. It's good to check along the way to see if your studying is paying off. One of the most effective ways to do this is by taking practice tests to evaluate your progress. Practice tests are useful because they show exactly where you need to improve. Every time you take a practice test, pay special attention to these three groups of questions:

- The questions you got wrong
- The questions you had to guess on, even if you guessed right
- The questions you found difficult or slow to work through

This will show you exactly what your weak areas are, and where you need to devote more study time. Ask yourself why each of these questions gave you trouble. Was it because you didn't understand the material? Was it because you didn't remember the vocabulary? Do you need more repetitions on this type of question to build speed and confidence? Dig into those questions and figure out how you can strengthen your weak areas as you go back to review the material.

Additionally, many practice tests have a section explaining the answer choices. It can be tempting to read the explanation and think that you now have a good understanding of the concept. However, an explanation likely only covers part of the question's broader context. Even if the explanation makes perfect sense, **go back and investigate** every concept related to the question until you're positive you have a thorough understanding.

As you go along, keep in mind that the practice test is just that: practice. Memorizing these questions and answers will not be very helpful on the actual test because it is unlikely to have any of the same exact questions. If you only know the right answers to the sample questions, you won't be prepared for the real thing. **Study the concepts** until you understand them fully, and then you'll be able to answer any question that shows up on the test.

It's important to wait on the practice tests until you're ready. If you take a test on your first day of study, you may be overwhelmed by the amount of material covered and how much you need to learn. Work up to it gradually.

On test day, you'll need to be prepared for answering questions, managing your time, and using the test-taking strategies you've learned. It's a lot to balance, like a mental marathon that will have a big impact on your future. Like training for a marathon, you'll need to start slowly and work your way up. When test day arrives, you'll be ready.

Start with the strategies you've read in the first two Secret Keys—plan your course and study in the way that works best for you. If you have time, consider using multiple study resources to get different approaches to the same concepts. It can be helpful to see difficult concepts from more than one angle. Then find a good source for practice tests. Many times, the test website will suggest potential study resources or provide sample tests.

Practice Test Strategy

If you're able to find at least three practice tests, we recommend this strategy:

Untimed and Open-Book Practice

Take the first test with no time constraints and with your notes and study guide handy. Take your time and focus on applying the strategies you've learned.

Timed and Open-Book Practice

Take the second practice test open-book as well, but set a timer and practice pacing yourself to finish in time.

Timed and Closed-Book Practice

Take any other practice tests as if it were test day. Set a timer and put away your study materials. Sit at a table or desk in a quiet room, imagine yourself at the testing center, and answer questions as quickly and accurately as possible.

Keep repeating timed and closed-book tests on a regular basis until you run out of practice tests or it's time for the actual test. Your mind will be ready for the schedule and stress of test day, and you'll be able to focus on recalling the material you've learned.

Secret Key #4 – Pace Yourself

Once you're fully prepared for the material on the test, your biggest challenge on test day will be managing your time. Just knowing that the clock is ticking can make you panic even if you have plenty of time left. Work on pacing yourself so you can build confidence against the time constraints of the exam. Pacing is a difficult skill to master, especially in a high-pressure environment, so **practice is vital**.

Set time expectations for your pace based on how much time is available. For example, if a section has 60 questions and the time limit is 30 minutes, you know you have to average 30 seconds or less per question in order to answer them all. Although 30 seconds is the hard limit, set 25 seconds per question as your goal, so you reserve extra time to spend on harder questions. When you budget extra time for the harder questions, you no longer have any reason to stress when those questions take longer to answer.

Don't let this time expectation distract you from working through the test at a calm, steady pace, but keep it in mind so you don't spend too much time on any one question. Recognize that taking extra time on one question you don't understand may keep you from answering two that you do understand later in the test. If your time limit for a question is up and you're still not sure of the answer, mark it and move on, and come back to it later if the time and the test format allow. If the testing format doesn't allow you to return to earlier questions, just make an educated guess; then put it out of your mind and move on.

On the easier questions, be careful not to rush. It may seem wise to hurry through them so you have more time for the challenging ones, but it's not worth missing one if you know the concept and just didn't take the time to read the question fully. Work efficiently but make sure you understand the question and have looked at all of the answer choices, since more than one may seem right at first.

Even if you're paying attention to the time, you may find yourself a little behind at some point. You should speed up to get back on track, but do so wisely. Don't panic; just take a few seconds less on each question until you're caught up. Don't guess without thinking, but do look through the answer choices and eliminate any you know are wrong. If you can get down to two choices, it is often worthwhile to guess from those. Once you've chosen an answer, move on and don't dwell on any that you skipped or had to hurry through. If a question was taking too long, chances are it was one of the harder ones, so you weren't as likely to get it right anyway.

On the other hand, if you find yourself getting ahead of schedule, it may be beneficial to slow down a little. The more quickly you work, the more likely you are to make a careless mistake that will affect your score. You've budgeted time for each question, so don't be afraid to spend that time. Practice an efficient but careful pace to get the most out of the time you have.

Secret Key #5 – Have a Plan for Guessing

When you're taking the test, you may find yourself stuck on a question. Some of the answer choices seem better than others, but you don't see the one answer choice that is obviously correct. What do you do?

The scenario described above is very common, yet most test takers have not effectively prepared for it. Developing and practicing a plan for guessing may be one of the single most effective uses of your time as you get ready for the exam.

In developing your plan for guessing, there are three questions to address:

- When should you start the guessing process?
- How should you narrow down the choices?
- Which answer should you choose?

When to Start the Guessing Process

Unless your plan for guessing is to select C every time (which, despite its merits, is not what we recommend), you need to leave yourself enough time to apply your answer elimination strategies. Since you have a limited amount of time for each question, that means that if you're going to give yourself the best shot at guessing correctly, you have to decide quickly whether or not you will guess.

Of course, the best-case scenario is that you don't have to guess at all, so first, see if you can answer the question based on your knowledge of the subject and basic reasoning skills. Focus on the key words in the question and try to jog your memory of related topics. Give yourself a chance to bring the knowledge to mind, but once you realize that you don't have (or you can't access) the knowledge you need to answer the question, it's time to start the guessing process.

It's almost always better to start the guessing process too early than too late. It only takes a few seconds to remember something and answer the question from knowledge. Carefully eliminating wrong answer choices takes longer. Plus, going through the process of eliminating answer choices can actually help jog your memory.

Summary: Start the guessing process as soon as you decide that you can't answer the question based on your knowledge.

How to Narrow Down the Choices

The next chapter in this book (**Test-Taking Strategies**) includes a wide range of strategies for how to approach questions and how to look for answer choices to eliminate. You will definitely want to read those carefully, practice them, and figure out which ones work best for you. Here though, we're going to address a mindset rather than a particular strategy.

Your odds of guessing an answer correctly depend on how many options you are choosing from.

Number of options left	5	4	3	2	1
Odds of guessing correctly	20%	25%	33%	50%	100%

You can see from this chart just how valuable it is to be able to eliminate incorrect answers and make an educated guess, but there are two things that many test takers do that cause them to miss out on the benefits of guessing:

- Accidentally eliminating the correct answer
- Selecting an answer based on an impression

We'll look at the first one here, and the second one in the next section.

To avoid accidentally eliminating the correct answer, we recommend a thought exercise called **the $5 challenge**. In this challenge, you only eliminate an answer choice from contention if you are willing to bet $5 on it being wrong. Why $5? Five dollars is a small but not insignificant amount of money. It's an amount you could afford to lose but wouldn't want to throw away. And while losing $5 once might not hurt too much, doing it twenty times will set you back $100. In the same way, each small decision you make—eliminating a choice here, guessing on a question there—won't by itself impact your score very much, but when you put them all together, they can make a big difference. By holding each answer choice elimination decision to a higher standard, you can reduce the risk of accidentally eliminating the correct answer.

The $5 challenge can also be applied in a positive sense: If you are willing to bet $5 that an answer choice *is* correct, go ahead and mark it as correct.

Summary: Only eliminate an answer choice if you are willing to bet $5 that it is wrong.

Which Answer to Choose

You're taking the test. You've run into a hard question and decided you'll have to guess. You've eliminated all the answer choices you're willing to bet $5 on. Now you have to pick an answer. Why do we even need to talk about this? Why can't you just pick whichever one you feel like when the time comes?

The answer to these questions is that if you don't come into the test with a plan, you'll rely on your impression to select an answer choice, and if you do that, you risk falling into a trap. The test writers know that everyone who takes their test will be guessing on some of the questions, so they intentionally write wrong answer choices to seem plausible. You still have to pick an answer though, and if the wrong answer choices are designed to look right, how can you ever be sure that you're not falling for their trap? The best solution we've found to this dilemma is to take the decision out of your hands entirely. Here is the process we recommend:

Once you've eliminated any choices that you are confident (willing to bet $5) are wrong, select the first remaining choice as your answer.

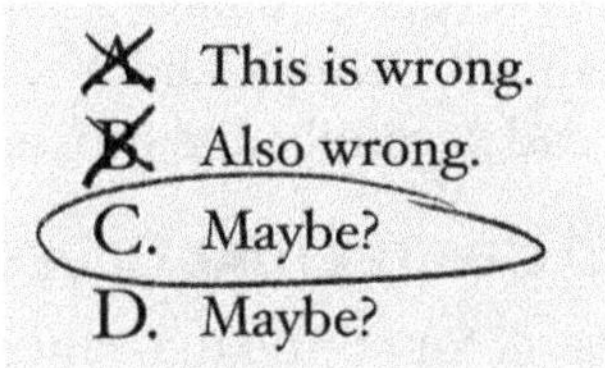

Whether you choose to select the first remaining choice, the second, or the last, the important thing is that you use some preselected standard. Using this approach guarantees that you will not be enticed into selecting an answer choice that looks right, because you are not basing your decision on how the answer choices look.

This is not meant to make you question your knowledge. Instead, it is to help you recognize the difference between your knowledge and your impressions. There's a huge difference between thinking an answer is right because of what you know, and thinking an answer is right because it looks or sounds like it should be right.

Summary: To ensure that your selection is appropriately random, make a predetermined selection from among all answer choices you have not eliminated.

Test-Taking Strategies

This section contains a list of test-taking strategies that you may find helpful as you work through the test. By taking what you know and applying logical thought, you can maximize your chances of answering any question correctly!

It is very important to realize that every question is different and every person is different: no single strategy will work on every question, and no single strategy will work for every person. That's why we've included all of them here, so you can try them out and determine which ones work best for different types of questions and which ones work best for you.

Question Strategies

⊘ Read Carefully

Read the question and the answer choices carefully. Don't miss the question because you misread the terms. You have plenty of time to read each question thoroughly and make sure you understand what is being asked. Yet a happy medium must be attained, so don't waste too much time. You must read carefully and efficiently.

⊘ Contextual Clues

Look for contextual clues. If the question includes a word you are not familiar with, look at the immediate context for some indication of what the word might mean. Contextual clues can often give you all the information you need to decipher the meaning of an unfamiliar word. Even if you can't determine the meaning, you may be able to narrow down the possibilities enough to make a solid guess at the answer to the question.

⊘ Prefixes

If you're having trouble with a word in the question or answer choices, try dissecting it. Take advantage of every clue that the word might include. Prefixes can be a huge help. Usually, they allow you to determine a basic meaning. *Pre-* means before, *post-* means after, *pro-* is positive, *de-* is negative. From prefixes, you can get an idea of the general meaning of the word and try to put it into context.

⊘ Hedge Words

Watch out for critical hedge words, such as *likely, may, can, often, almost, mostly, usually, generally, rarely*, and *sometimes.* Question writers insert these hedge phrases to cover every possibility. Often an answer choice will be wrong simply because it leaves no room for exception. Be on guard for answer choices that have definitive words such as *exactly* and *always.*

⊘ Switchback Words

Stay alert for *switchbacks.* These are the words and phrases frequently used to alert you to shifts in thought. The most common switchback words are *but, although*, and *however.* Others include *nevertheless, on the other hand, even though, while, in spite of, despite*, and *regardless of.* Switchback words are important to catch because they can change the direction of the question or an answer choice.

⊘ FACE VALUE

When in doubt, use common sense. Accept the situation in the problem at face value. Don't read too much into it. These problems will not require you to make wild assumptions. If you have to go beyond creativity and warp time or space in order to have an answer choice fit the question, then you should move on and consider the other answer choices. These are normal problems rooted in reality. The applicable relationship or explanation may not be readily apparent, but it is there for you to figure out. Use your common sense to interpret anything that isn't clear.

Answer Choice Strategies

⊘ ANSWER SELECTION

The most thorough way to pick an answer choice is to identify and eliminate wrong answers until only one is left, then confirm it is the correct answer. Sometimes an answer choice may immediately seem right, but be careful. The test writers will usually put more than one reasonable answer choice on each question, so take a second to read all of them and make sure that the other choices are not equally obvious. As long as you have time left, it is better to read every answer choice than to pick the first one that looks right without checking the others.

⊘ ANSWER CHOICE FAMILIES

An answer choice family consists of two (in rare cases, three) answer choices that are very similar in construction and cannot all be true at the same time. If you see two answer choices that are direct opposites or parallels, one of them is usually the correct answer. For instance, if one answer choice says that quantity *x* increases and another either says that quantity *x* decreases (opposite) or says that quantity *y* increases (parallel), then those answer choices would fall into the same family. An answer choice that doesn't match the construction of the answer choice family is more likely to be incorrect. Most questions will not have answer choice families, but when they do appear, you should be prepared to recognize them.

⊘ ELIMINATE ANSWERS

Eliminate answer choices as soon as you realize they are wrong, but make sure you consider all possibilities. If you are eliminating answer choices and realize that the last one you are left with is also wrong, don't panic. Start over and consider each choice again. There may be something you missed the first time that you will realize on the second pass.

⊘ AVOID FACT TRAPS

Don't be distracted by an answer choice that is factually true but doesn't answer the question. You are looking for the choice that answers the question. Stay focused on what the question is asking for so you don't accidentally pick an answer that is true but incorrect. Always go back to the question and make sure the answer choice you've selected actually answers the question and is not merely a true statement.

⊘ EXTREME STATEMENTS

In general, you should avoid answers that put forth extreme actions as standard practice or proclaim controversial ideas as established fact. An answer choice that states the "process should be used in certain situations, if..." is much more likely to be correct than one that states the "process should be discontinued completely." The first is a calm rational statement and doesn't even make a definitive, uncompromising stance, using a hedge word *if* to provide wiggle room, whereas the second choice is far more extreme.

✓ BENCHMARK

As you read through the answer choices and you come across one that seems to answer the question well, mentally select that answer choice. This is not your final answer, but it's the one that will help you evaluate the other answer choices. The one that you selected is your benchmark or standard for judging each of the other answer choices. Every other answer choice must be compared to your benchmark. That choice is correct until proven otherwise by another answer choice beating it. If you find a better answer, then that one becomes your new benchmark. Once you've decided that no other choice answers the question as well as your benchmark, you have your final answer.

✓ PREDICT THE ANSWER

Before you even start looking at the answer choices, it is often best to try to predict the answer. When you come up with the answer on your own, it is easier to avoid distractions and traps because you will know exactly what to look for. The right answer choice is unlikely to be word-for-word what you came up with, but it should be a close match. Even if you are confident that you have the right answer, you should still take the time to read each option before moving on.

General Strategies

✓ TOUGH QUESTIONS

If you are stumped on a problem or it appears too hard or too difficult, don't waste time. Move on! Remember though, if you can quickly check for obviously incorrect answer choices, your chances of guessing correctly are greatly improved. Before you completely give up, at least try to knock out a couple of possible answers. Eliminate what you can and then guess at the remaining answer choices before moving on.

✓ CHECK YOUR WORK

Since you will probably not know every term listed and the answer to every question, it is important that you get credit for the ones that you do know. Don't miss any questions through careless mistakes. If at all possible, try to take a second to look back over your answer selection and make sure you've selected the correct answer choice and haven't made a costly careless mistake (such as marking an answer choice that you didn't mean to mark). This quick double check should more than pay for itself in caught mistakes for the time it costs.

✓ PACE YOURSELF

It's easy to be overwhelmed when you're looking at a page full of questions; your mind is confused and full of random thoughts, and the clock is ticking down faster than you would like. Calm down and maintain the pace that you have set for yourself. Especially as you get down to the last few minutes of the test, don't let the small numbers on the clock make you panic. As long as you are on track by monitoring your pace, you are guaranteed to have time for each question.

✓ DON'T RUSH

It is very easy to make errors when you are in a hurry. Maintaining a fast pace in answering questions is pointless if it makes you miss questions that you would have gotten right otherwise. Test writers like to include distracting information and wrong answers that seem right. Taking a little extra time to avoid careless mistakes can make all the difference in your test score. Find a pace that allows you to be confident in the answers that you select.

⊘ Keep Moving

Panicking will not help you pass the test, so do your best to stay calm and keep moving. Taking deep breaths and going through the answer elimination steps you practiced can help to break through a stress barrier and keep your pace.

Final Notes

The combination of a solid foundation of content knowledge and the confidence that comes from practicing your plan for applying that knowledge is the key to maximizing your performance on test day. As your foundation of content knowledge is built up and strengthened, you'll find that the strategies included in this chapter become more and more effective in helping you quickly sift through the distractions and traps of the test to isolate the correct answer.

Now that you're preparing to move forward into the test content chapters of this book, be sure to keep your goal in mind. As you read, think about how you will be able to apply this information on the test. If you've already seen sample questions for the test and you have an idea of the question format and style, try to come up with questions of your own that you can answer based on what you're reading. This will give you valuable practice applying your knowledge in the same ways you can expect to on test day.

Good luck and good studying!

Three-Week Study Plan

On the next few pages, we've provided an optional study plan to help you use this study guide to its fullest potential over the course of three weeks. If you have six weeks available and want to spread it out more, spend two weeks on each section of the plan.

Below is a quick summary of the subjects covered in each week of the plan.

- Week 1: Hazard Identification and Control
- Week 2: Emergency Preparedness and Fire Prevention, Safety Program Development and Implementation, & Leadership, Communication, and Training
- Week 3: Practice Tests

Please note that not all subjects will take the same amount of time to work through.

Two full-length practice tests are included in this study guide. We recommend saving any additional practice tests until after you've completed the study plan. Take these practice tests without any reference materials a day or two before the real thing as practice runs to get you in the mode of answering questions at a good pace.

Week 1: Hazard and Risk Identification and Control

Instructional Content

First, read carefully through the Hazard and Risk Identification and Control chapter in this book, checking off your progress as you go:

- ❑ Common Hazards and Controls Associated with Hot Work
- ❑ Common Electrical Hazards and Controls
- ❑ Common Hazards and Controls Associated with Excavation
- ❑ Common Hazards and Controls Associated with Working at Heights
- ❑ Common Hazards and Controls Associated with Working in Confined Spaces
- ❑ Common Struck by Hazards and Associated Controls
- ❑ Common Hazards of Mechanized Equipment and Associated Controls
- ❑ Common Hazards and Controls Associated with Hoisting and Rigging
- ❑ Common Hazards and Controls Associated with Crane Operations
- ❑ Common Hazards and Controls Associated with Material Handling
- ❑ Common Hazards and Controls Associated with Material Storage
- ❑ Common Hazards and Controls Associated with Housekeeping
- ❑ Common Hazards and Controls Associated with Powder Actuated Tools
- ❑ Common Hazards and Controls Associated with Hand and Power Tools
- ❑ Common Hazards and Controls Associated with Asbestos Exposure
- ❑ Common Hazards and Controls Associated with Lead Exposure
- ❑ Common Hazards and Controls Associated with Noise Exposure
- ❑ Common Hazards and Controls Associated with Radiation Exposure
- ❑ Common Hazards and Controls Associated with Silica Exposure
- ❑ Common Hazards and Controls Associated with Chemical Exposure
- ❑ Common Hazards and Controls Associated with Working in Extreme Temperatures
- ❑ Common Hazards and Controls Associated with Vibration and Impact Exposures
- ❑ Common Biological Hazards and Controls
- ❑ The Globally Harmonized System of Classification and Labeling of Chemicals (GHS)
- ❑ Basic Safety Through Design
- ❑ Risks Associated with Multiple Trades Working Simultaneously in Work Area
- ❑ Principles of Ergonomics as Applied to Construction Practices and Material Handling
- ❑ Requirements, Usage, and Limitations of Personal Protective Equipment
- ❑ Basic Testing and Monitoring Equipment

As you read, do the following:

- Highlight any sections, terms, or concepts you think are important
- Draw an asterisk (*) next to any areas you are struggling with
- Take notes in your notebook or in the margins of this book

After you've read through everything, go back and review any sections that you highlighted or that you drew an asterisk next to, referencing your notes along the way.

Week 2: Emergency Preparedness, Incident Investigation, and Response; Safety Program Development, Implementation, and Sustainment; & Leadership, Communication, and Training

Instructional Content

First, read carefully through the Emergency Preparedness, Incident Investigation, and Response; Safety Program Development, Implementation, and Sustainment; & Leadership, Communication, and Training chapters in this book, checking off your progress as you go:

- ❑ Proper Fire Protection and Prevention Methods
- ❑ Components of Emergency Action Plans
- ❑ Common Elements of Response Plans for Environmental Hazards
- ❑ Emergency Response System
- ❑ Potential First Aid or Medical Needs
- ❑ Universal Precautions
- ❑ Applicable Health and Safety Standards and Best Practices
- ❑ Common Components of Site-Specific Safety Plans
- ❑ Worksite Assessment or Audit Processes
- ❑ Roles, Responsibilities, and Lines of Authority as They Relate to Safety Management
- ❑ Recommended Equipment Inspection Records or Logs
- ❑ Basic Risk Management Concepts
- ❑ Data Gathering Techniques and Procedures Used in Incident Investigation
- ❑ Recordkeeping
- ❑ Techniques for Determining the Root Cause of Accidents or Incidents
- ❑ Post-Incident/Accident Reporting and Follow-Up Procedures
- ❑ Documentation Requirements of Occupational Injuries and Illness
- ❑ Available Training Delivery Methods and Instructional Materials
- ❑ Appropriate Human Behavior Motivation Methods and Techniques
- ❑ Communication Strategies
- ❑ When to Consult with Equipment Manufacturers, Suppliers, or Subject Matter Experts
- ❑ Information Confidentiality Requirements
- ❑ Leadership

As you read, do the following:

- Highlight any sections, terms, or concepts you think are important
- Draw an asterisk (*) next to any areas you are struggling with
- Take notes in your notebook or in the margins of this book

After you've read through everything, go back and review any sections that you highlighted or that you drew an asterisk next to, referencing your notes along the way.

Week 3: Practice Tests

Your success on test day depends not only on how many hours you put into preparing, but also on whether you prepared the right way. It's good to check along the way to see if your studying is paying off. One of the most effective ways to do this is by taking practice tests to evaluate your progress. Practice tests are useful because they show exactly where you need to improve. Every time you take a practice test, pay special attention to these three groups of questions:

- The questions you got wrong
- The questions you had to guess on, even if you guessed right
- The questions you found difficult or slow to work through

This will show you exactly what your weak areas are, and where you need to devote more study time. Ask yourself why each of these questions gave you trouble. Was it because you didn't understand the material? Was it because you didn't remember the vocabulary? Do you need more repetitions on this type of question to build speed and confidence? Dig into those questions and figure out how you can strengthen your weak areas as you go back to review the material.

Practice Test #1

Now that you've read over the instructional content, it's time to take a practice test. Complete Practice Test #1. Take this test with **no time constraints**, and feel free to reference the applicable sections of this guide as you go. Once you've finished, check your answers against the provided answer key. For any questions you answered incorrectly, review the answer rationale, and then **go back and review** the applicable sections of the book. The goal in this stage is to understand why you answered the question incorrectly, and make sure that the next time you see a similar question, you will get it right.

Practice Test #2

Next, complete Practice Test #2. This time, give yourself **4 hours** to complete all of the questions. You should again feel free to reference the guide and your notes, but be mindful of the clock. If you run out of time before you finish all of the questions, mark where you were when time expired, but go ahead and finish taking the practice test. Once you've finished, check your answers against the provided answer key, and as before, review the answer rationale for any that you answered incorrectly and then go back and review the associated instructional content. Your goal is still to increase understanding of the content but also to get used to the time constraints you will face on the test.

As you go along, keep in mind that the practice test is just that: practice. Memorizing these questions and answers will not be very helpful on the actual test because it is unlikely to have any of the same exact questions. If you only know the right answers to the sample questions, you won't be prepared for the real thing. **Study the concepts** until you understand them fully, and then you'll be able to answer any question that shows up on the test.

Hazard and Risk Identification and Control

Transform passive reading into active learning! After immersing yourself in this chapter, put your comprehension to the test by taking a quiz. The insights you gained will stay with you longer this way. Scan the QR code to go directly to the chapter quiz interface for this study guide. If you're using a computer, simply visit the online resources page at **mometrix.com/resources719/chst** and click the Chapter Quizzes link.

Common Hazards and Controls Associated with Hot Work

Hot work refers to work with hot metal such as brazing, welding, soldering, cutting with a torch, and drilling or grinding that potentially creates a fire hazard. Employees engaged in hot work must be trained in the hazards posed by the work and how to mitigate them. A hot work program must include assessment of industrial hygiene hazards (exposure to metal dusts and fumes), assessment of noise hazards, assessment of proper PPE (eye protection, proper gloves, and respiratory protection if required), and assessment of fire hazards posed by the hot work.

Hot Work Permit Systems

A hot work permit ensures that safety precautions have been taken for welding or torch-cutting activities. The permit is designed as a job aid to check that the work area has been prepared for flying sparks by removing **flammable debris** in a sufficient radius around the work area. The work permit should prompt a check that the **fire sprinkler protection system** is in place and functional, and that fire extinguishers of the correct type are staged and ready. The permit should include a provision to test the work environment to ensure that there is no **explosive atmosphere**. The permit should document that the employees have received the required training and have the proper PPE available. The permit should include a provision to have a **fire watch** in place for at least thirty minutes after the hot work has been completed, and for the area to be periodically monitored for at least six hours after the work is completed.

Supervisor's Duties for Hot Work

Supervisors are responsible for planning the work and ensuring that the hot work permit system is implemented. The use of the hot work permit checklist is a means of ensuring that the proper safety precautions are taken and that planning is completed prior to the hot work beginning. The supervisor is also responsible for ensuring that employees have the proper PPE on hand, providing the proper manpower and personnel to use a designated fire watch employee if necessary, and ensuring that employees have received the required safety and operational training to conduct the hot work.

Fire Watch

Proper planning for hot work or welding includes having a person to act as "fire watch." This is necessary for any activities that have a fire risk. The fire watch person should be trained in the duties and should understand the requirements of the **hot work permit**. The person should be stationed in the area and remain alert for sparks or embers emitted during the work. The person must be prepared to use a fire extinguisher and to alert other workers and authorities in the event of an emergency. Fire watch duties must extend after the hot work is finished until there is absolutely no chance that any embers remain that could cause a fire. If the fire watch must leave the work area at any time during work, work must be stopped until he or she returns.

Welding Project Hazards and Controls

There are several hazards to consider when planning a welding project. The safe use of the compressed gases used must be considered. The types of metals to be welded should be considered to determine whether hazardous metal fumes or dusts will be generated that will require respiratory protection (and specialized equipment that will fit under a welding helmet). The eye hazard posed by exposure to the wavelengths of light emitted by the welding process must be considered and the appropriate welding glasses or face shield obtained. The risk of fire and sparks must be considered, and appropriate protective clothing should be worn.

Welders will be exposed to sparks when welding. If the welder is wearing the wrong clothing or gloves, these sparks can **ignite** the clothing and burn the worker. Proper protection for the eyes and face includes a **welding helmet** with face shield or goggles. The work uniform should be of **heavy-duty cotton** rather than synthetic materials because synthetic materials such as polyester have a lower ignition temperature. Clothing can also be manufactured from flame-resistant materials such as Nomex®. For additional protection, welders can wear a leather **apron** that will protect the chest and abdominal area.

Common Electrical Hazards and Controls

The Occupational Safety and Health Administration (OSHA) Electrical Safety Standard (29 CFR 1910.303) considers that guarding and personal protective equipment (electrical safety gloves or other measures) must be taken when employees are exposed to live electrical wires greater than 50 volts. Many companies have adopted more stringent requirements that require protection if exposure levels are greater than 30 volts. Electrical safety gloves must be worn, and untrained and unauthorized employees must stay at least 10 feet from the energized parts. There must be a means of triggering an emergency stop available to the employee, and there must be personnel on site who are trained in first aid measures for electric shock if employees will be exposed to electricity above these threshold levels.

Electrical Shock

An electrical shock occurs when the human body unintentionally becomes part of an electrical circuit. Electric shock occurs when:

- A person's body becomes part of the **conduction circuit or path** followed by electrical energy.
- A person becomes part of the circuit by holding both **positive and negative wires** of an electrical circuit.
- An otherwise closed electrical circuit loses or has a break in its **insulation**. Any person holding the line then becomes part of the circuit.
- Standing in or touching another **conductor carrying a current**. This is often caused by persons standing in water and making contact with electrical lines.

Impact of Electrical Current Flowing Through the Human Body

Depending on the strength of an electrical current flowing through the human body, the impact can be minimal to fatal. The factors that determine the **severity of impact** are:

- The actual amount of current flowing through the body
- The path that the electric current takes as it passes through the body
- The length of contact the human body has with the sources of the electrical current
- The frequency of the electrical current flowing through the body

Assuming that the length of contact with an electrical source is at one second, an exposure of up to **5 milliamperes of current** can be experienced with only minor discomfort. As the current increases beyond 5 milliamperes for one second exposures, the exposures become gradually more dangerous.

Energy Control Systems

Workers are often injured when performing service or maintenance operations on equipment operated by electric or other motorized power. The equipment being serviced must be secured against possible accidents by an energy control program. The employer's energy control program should train employees in safe procedures for servicing machines and equipment which might be subject to unexpected startup or operation. Tagout and Lockout devices are most often used to prevent this type of accident.

- **Tagout** refers to the placement of a tag or warning label which informs workers that a machine cannot be operated until the tagout device is removed by the health and safety officer.
- A **Lockout Device** is any mechanical method of preventing a machine from releasing energy or starting up operations while it is being serviced or maintained. Simple locks and keys may be sufficient. Other situations may require blocks, flanges, and bolted slip blinds.

Training Requirements

The OSHA lockout/tagout regulation lists specific training requirements for employees, both when there is a change to the equipment used or the circumstances of its use and on an annual basis. The training must cover the types of hazardous energy that can/should be controlled (not just electrical energy), the definitions and identities of authorized and affected employees, the specific lockout procedures for equipment they will be locking out, the difference between lockout and tagout, why tagout alone is not an approved control of hazardous energy, and the process to commence safe start-up of equipment after lockout.

Authorized and Affected Employees

An **authorized employee** under the lockout/tagout regulations is an employee who has been thoroughly trained in the reasons for lockout and the methods to lock out equipment, and has been issued a lock to use. The lock should be personalized by color coding and/or the use of personalized tags to notify others who is responsible for a particular lockout.

An **affected employee** is one that has a general awareness of what lockout/tagout is, and works near machinery that will be locked out from time to time. However, these employees are not responsible themselves for performing the lockout or for following the steps to safe start-up of equipment after lockout.

GFCI

A **Ground Fault Circuit Interrupter** (**GFCI**) protects against the most common type of electrical shock, a ground fault. A ground fault occurs when an electrical current finds an alternate "path of least resistance" to the ground instead of the ground wire. A person receives a shock when this path of least resistance is through their body. GFCIs are commonly used and required in wet areas such as bathrooms or kitchens, since the presence of water increases the chance that electrical current will experience a ground fault. Water is very conductive and provides an easy path for electrical current. The GFCI works by comparing the current that exits the unit to the current returning to it; if a significant difference is detected, it immediately cuts off the electricity flow through the circuit and protects from the risk of electrical shock.

Employer Responsibilities Regarding the Use of Ground Fault Circuit Interrupters

Employers have a responsibility to provide appropriate ground fault circuit interrupters for power tools and electrical equipment at construction sites.

- While the program is the direct responsibility of the employer, the employer may delegate the **implementation** of it to an individual designated as a "competent person."
- **120 Volt receptacle single phase outlets** on construction sites which operate at 15 and 20 amps, and which are not part of permanent building wiring, must be equipped with GFCIs for the protection of workers.
- The provision requiring adequate **ground and circuit interruption** must apply to all equipment available for use by workers. It applies to all cord sets and receptacles not part of the permanent building structure.
- The employer must maintain a written **description of the provisions** of the electrical safety program on site. The written documentation of safety provisions must be available for inspection by any regulatory or safety official and by employers who are affected by the provisions of it.

Exceptions to GFCI Requirements for Electrical Generators

There are **exceptions** to the OSHA provisions regarding GFCI circuit protection for non-permanent electrical circuits. When electricity is being delivered to small tools via portable generators, the frame of the generator may in itself serve as the grounding electrode mechanism for the electric circuit supplied by the generator. This is true when:

- The **portable generator** is the supplier of electricity for equipment connected directly to the generator by cords, plugs, or receptacles mounted on the generator expressly.
- The extraneous metal parts of the equipment (the parts not designed to conduct electricity) and the equipment grounding terminals of the receptacles are bonded to the frame of the generator.

The exceptions apply to both dedicated portable generators and vehicle-mounted generators. With regard to vehicle-mounted generators:

- The **frame of a motor vehicle** is considered sufficient grounding for vehicle-mounted generators incorporated into the vehicle's design and construction.
- OSHA regulations state that, in this instance, the generator must be **bonded** to the vehicle frame.
- OSHA regulations state that a vehicle-mounted generator grounded through bonding with the vehicle frame can only supply equipment located **on the vehicle**. Other means of permissible connection, in this usage, are through connection to mounted receptacles on the vehicle.

Onsite Electricity Grounding Requirements

The **Assured Equipment Grounding Conductor Program** is a written protocol which covers cord sets and receptacles which are not a permanent part of a building structure's electrical configuration. The safety program applies to cords, plugs, connectors, and electric-powered tools.

- The written program and the name of the designated "competent person" must be **available for review** by any affected employee or by OSHA officials.
- **Tests** are routinely required to assure workers that equipment and non-permanent electrical lines are adequately grounded. Required tests must be recorded and maintained until the next testing period.
- Two types of testing are required as part of the Assured Equipment Grounding Conductor Program:
    - A **continuity test** is needed to ensure that equipment grounding conductors are electrically continuous.
    - Another test must be done to ensure that the receptacles and plugs have the ground circuits connected to the **proper terminal**.

Temporary Power Cord Safety

Temporary power cords and extension cords should only be used on a short-term basis. They should not be installed as substitutes for permanent wiring. Any actual wiring work must be performed by a qualified electrical worker that has been trained under OSHA requirements and NFPA 70E. Extension cords should always be examined before use to ensure they are not frayed or have insulation missing and that the third ground prong is in place and not bent. Care must be taken to install the cords in such a way that they do not pose a trip hazard; for example, a cord protector can be used to prevent people from catching their feet on the cord and tripping. If the cord will be used for electronic equipment or computers, it should have a surge protector in it to protect the electronic equipment against power surges.

Testing Temporary Electrical Circuitry on the Construction Site

Inspection of non-permanent electrical lines and equipment should be a routine part of the daily round on a construction site.

- All required tests must be carried out **before** any work is performed when equipment is first used.
- Equipment should be tested again after it has been **repaired** and returned to the worksite. This is especially important when electrical parts are known to have been damaged.
- Routine inspections of electrical equipment must be re-tested within a **three-month period** unless cord sets and/or receptacles are part of a fix-wire system (in which case inspections must be within six months).

Test records must be maintained onsite and available for inspection. A record date of all inspections (and identification of the item being inspected) is required.

Electrical Continuity and Impedance Testing

On construction sites, all power tools and electrical lines which are not part of a building's permanent structure must be tested for **grounding and continuity of circuitry**. The types of equipment testing equipment which may be encountered for these purposes are:

- **Ground loop impedance tester**: This measures the amount of resistance or impedance in the ground terminal of power equipment used for drilling, sawing, lighting, etc.
- A **volt-ohm meter** would also measure the amount of impedance but the **impedance meter** provides data regarding both low and high ground faults.
- **Lamp and battery, battery and buzzer/bell**, and **neon lamp testing equipment** are used to measure continuity in circuitry but not the amount of resistance.

Arc Flash

An arc flash is an event in which electrical current jumps (or flows) through an air gap between two conductors; for example, broken or torn wire insulation can cause conditions that promote arc flash, as can dust buildup in an electrical panel. Arc flash is an extremely dangerous condition due to the tremendous energy released during the event. The arc flash explosion can cause severe burns, and the explosive concussion can throw an employee some distance from the event. It can also cause hearing damage. Unfortunately, the susceptibility of a given electrical panel to arc flash is not apparent to the naked eye, making an arc flash assessment and electrical panel labeling program important.

All electrical workers must be trained in the basic elements of arc flash safety. All electrical panels must be evaluated by qualified personnel to determine the arc flash potential. This evaluation considers the potential release of energy that would occur (based on voltage and load) were there to be an arc flash in that panel. From this, the panel is rated on a scale of 0 to 4. Personnel that will be working in the panel must wear personal protection suits rated for the appropriate arc flash hazard level. In addition, workers must be trained to set up appropriate exclusion zones around areas where work will occur on electrical panels. Workers must be trained in emergency procedures and first aid procedures for victims of electrical shock and burns.

Common Hazards and Controls Associated with Excavations

Underground Excavations

A competent and knowledgeable person must assure that workers receive proper training when working in **underground areas**. The requisite areas of required training demand that all underground workers be taught to recognize and avoid hazards which are peculiar to underground construction. Those areas where training must be applied are:

- Employees must be taught the methods of **air monitoring** and to recognize potential indications of contaminants.
- Workers should know the air **ventilation standards** which apply and the required amounts of **illumination**.
- The use of **communication equipment** and appropriate communication methods must be taught to workers. Effective means of communication from excavation to the above-ground area must be provided.
- Employers must be trained in **flood control**, **fire control**, and **fire prevention**.

- **Personal protective equipment** training is of vital importance. Familiarity with emergency breathing apparatuses is part of this training.
- Workers must be fully informed regarding **check-in and check-out methods** as well as **evacuation plans**.

The employer is required to take swift action when a competent person has determined the presence of hazardous gases in an underground excavation:

- **Notices** of the gaseous condition must be posted at all entrances to the excavation.
- If the hazardous gases amount to at least **five percent** of the Lower Explosive Limit (LEL), **ventilation air volume** must be immediately increased to reduce the concentration to an amount less than five percent of the LEL. There is an exception when operating under gassy/potentially gassy requirements.
- Work must be entirely **stopped** when welding, cutting or other hot work is being done if the concentration is **ten percent** or more of the LEL. Concentrations must be reduced by ventilation until the concentration is below ten percent of LEL.

Underground Tunnels, Shafts, or Underground Cavities

Underground work has a wide array of dangerous conditions which must be addressed in worker training. While some of the areas of instruction may be specialized, as in the use of electronic testing equipment, all workers should be informed of several general hazards:

- **Illumination**: Workers should be informed of appropriate light levels when working underground.
- **Communications**: A clear understanding of the communications systems to be used in both routine work and/or emergency conditions is required.
- **PPE**: The use of personal protective equipment (PPE) is necessary in many cases. Workers must be trained in the potential hazards requiring respiratory protections.
- **Fire Prevention and Protection**: Whether the activity is routine or specialized, such as in the use of explosives, workers should be trained sufficiently in these areas so as to contribute to overall safety of all workers.
- **Air monitoring**: Air quality cannot always be detected without instruments but workers should be trained to recognize certain indications of developing hazardous conditions.
- **Ventilation**: Workers should be educated to understand individual requirements and the methods used to maintain proper ventilation.
- **Floor Control**: Water seepage into shafts, tunnels, and excavations is a common risk factor. Workers must be trained in emergency evacuation procedures.

Regulations Designed to Prevent Injury When Workers Are Required to Work Underground

The OSHA provisions in **1926.800** provide the basis for worksite safety requirements of underground workers employed in the excavation of underground cavities, tunnels, shafts or chambers. The general provisions of those requirements are:

- Underground areas must be properly **vented** to prevent accumulation of dusts, fumes, vapors, and gases.
- The number of employees working underground should determine the minimum capacity of ventilation requirements. **Two hundred (200) cubic feet per minute of fresh air** is required for each employee.

- **Mechanical ventilation** is a must in most underground circumstances. The velocity of air must be regulated to flow at least 30 feet per minute in underground areas where conditions are likely to create dust, fumes, or gas vapors.
- **Air flow direction** should be reversible.
- If **blasting** is to be conducted in an underground tunnel or cavity, workers cannot return to work until the ventilation system has cleared the area of all exhaust and fumes.
- **Ventilation** doors must be designed to remain closed when in use regardless of the direction of air flow. Ventilation doors are designed to provide a barrier to dangerous fumes should they build up in adjacent areas.
- If the ventilation system has been shut down for any purpose, **testing** must be done by a competent person to determine whether the area is safe for reoccupation by workers.
- **Entry/exit openings** must be strictly controlled and posted with appropriate safety signs.
- Whenever any employee is working underground, a "**designated person**" must be present above ground. The "designated person" is required to keep a running account of the number of underground workers and to secure and communicate a rescue response when necessary.
- It is the responsibility of the employer to **communicate and coordinate** with other workers and/or job operations which may impact the safety of underground workers. Movement of heavy equipment is an example of other worksite activity which may impact underground workers.
- OSHA section 1926.800 mandates that **rescue plans** must be established in the event of emergency. Specific requirements are based on the number of workers employed underground.

OSHA section **1926.800** requires contingency plans for the emergency use of rescue personnel on construction job sites.

- Whenever employees working underground number 25 or more, regulations require that the employer make **contingency plans** with locally available trained rescue personnel.
- When 25 or more employees are working underground, contingency must be made for **two rescue teams**. One team must be located within a half-hour traveling distance; the other must be within a two-hour travel time period.
- If less than 25 persons are working underground, contingency plans may involve a **single rescue crew** which must be positioned within one-half hour travel time from the job site.

Proper contingency planning requires that rescue personnel be made familiar in advance of conditions at the job site.

Construction Trenching

When a **trenching system** is to be dug to a depth of twenty feet or more, a professional engineer is required to design a safe, strong, and reliable method of shoring up the excavation. The CHST must be familiar with ground materials and slope requirements for trenches less than 20 feet deep. Soil types are categorized as Types A, B, or C. The maximum angle of a trench wall depends on the category type into which the excavation has been made. Clay is more stable than other ground materials like sand or silt and is considered a Type A material. The maximum slope of an excavation into Type A soil is 53 degrees. Angular gravel and silt are Type B materials. The maximum height to depth ratio for trenching in Type B soil is 1:1, yielding a maximum 45-degree angle. Type C is all other gravel and sand, which are very unstable for trenching. The slope angle for a trench into Type C soil can be no more than 34 degrees. The only material in which vertical trench walls are allowed is solid rock.

If sloping is undesirable, **shielding** may be used instead. Various types of shield systems may be designed to protect workers. These may range from portable systems which may be moved from one site to another to systems manufactured on the spot according to OSHA specifications. "Trench boxes" or "trench shields" are terms which may be used to describe shielding used to protect workers in trenches.

Physical Conditions Which Apply When a Soil or Earth Is Reported as Distressed

A **distressed soil** or area of earth and rock is one that poses a special danger to workers in the vicinity. There is an imminent and greater potential for a collapse and injury to workers occupied in an excavation of distressed soil. Distressed soils can be recognized by:

- **Fissures or cracks** are apparent in the face of or on the sides of an open excavation.
- When the edges of an excavation show signs of **subsidence or sinking**.
- When the bottom of an excavation shows signs of **bulging or separating**.
- **Spalling** means that a material is breaking up into small chips, flakes, or splinters. It is an indicator of distressed earthen materials.
- Pebbles and little clumps of material **separate** from the face of the side cut and roll down into the excavation.

Conducting the Required Daily Investigation of an Ongoing Excavation Site

Daily inspections of excavation sites are required as a minimum but inspections can be made any time a change in work, weather, or environmental conditions makes more frequent inspections desirable. The safety inspector must be alert to all hazardous conditions affecting worker safety but there are certain commonalities of concern:

- Examine for failure of protective systems, hazardous atmospheres, water accumulation, and the ground signs which may indicate a potential collapse of excavation walls.
- Means of **exit or egress** should be inspected. Stairways, ladders, ramps or other safe means of evacuation from trenches are required in ditches or trenches that are 4 feet or more in depth. A worker should not have to travel more than 25 feet to reach a means of safe exit.
- **Stability** of adjoining walls, buildings, or other nearby structures should be inspected for settling, leaning, or other signs of tipping.

Above-Ground Utilities Likely to Be Affected by Construction Site Excavation

Above ground utilities are usually visible and are easily mapped in preparation for site excavations. Though they are clearly visible in most instances, above ground installations of utilities require caution and preparation on the part of excavation planners:

- **Trucks and other vehicle drivers** passing through the site must be adequately warned of the hazards of above-ground utility lines. **Operators** must be briefed beforehand. Caution signs at planned traffic locations increase safety significantly. Drivers should be forewarned as to the height of their vehicles or excavating equipment.
- In the planning stage, great care should be given to the **route of traffic** which must bypass the planned excavation. Traffic from heavy equipment must not be allowed to pass in such proximity to the excavation that might cause undue pressures upon the walls of the excavation and trigger cave-ins.

Underground Utilities Likely to Be Affected by Construction Site Excavation

Preparation for site excavation in areas where underground sewer, power, phone, or gas lines may be buried requires:

- Individual **utility companies** must be contacted and apprised of the nature and specific location of the proposed work. Requests should be made for information pertaining to the specific location of underground installations.
- If the individual utility does not respond to requests within a **24-hour period**, the construction manager may begin work on the excavation, exercising all caution. The exercise of caution means that detection equipment to locate lines is routinely used, along with other acceptable measures.
- **Mapping**: The estimated location of all underground utilities should become part of the site map drawn up prior to beginning the excavation. Sufficient numbers of copies should be distributed at briefings to managers, equipment operators, and laborers.

Training Requirements

As part of a comprehensive injury prevention plan and hazard communication plan, workers who work in an excavation site must be trained on the hazards of an excavation and how they are protected from injury. They should be trained in site-specific emergency procedures. If wearing a harness, the employee must be trained in how to use the harness and how to inspect it to ensure it is working optimally.

Safety Requirements

The risk of a cave-in is the greatest risk at an excavation site. Any excavation of greater than five feet deep must have a protective system in place to prevent cave-ins (barricades, sloping side, benching, etc.). If the excavation is greater than twenty feet deep, this protective system must be designed by a registered professional engineer. There are several types of cave-in protection available, such as shoring the sides, sloping the sides to reduce the vertical edge, benching the sides, and shielding workers from the sides. A qualified individual should be consulted when choosing cave-in protection, as site conditions such as soil type, moisture level, and the amount of activity around the excavation must be taken into consideration. Access and egress to the excavation area and the excavation itself must be controlled with barricades and work practice controls. A ladder or ramp must be provided as a means of egress if the depth of the excavation is four feet or greater. The route to the means of egress cannot be more than 25 feet for any employee who may be working in the excavation. The spoil pile (the soil removed from the pit) must be set back from the edge of the excavation at least two feet. This prevents material falling on workers and makes an even walking surface at the edge of the excavation.

Common Hazards and Controls Associated with Working at Heights

Safety Nets

According to the provisions of OSHA 1926.502, **safety net systems** should be constructed with the following considerations:

- Safety nets should not be positioned more than **30 feet below the working level**. Nets should be placed as close to the work area as the activity permits as long as they are unobstructed.
- Safety nets should be positioned so as to extend **beyond the vertical plane** of the working surface at a distance of 8 to 13 feet, depending on the distance of the net from the working area.
- Safety nets should have sufficient **"ground" clearance** so as not to strike surfaces below the working level. This precaution is determined by **drop testing** and is the responsibility of the employer. Drop testing is a significant factor in determining the strength of the net and its positioning.
- Drop testing consists of a **400-pound sandbag** dropped from the highest working surface to which workers are exposed.
- **Scrap materials and/or tools** which have fallen into the safety net must be removed expeditiously and certainly no later than the start of the next work shift.

Scaffolding

There are many different types of scaffolding designed around specific construction uses. The terminology used to describe scaffolding is varied. Therefore, the CHST should know both slang terms for the various types as well as the technically correct names provided by the manufacturer.

- **Horse Scaffold**: This is merely a platform supported on each end by construction "horses" or "saw horses." When made of metal, these are sometimes called trestle scaffolds.
- **Chimney Hoist**: The technical name for a "chimney hoist" is multi-point adjustable suspension scaffold. This type of scaffold is used to provide access to the inside of large chimneys.

The **fabricated frame scaffold** is a common type made of tubular steel and welded at the assembly joints. Aside from the mainframe, the fabricated frame scaffold consists of other components like horizontal bearers and intermediate bearing braces or members. **Bricklayers' square scaffolding** is constructed from a system of framed squares which provide solid, wide, and stable footing for masons and tenders. A **carpenters' bracket scaffold** attaches to the building or the stable vertical walls of a building under construction. A **catenary scaffold** is a standing platform suspended from parallel horizontal ropes or cables attached to beams or other solidly supported structures. Very often, the catenary scaffold is supported by additional vertical hoist and pulley system. A **hoist** is a manually operated or power-driven device which attaches to a **suspension scaffold**. It is the mechanism which allows the suspended platform to be pulled up or "hoisted" aloft. There are two common safety devices which are designed to work in conjunction with a hoist to augment worker safety:

- **Lifeline**: Scaffolds and suspension devices must be anchored to solid structures to prevent swaying, settling, or otherwise collapsing. The lifeline is a rope, chain, or cable that ties a platform to strong and stable anchor points. A lifeline may be positioned vertically or horizontally and fastened at several points.
- **Decelerating Devices** are mechanisms specifically designed to augment scaffold safety systems. Their chief purpose is to dissipate energy which would harm a falling worker attached by a harness or belt. A decelerating device protects against the shock or whip action imparted to a falling body as it strikes the end of a safety line or other implement. It slows and gentles the falling motion.

The terms refer to safety devices or components associated with scaffolding or suspension mechanisms which allow workers to conduct safer activities alongside or upon the heights of buildings.

- **Stall Load**: The stall load refers to the point at which a power-operated hoist is automatically disconnected for worker safety. Stall load is designed into the mechanism to prevent overload. Should a worker overload a suspension scaffold, the operating mechanism disconnects until a proper load weight is restored and amendments are made.
- **Rated Load**: The rated load is the maximum load for which a hoist or scaffold is designed to accommodate. The rated load of a scaffolding mechanism can be obtained by reading the manufacturer specifications.

Reviewing and Inspecting Safe Upright Standing Scaffolding Procedures

Since scaffolding is commonly used, and because there are so many types of scaffolding, the CHST has a long checklist of safety items to review when assessing **worker safety**:

- All scaffolds must be **assembled** by a competent person capable of supervising the placement, disassembly, and moving of scaffolding.
- The competent person must have knowledge of **guardrail specifications** for tubular welded frame scaffolds and for the manner in which they are to be braced.
- Mobile scaffolds must be tightly planked with **scaffold grade planking**. Planking which extends beyond scaffold supports must meet additional OSHA specifications.
- The base of any scaffolding must be set upon a **sound and stable base**. The positioning of upright supports must be plumb and the joints securely locked.
- Safe access by means of a **secure ladder or permanent stairway** must be provided. Ladders must be affixed to the scaffold frame. The scaffold frame is to be anchored into solid and stable side structures at points along its heights.

Circumstances Under Which Ladders Can Be Used Upon Scaffolding Platforms

As a general rule of thumb, ladders should not be used to increase the working height of employees on elevated scaffolding. There are exceptions, however:

- Where there is a large area scaffold designed to be used as a temporary "floor" and wide enough to provide security, a ladder may be used to elevate the working height.
- When ladders are used upon scaffolding, the structure is subject to a sideways thrust. Under these circumstances, the main body of the scaffolding must be firmly anchored to walls or secure beams.

- The ladder's legs must be solidly planted and blocked to prevent skidding or slipping.
- All other moving parts of the scaffold must be secured against movement or shifting.

Measurement Rules for Setting Ladders Against the Walls of a Work Area

There are three elements to be considered when observing the regulations regarding setting a straight ladder against the walls of a work area.

- The **base** of the ladder refers to the distance between the wall and the bottom of the ladder. This distance should be no more nor less than 1/4th of the length of the ladder from floor to top point.
- The **length** of the ladder in this application does not mean the entire length of the ladder but refers to the distance from the point where it touches the floor to the point where it is supported by the wall.
- Ladders should be firmly **anchored** with stops in place and secured with blocking.

Load Ratios for Scaffolding and Suspension Lines or Ropes

Scaffolding load ratios require that scaffolds be constructed with the capability of supporting many times their intended load.

- **Scaffolding and planking** must be able to support at least four times the maximum intended load.
- **Suspension ropes** must be capable of supporting at least six times the maximum intended load.

There is good reason for the OSHA requirement that scaffolds and ropes be able to support far more than their intended load:

- In addition to the workers using the scaffold, there is the added weight of **tools and equipment** to consider.
- The effect of the **elements** like snow and wind can have an impact on the carrying capacity of scaffolding.
- **Planking** may be weaker than the specifications would indicate, particularly if the lumber is of lower quality and contains knots or fissures.

Personal Fall Arrest System

A personal fall arrest system is a method of stopping an employee from falling from the heights at which he is working. Some personal fall arrest systems involve harnesses, hooks, belts, and loops of rope or other suspension mechanisms designed to catch a falling worker. Fall arrest systems must:

- Not subject an employee to a force greater than 900 pounds if a **body belt** is used
- Not subject an employee to a force greater than 1800 pounds if a **body harness** is used
- Not allow an employee to fall more than **six feet** nor come in contact with the **floors or objects below**

Personal fall arrest systems must also be able to sustain two times the impact caused by an employee falling six feet. The deceleration distance of travel can be no longer than 3.5 feet.

Positioning Device Systems

Several types of construction activity require the use of positioning devices which allow the worker to hang suspended by a harness or belt upon the area being worked on. OSHA regulations require that:

- The **rigging** for the positioning device be designed such that an employee cannot fall more than two feet.
- The point to which the positioning device is **anchored** must be stable and capable of supporting the impact of a 3,000-pound inertial force.
- Connectors should be of **drop forged** or formed steel or similarly strong materials. They should be corrosion resistant and smooth with a tensile strength of at least 5000 pounds.
- D-rings and snap hooks must have a minimum **tensile strength** of 3600 pounds.

Positioning device systems must be inspected for wear prior to use. Equipment showing signs of wear or damage must be replaced immediately.

Height at Which Fall Protection Systems Are Required

Personal fall arrest systems (harnesses, for example), guardrails, and safety net systems must be used whenever workers are employed in construction or walking in work areas six feet or more from the lower level.

OSHA regulations require that the "six foot" rule be applied to the following situations:

- Workers must be protected from falling through **holes or floor openings** like skylights or stairwells. Skylights may be covered with solid structural materials. Another method may be to erect a guardrail system around floor openings or skylights.

- Workers employed in **framework or reinforcing steel frame construction** must be protected by personal arrest systems and/or safety nets.
- Workers who must walk through or work near **trenching areas or excavations** must be protected by fences, barricades, or guardrail systems if the excavation is six feet or more.

Providing Greater Fall Protection to Construction Site Workers

Common errors are often repeated around construction sites so it is important to recognize the following **fall protection strategies**:

- The assignment of a guard for **floor holes** is necessary, especially when an area is being used as a walkway by materials handlers who may not be able to see over the materials they are carrying.
- Body belts, lanyards, and fall-breaking mechanisms must be employed while working from **aerial lifts**.
- **Wall openings** must either be guarded or securely barricaded. The same strategy may be employed in preventing falls from open-sided floors or platforms.
- When no other practical means of fall protections is practical, **safety nets** must be securely arranged to protect workers at heights of 25 feet or more.

Safe Practices for Using Fall Protection Systems

For effective fall protection, OSHA recommends that companies adhere to the following practices:

- The company should have a written fall protection plan as part of its overall health and safety plan. The plan should include company rules for how and when to use fall protection equipment.
- The company should follow standard fall protection requirements when fall protection equipment must be used, usually when an employee in a general industry is four feet above the floor, when an employee of a construction company is six feet above the ground, or when an employee is on scaffolding 10 feet above the ground.
- The company should provide correct fall protection equipment and ensure that it is not only used, but used properly.
- The company should inspect, maintain, repair, and replace fall protection equipment regularly.
- The company should provide supervisors and workers with training on how to recognize fall-related hazards and how and when to use fall protection equipment.

A fall protection system can limit or prevent falls. A fall protection system can include safety belts, safety harnesses, lanyards, hardware, grabbing devices, lifelines, fall arrestors, climbing safety systems, and safety nets. Most of these elements stop a fall that has already started and must meet specific standards. Safety belts are worn around the waist while harnesses fit around the chest and shoulders and occasionally the upper legs. Safety harnesses lessen the number and severity of injuries when they arrest a fall because the force is distributed over a larger part of the body. Lanyards and lifelines connect safety harnesses to an anchoring point while grabbing devices connect lanyards to a lifeline. Lanyards absorb energy, so they reduce the impact load on a person when the fall is arrested.

Safe Work Practices When Working at Heights or on Elevated Work Platforms

Elevated work platforms or working at heights pose a fall hazard that can be fatal if an employee falls on his or her head. When using a platform, there should be guardrails that conform to Occupational Safety and Health Administration (OSHA) guardrail guidelines. Employees should be trained in and use fall protection harnesses; the harness must be tied to a place that will safely arrest the fall and not tip over any equipment. If the height is accessible using a ladder, make sure the ladder is on a stable surface and that a coworker is available to hold the ladder and position it so that reaching from the ladder is not necessary. Never stand on the top step of the ladder.

Aerial Lifts and Scissor Lifts

An **aerial lift** is a basket on an extended arm for lifting workers to heights. A **scissor lift** accomplishes the same thing but is conveyed upward vertically. Both have guard rails around the platform. Per 29 CFR 1926.453(b)(2)(v), OSHA requires that a body belt must be worn and a lanyard must be attached to the boom or basket when working from an aerial lift.

Common Hazards and Controls Associated with Working in Confined Spaces

Permit-Required and Non-Permit-Required Confined Spaces

There is a significant difference between a permit-required and non-permit required confined space. **Confined spaces** are areas not designed for continuous employee occupancy, and which have limited means of egress. The definition includes spaces that even arms are placed into, not just spaces that can accommodate the entire body. They may also have the potential for a hazardous atmosphere, either oxygen deficiency, presence of chemical vapors, or extreme temperatures. For this reason, the Occupational Safety and Health Administration (OSHA) requires assessments and inventories of potential confined spaces to determine appropriate entry procedures. Entry into a confined space using a permit system is required when there is a potential for oxygen deficiency, explosive atmosphere, and/or chemical vapor exposures. The permit provides a mechanism to track entries and to document that the proper pre-entry procedures have been followed, such as measuring oxygen levels and measuring chemical vapors. Employees must also have an attendant outside the confined space at all times and emergency equipment on hand in case of emergency.

Permits issued for this purpose must:

- Identify the space to be entered
- Identify the type and nature of the work to be performed and provide substantial justification for issuing a permit
- Be date stamped with the date of authorization and the number of hours or days in which confined-space work has been authorized
- Identify the person or persons who are authorized to enter the confined space. This section of the permit should specify the type of tracking and monitoring system used to control entrance and egress

Working in Permit-Required Confined Spaces

A confined space has the following characteristics:

- An area sufficient in size to accommodate a worker of normal size to enter and perform assigned work with necessary tools and uniform.
- A confined space has limited entry and/or egress points which are so constructed as to limit the free passage of employees or rescuers into the area. Holding tanks, grain silos, manure pits, and entry passages to sewage tunnels are confined spaces requiring special methods of access and preparatory measures taken to ensure health and safety under poor atmospheric conditions.

- A confined space is one in which continuous employee occupancy for work purposes cannot be expected to last for an entire 8-hour period of time as would be the case under normal working conditions. OSHA limits the amount of time a worker may be exposed to confined spaces with limited or extremely hazardous atmospheres.

Confined Space Program

A confined space program must contain an inventory of all confined spaces at a site, a formal documented assessment of the potential hazards posed by the space (e.g., Is there a danger of oxygen deficiency? Are there potentially hazardous vapors or fumes present? Is there a risk of high or low temperature environment?). A confined space entry permit must be developed and required to be used for all permit-required confined spaces. All confined spaces must be labeled as such, and the label should identify whether the space is permit required or not. The employees that will enter the confined space must receive appropriate training on entry procedures and appropriate emergency procedures.

To fully implement a confined space safety program, one must be able to evaluate whether a hazardous atmosphere exists inside the confined space. The first assessment that must be made is to determine whether there is an oxygen-deficient atmosphere. For this, an oxygen level meter is needed to confirm that the oxygen content of the air inside the confined space is at least 19.5 percent. Second, if there is any possibility that flammable dust or vapors exist, the air must be analyzed for either dust or organic vapors. Organic vapors can be analyzed using a handheld organic vapor analyzer, and the results can be used to establish whether respiratory protection is necessary. The confined space entry plan and testing plan should ideally be overseen by an industrial hygienist to ensure that no potential hazards are overlooked.

Controls for Working in Confined Spaces

Confined spaces include such work areas as tank cars, boilers, silos, underground tunnels, and railroad boxcars. All these spaces have limited entrances and exits and require specific controls to ensure worker safety. Hazards that workers in confined spaces face include toxicity, potential oxygen deficiency, and fire or explosion from flammable or combustible gases or dust. To protect workers, the following actions should be taken:

- Always evaluate a confined space for hazards before workers enter.
- Ensure that the confined space has adequate ventilation.
- Include equipment for suppressing fires and removing smoke and fumes.
- Train workers on safety procedures they need to follow when working in a confined space.
- Institute a buddy system for confined spaces so two workers are always present.

Common Struck by Hazards and Associated Controls

OSHA defines 'Struck by' as an injury caused by the contact or impact of a piece of equipment or object and the injured person. The impact of the object with the person is the cause of the injury. Types of 'Struck by' hazards include: flying objects, falling objects, swinging objects, and rolling objects. These can cause severe injury. Control measures to prevent struck by injuries include enforcing height limits on pallet stacking, toe boards and other barriers, proper training for those operating various equipment and vehicles, etc. Finally, proper PPE is critical, especially hard hats. The hard hat is the best piece of PPE to prevent 'Struck by' injuries due to falling objects.

Flying Objects

Flying objects are a common occurrence at a construction site. Activities that may cause flying objects include **demolition activities** such as wall removals or removal of other infrastructure. Flying objects can also occur if **compressed air** is used to clean up the work site; for example, nails, wood, and concrete debris can become airborne if compressed air is used. Best practices for preventing injuries from flying objects include mandating that all workers wear **eye protection** (safety glasses or goggles) on the work site, along with a **hard hat**. High-risk activities should require the worker to wear a **face shield**. Compressed air should not be used as a method of clean up due to the risk of flying objects and dust. Instead, use vacuums and sweeping.

Falling Objects

Overhead work is common at a construction site. Overhead work could include work on ceilings or roofs, or having materials transported overhead by cranes or hoists. Measures that can be taken to reduce the risk of injury from being struck by a falling object are to **secure** any tools and raw materials by means of fastening them to a fixed object (such as scaffolding) to prevent them from falling. **Debris nets** can also be used beneath the work area to catch anything that might fall. Additional measures include providing **barricades** (safety cones, caution tape, warning signs) to restrict entry beneath and around the work zone. **Personal protective equipment** (hard hats) should always be worn on a construction site.

Swinging Objects

Swinging objects can be found on a construction site when transporting materials overhead using a **crane** or **hoist**. For example, beams or joists must be transported overhead during installation. Best practices are to first **plan the overhead move** and ensure there are no other activities occurring in the immediate vicinity at the same time. All job site participants need to be aware of the overhead activity and the potential for a swinging object. The immediate area under and around the overhead activity should be barricaded off to restrict entry beneath the crane or hoist move. Ensure that all inspections are done on the crane or hoist prior to doing the overhead lift. Also ensure that the movements are done in a slow and controlled fashion to prevent excessive movement of the load.

Rolling Objects

Rolling objects at a construction site can include items such as roofing materials that are supplied by the roll, or a vehicle that escapes control. Controls for being struck by rolling objects include being **aware** of the possibility and ensuring that movements of these items are **planned and controlled**, and discussing the potential for being struck by these objects at daily safety tailgate talks. Ensure that large rolls of material are secured before moving them, and ensure that there are no workers in the movement path. Workers should not walk in front of moving vehicles but should follow behind. When unattended, vehicles should be parked on flat ground, not a slope, with the parking brake engaged.

PPE for Struck-by Hazards

The best practice at a construction site is for all employees and visitors to wear **hard hats** that are approved by the **American National Standards Institute (ANSI)**. OSHA requires hard hats to be worn whenever there is a risk of being struck by falling objects. Two types of hard hats are available: Type I and Type II. **Type I hard hats** are designed to protect from the full force of the impact directly on the top of the head, whereas **Type II** will also protect from a blow that is received on the side of the head or off-center, in addition to one received on the top. Be aware that the suspension of the hard hat should be replaced annually to ensure top performance.

Planning for Overhead Lifts

Using a crane or hoist to perform an overhead lift requires planning. Many factors must be taken into account. For example, what other activities are occurring at the same time? Is any accommodation or rescheduling needed to ensure that the lift doesn't cause additional hazard? Roles and responsibilities for the lift need to be planned, as well as how to communicate during the lift. The capacity of the crane or hoist to carry the weight of the load needs to be verified. The stability of the crane or hoist needs to be ensured. The path the crane or hoist will take needs to be planned out to ensure there are no structures in the path that will be impacted. The weather needs to be considered; if there are high winds, it needs to be determined whether this will cause unintentional swinging of the load.

Caught-in or Caught-Between Hazards and Associated Controls

OSHA defines 'caught-in or between' injuries as harm caused by a person being squeezed, caught, crushed, pinched, or compressed between machinery or parts of equipment. This type of hazard and injury can cause broken bones, amputations, strangulation, or internal injuries. The best control measure for this type of risk is to construct physical machine guards that prevent any body parts from insertion in or near the equipment. Also, workers can help mitigate potential harm by not wearing jewelry (necklaces, and other things that dangle) or loose clothing (un-tucked shirts, etc.).

Cave-in Protection

Trenches can be dangerous due to the possibility of cave-ins. If the trench is **greater than five feet deep**, it is required to be shored up by an engineered protective system unless it is constructed solely of stable rock. Trenches can also be protected from cave-ins by sloping the sides to reduce risk of the walls collapsing. Devices called **trench boxes** can be used to shore up the sides and stabilize them. Any shoring or sloping of a trench must be overseen by a competent and trained person. Trenches must be inspected daily for signs that the walls are stable, and all workers should be informed of trench safety practices. Heavy equipment should not be operated near the open edge of the trench due to the possibility of weakening the walls and causing a cave-in.

Control and Inspection Measures for Scaffolding

Scaffolding is commonly used in the construction industry to access elevated parts of buildings and to provide a work platform for workers and tools. Scaffolds must be constructed to hold at least **four times the expected weight** on the platform. Boards must be of **proper grade** (for example, no plywood) and must not overlap with one another; that is, each board must rest on its own support rail. **Guardrails** must be in place, and **fall protection** (harness) must be used if the scaffold is more than ten feet high. The scaffolding should be inspected daily to ensure that the construction is still sound, the boards do not show signs of wear, and there is no debris or ice on the scaffolding.

Machine Guarding

Machine guarding is the practice of placing **physical barriers** on tools or machinery to prevent human contact with moving parts that can damage body parts. An important part of a **tool inspection program** is examining machine guards for their presence and ensuring they are not removed. On construction sites, commonly encountered machine guards include **blade protectors** on circular saws and **guards** on grinding wheels. Guards must also be in place when there is a risk of exposure to rotating shafts, belts, or gears.

Common Hazards of Mechanized Equipment and Associated Controls

Heavy equipment includes any large piece of mechanized equipment, such as an excavator, loader, backhoe, grader, or roller. Due to its size, power, and design, heavy equipment can create various hazards at a job site.

The heavy equipment driver may have **limited visibility**, particularly when backing up, which can lead to the equipment running into structures, staged material, other equipment, or pedestrians. Pedestrians near moving equipment can get **caught between** the equipment and other objects, resulting in injury or even death. Due to its weight and design, heavy equipment that is used on unstable or uneven ground can tip over, damaging the equipment and potentially injuring the driver. Equipment that digs can strike **buried utilities**, which can cause an electric shock or puncture natural gas lines, creating a flammability hazard. Tall equipment or equipment with extendable buckets or loaders may have clearance issues causing it to strike **overhead objects**, such as structures, trees, or power lines. Finally, there is a potential for **operational failure** where the lifting mechanism or another working piece fails, resulting in a dropped load or falling equipment.

Distracted Driving

Distracted driving is the operation of a motor vehicle while performing any other activity that diverts attention from the driving task. When the driver's focus is not entirely on operating the vehicle, the risk for an **accident** increases significantly. Driver attention can be **diverted** by occupants, navigating unfamiliar territory, eating, drinking, other drivers, or mobile devices. Research has shown that drivers distracted by mobile devices can miss up to 50% of the information in the driving environment. In 2019, distracted driving contributed to 15% of all vehicle crashes in the United States, accounting for over 3,000 deaths.

Protective Measures for Workers Near Heavy Equipment

Due to the size of the equipment and the limited visibility for the driver, especially while backing up, heavy equipment can pose a hazard to nearby workers. Employees who will be on a job site with heavy equipment should be issued **high-visibility clothing** to make them more noticeable to the driver. Vehicles should only reverse if the driver has a clear view of what is behind them, if they use a **spotter**, or if the vehicle is outfitted with a **reverse indicator** to warn nearby pedestrians of the vehicle's movement. If equipment has a cracked or broken windshield that can impact their ability to see nearby workers, the windshield should be repaired or replaced. Equipment lights, including headlights, taillights, and brake lights, should all be functioning to increase the visibility of the equipment to others on the job site. Employees should be instructed on precautions that are necessary when working or walking near heavy equipment.

Controls to Protect Drivers of Heavy Equipment

Whenever heavy equipment is to be on a job site, the employer should make sure that all roadways are safe for travel to prevent tipping. Heavy equipment must be outfitted with a **roll-over protection system (ROPS)** to provide the driver additional protections in the event of a tip. Vehicles that are top loaded (such as dump trucks) must have **canopies** or **cab shields** to protect the driver from material being loaded. All vehicles must have seat belts, which should be used whenever the vehicle is in motion. The number of occupants of the vehicle should never exceed the number of seat belts in the cab. All drivers must be trained on load limits, proper loading, safe operation, pre-shift inspections, and all safety features for each piece of equipment they will be using.

Precautions Around Overhead Powerlines

Whenever there are overhead powerlines present and heavy equipment will be used, signs should be installed to indicate the presence of the lines. Regarding 1926.1411, when equipment is moving beneath power lines of 50,000 volts or less with no load and the boom lowered, the **minimum clearance** between the cab and the lines is 6 feet. For lines of 50,000 to 345,000 volts, the minimum clearance is 10 feet. For lines of 345,000 volts or greater, the minimum clearance is 16 feet. A spotter must be used to ensure adequate clearance between the lines and the equipment. Boom cage guards, insulating links, and proximity warning devices may be used, but the minimum distance must be maintained.

Common Hazards and Controls Associated with Hoisting and Rigging

Personnel Hoisting

When conventional means of moving workers to higher levels are not feasible, a crane can be used to hoist personnel, so long as the following cautionary procedures are observed:

- The crane must be on level ground. The angle between the crane base and the ground cannot exceed **one percent**.
- Total weight of personnel hoisted plus the weight of the platform itself cannot exceed **fifty percent of the rated capacity** of the crane or derrick.
- If workers must be hoisted aloft by crane, the **platform** should be moved in a slow, cautious and controlled manner.
- **Carrying load lines** must be capable of supporting seven times the weight of the load. If anti-rotation rope is used, it should be of a type capable of supporting ten times the intended carry weight.
- All **brakes and locking devices** on the crane must be engaged when the personnel platform is stopped.

Personnel Platforms

When conventional means of hoisting personnel to higher building levels are unavailable or infeasible, a **personnel platform** may be used. OSHA regulations require that such platforms be designed by a qualified engineer or another qualified competent person.

- The platform must be designed to **minimize tipping** when workers move or shift their weight on the platform.
- The platform itself (without guardrails or fall arrest systems attached) must be able to support **five times the maximum intended load**.
- **Guardrails and grab rails** should be installed around the entire perimeter of the platform.
- **Gates** for access must not be outward swinging. A fail-safe device must be installed to prevent accidental opening of the gate or entrance.
- The platform or cage must be of a height permitting employees to stand **upright**.
- The platform or cage must be provided with **overhead protection** to prevent injury from falling objects.

Personnel platforms are designed by an engineer or another qualified person to meet specific OSHA requirements. The load rating must be specified on a plate attached to the platform itself. However, these precautions must be augmented by an attempted lift or "trial lift" at the rated load capacity.

- The platform must be loaded to the **weighted load capacity** (without people) and lifted from the ground.
- During this trial lift, the hoist operator must **test** all controls, systems, and safety devices to see if they are properly functioning.
- A single **load test or trial lift** is sufficient so long as all systems are properly functioning and the crane or hoist is not moved.
- If the crane or hoist must be moved and used in a different area, a **new trial lift** must be conducted at the new location.

After the trial lift of a personnel hoisting platform, a competent person must immediately conduct a **visual inspection** of the crane or hoist to see if any defects resulted from the lift. Among the areas for inspection are:

- The **platform** must be examined to see if it is properly balanced and does not tilt to one side.
- **Hoist ropes** must not be twisted, frayed, or entangled.
- The **linking mechanism** between the hoisting lines and the platform itself should be centrally positioned so as to prevent tilting.
- **Hoisting lines** should be inspected for slackness. Slack lines may indicate that hoist ropes or cables are caught or hung up in drums.

A good safety and health monitoring program at a construction worksite requires constant attention. There are many events which might cause personnel hoist operators to lose focus. Some of the **unsafe conditions** a CHST might observe are:

- A crane operator leaves the cab of his equipment to talk to a co-worker while workers are suspended in the platform.
- A harried site manager urges the crane operator to continue hoisting personnel aloft even when a dangerous wind has begun to blow.
- The crane operator loses communication and visual contact with the workers in the hoist.
- A group of workers takes a shortcut beneath the crane and platform boom in order to reach the area where they've parked their cars.

Safety in the use of personnel hoisting systems is a combination of **preparatory precautions** (proper construction of the platform) and **operating precautions**. With regard to operating infractions, the CHST might notice the following unsafe practices during crane or hoist operation:

- An employee reaches his arms outside the perimeter of the platform or cage during raising and lowering. Employees must remain entirely inside of platforms during operations.
- An employee tries to open the gate or access door to the platform before the platform is landed. The platform must be properly landed and secured to the structure before exit operations.
- Tag lines are not being used.

Pre-Job Inspection of Rigging Equipment

A pre-job safety inspection must be conducted before any rigging job. Ensure all lines to be used are not frayed or broken. Ensure all employees are aware of how the job is planned to be conducted and are aware of their respective roles. Ensure the controls on the equipment operate as planned and that the equipment is parked in a stable surface and will not rock once the load is engaged and lifted. Inspect hydraulic lines for leaks and ensure there are proper levels of hydraulic fluid. Evaluate the need for tag lines and employ them if necessary.

Tag Lines

Tag lines are used in the use of cranes and hoisting equipment to attach to and stabilize the object being lifted. Lifting and moving an object with a crane will cause the load to swing on the end of the cable like a pendulum, which is potentially dangerous to employees. The tag lines are affixed to the load and to the ground to stabilize the load and prevent it from swinging. The tag lines must be made of soft material such as nylon or sisal, not of wire rope. Tag lines are required when the risk to employees of being hit by a swinging load exists or when there are overhead electrical lines that may be hit by a swinging load or crane.

Wire Rope Used in Construction Rigging

The use of wire rope for construction rigging must meet OSHA specifications:

- There cannot be more than **three full tucks** in a wire rope. Other permitted types of connection may be used if they meet safety specifications.
- Wire ropes used for hoisting or lowering must be constructed of a **continuous length of rope** without knot or splice except for the eye splices at the end of the rope.
- The eyes in wire rope slings or bridles must not be comprised of **wire rope clips or knots**.
- Wire rope shouldn't be used if the total number of **visible broken wires** is in excess of ten percent of the total number of wires. Nor can wire rope be used if the rope shows other signs of excessive corrosion or wear.

Hand Signals for Communicating in a Rigging and Hoisting Job

Hand signals should be used whenever the crane operator cannot see the load, landing area, or path of travel or when the load is close to power lines. Signals should be communicated bare-handed. The common signals are as follows:

- HOIST—Raise the arm at the elbow with the index finger pointing up, then move hand in a horizontal circle.
- LOWER—Lower the arm straight down with the index finger down, then move the hand in a horizontal circle, pointing down.
- BOOM UP—Extend right arm to the side with fingers closed and thumb pointing up.
- BOOM DOWN— Extend right arm to the side with fingers closed and thumb pointing down.
- TRAVEL FORWARD—With palm up, move the thumb in the direction of travel.
- STOP—Extend the arm straight out with palm down.
- EMERGENCY STOP—With the arm extended straight out, palm down, move the hand rapidly right and left.

HAZARDS AND CONTROLS FOR HOISTING APPARATUS AND ROPES, CHAINS, AND SLINGS

Hoisting apparatus can range from hand-operated derricks and winches to overhead cranes and aerial baskets. Hazards associated with **hoisting apparatus** include the following:

- Structural failure or tipping if the apparatus is overloaded or used on windy days
- Material falling on people or property below
- Limited visibility
- Use environment such as power lines.

To **reduce the hazards** associated with hoisting equipment, the following actions should be taken:

- Develop setup standards such as ensuring the setup site is level and away from power lines and that the apparatus is assembled properly.
- Ensure the apparatus is not overloaded. Train operators on both everyday and emergency procedures for using their equipment.

Ropes, **chains**, and **slings** are the rigging between the hoist and the load. Hazards associated with this rigging stem from the following:

- Overloading
- Deterioration
- Improper rigging leading to falls

To **reduce these hazards**, the following actions should be taken:

- Store rigging properly, out of sunlight and away from moisture and chemicals that can cause it to deteriorate.
- Inspect rigging regularly to ensure it is not deteriorating or wearing out.
- Follow load capacity charts for rigging to guard against overloading.

Common Hazards and Controls Associated with Crane Operations

Safety Operations Associated with Cranes and Lifting Devices

Construction cranes account for a high percentage of workplace injury and accidents on construction sites. The CHST can implement several strategies to minimize injury from unsafe conditions:

- Before operating, the crane's **hydraulic lift controls** should be inspected and determined to be in proper working order. The **outrigging hydraulic extenders** should be solidly placed and fully extended to guarantee utmost stability.
- The weight of intended loads must be within the crane's **specified load capacity**.
- **Human walking traffic** should be barricaded from the crane's area of operation. At no time should loads be moved over the heads of workers.
- Cranes must be situated and operated at safe distances from **electric power lines** or onsite **distributive power lines** needed for small equipment.
- **Rigging and brakes** must be examined and trial tested before use. Inspect cable, chains, and hook.

Daily Crane Inspection

A daily crane inspection program is an essential part of safe crane operation. The presence of appropriate personal protective equipment (PPE) should be confirmed, and it should be checked that access to the area that the crane will be used is controlled. Check the disconnect switch for correct operation, that the hoist is not loose or broken, that the wire rope is seated properly and not twisted or bent, that the push button controls have no damage, that the controls have the required American Nation Standards Institute (ANSI) tag, that the controls operate as they should (e.g., the crane goes up when that button is triggered), that hooks have no cracks or gouges, that the wire ropes are not frayed, and that the capacity plate is present and visible.

Monthly Crane Inspection

There are additional inspection elements for a monthly crane inspection above the daily requirements. All inspections must be conducted by qualified personnel. Items to be inspected include looking for deformed, cracked, or corroded members; loose bolts and rivets; cracked or worn sheaves and drums; worn bearings, pins, shafts, or gears; and excessive wear on brake system and parts. Check that the load indicators function correctly over the full operating range, and check for excessive wear on chain drive sprockets.

Crane Load Ratings

Manufacturers of cranes design them for a certain load capacity. The load capacity is the maximum that the crane can safely lift; however, it is not good practice to lift approaching the capacity limit unless the crane is designed for that type of use. Cranes are classified as follows:

- Class A: Standby service—crane on standby or used infrequently
- Class B Light service—used at slow speeds and infrequently
- Class C Moderate service—Average 50 percent of rated capacity, five to ten lifts per hour
- Class D Heavy service—Used at 50 percent of rated capacity constantly through the work day
- Class E Severe service—Used approaching capacity 20 times per hour
- Class F Continuous severe service—Used approaching capacity continuously throughout its life

Safe Operation of Mobile Cranes

In order to make a safe lift with a **mobile crane**, the operator needs to know the following:

- Whether the vehicle is level
- Whether the outriggers are extended or retracted
- If the outriggers are extended, whether they are supported by stable ground
- Whether the tires are fully inflated
- The angle of the boom
- The length of the boom and jibs
- The positions the boom will be in during the lift
- The weight of the load

With this information in hand, the operator can use a load chart to determine if the load is within the structural and stability limits of the crane.

Common Hazards and Controls Associated with Material Handling

Materials handling involves lifting, moving, and placing items, whether manually or with equipment. Equipment frequently used for materials handling includes jacks, hoists, backhoes, escalators, derricks, and cranes. As many as 20–25 percent of disabling work-related injuries involve materials handling. Hazards stemming from materials handling differ depending on the material being handled and on the equipment used. Materials may be heavy, toxic, radioactive, hazardous, or flammable: each involves a different set of hazards. Equipment used may be mobile, such as a forklift, and could hit someone. Other times, items can fall off a hoist or a crane can collapse. Electrically powered equipment poses electricity hazards such as shock. Moving items manually also poses hazards. For example, lifting too heavy a load or using improper lifting techniques can cause sprains or strains.

Materials Handling and Storage Safety

Storing and moving construction materials requires a great deal of twisting, turning, lifting, pushing, and bending. According to the National Safety Council, twenty percent of worker injuries are related to **back injury**. To minimize such injuries, OSHA recommends the following relatively simple preventatives:

- Workers should seek help when a load is so bulky or heavy that it cannot be safely moved by one person.
- Holders or handles should be attached to loads to make movement steadier.
- Steel-toed shoes and other protective devices should be worn to prevent injuries to feet from dropped materials.
- Gloves are obligatory to prevent cuts and cracks in the skin which invite injury.
- Walkways must be kept free of debris or extraneous stored materials, especially in areas where workers routinely carry and deliver objects needed in construction.
- Materials which are scheduled to be moved or are already moved should be stacked and racked with blocking which will prevent them from collapsing or rolling onto passing employees.

Preventing Accidents During Materials Handling

Developing and enforcing procedures for materials handling can help **reduce accidents**. Procedures should include how to do the following:

- Select the correct equipment for the job.
- Communicate during the materials transfer, whether using hand signals, two-way radios, or other forms of communication.
- Deal with problems that could occur during the materials handling activity.

The procedures should also include step-by-step instructions for actually completing the materials handling activity. Other actions that can help **prevent materials handling accidents** include the following:

- Create a safe materials handling environment with good lighting, wide aisles, and proper ventilation.
- Institute traffic controls to keep lift areas clear of people.
- Regularly maintain, inspect, and repair equipment used for materials handling.
- Train workers to properly use materials handling equipment, hand signals, and rigging.
- Train workers on safe techniques for manually moving materials, such as techniques for lifting heavy loads.
- Eliminate materials handling whenever possible.

OSHA has regulations which apply to construction and materials handling procedures, but employers or contractors set policy. A good management program designed to **limit worker material handling injury** would include:

- Training in the techniques of proper lifting and conditioning (warm-up and stretching) techniques.
- Training in the use of blocking, stacking, and proper storage of materials should be a requirement.
- Training in the observance of weight limits or maximum capacities for motorized or mechanical equipment should be implemented.

- Employers should set policies of minimum physical requirements which must be met by employees who will be occupied in the movement of heavy materials.
- Employees should be trained in and encouraged to use protective clothing and other equipment designed to prevent exposure to harmful substances.

Safe Lifting Techniques

Training workers on safe lifting techniques is important because improper lifting can cause sprain and strain injuries. As many as 25% of workers' compensation claims are for lower back injuries. The following steps should be performed to decrease the chance of injury when lifting:

- Check the weight of the item to be lifted. If it is higher than the RWL (recommended weight limit), take steps to lighten the load, whether by using equipment to pick up the item or by splitting the load into two or more loads.
- Ensure the floor is not slippery.
- Spread your feet for a more stable stance.
- Keep your back straight.
- Hold the load close to your body.
- If an item is on the floor, bend down, grasp it firmly, and then stand up slowly and steadily.

RWL and LI: RWL stands for **recommended weight limit**. The RWL is the weight that healthy workers could lift for up to eight hours without causing musculoskeletal injuries. To calculate the RWL, you need to multiply the following:

$$LC\ (load\ constant) \times HM\ (horizontal\ multiplier) \times VM\ (vertical\ multiplier) \\ \times DM\ (distance\ multiplier) \times AM\ (asymmetric\ multiplier) \\ \times FM\ (frequency\ multiplier) \times CM\ (coupling\ multiplier)$$

LI stands for **lifting index**. It measures the physical stress associated with lifting. As the LI goes up, the chance of injury also goes up. To calculate LI, you divide the load weight by the RWL. It is important to know the RWL and the LI because they can help you in the following ways:

- Identify tasks where injuries are more likely to occur.
- Help you develop procedures for safe lifting tasks.
- Identify which lifting tasks need to be redesigned first (generally those with higher LIs).

Ultimately, knowing the RWL and LI can lead to safer lifting procedures and fewer injuries. In order to determine the RWL, you must know the weight of the object. You must also know the following task variables:

- *H* (**horizontal distance**)—the distance of the hands away from the midpoint between the ankles.
- *V* (**vertical location**)—the distance of the hands above the floor.
- *D* (**vertical travel distance**)—the number of inches or centimeters the object is lifted.
- *A* (**asymmetry angle**)—the distance from 0 to 135 degrees that a worker turns during the lift.
- *F* (**lifting frequency**)—the average number of lifts per minute, measured over a 15-minute time span.
- *C* (**coupling classification**)—defined as good, fair, or poor depending on the type of grip (such as handles) and the type of container (box or bag; rigid or non-rigid, standard or irregular shape, etc.).

Handling of Drums and Containers for Hazardous Materials

The OSHA regulations in 1910.120 describe the special handling treatment of drums and containers of hazardous substances and contaminated liquids. Prior to the Environmental Protection Act and other legislation designed to protect workers and communities, it was fairly commonplace to store 55-gallon drums by burying. Regulations require that this situation be ameliorated by interventions.

- Whenever possible, drums and containers should be **inspected and tested for contamination** before being moved. If this is impractical, then the containers should be moved to an area which is more hospitable to inspection of contents.
- Currently, there are stringent regulations regarding **labeling and storage** of containers. If an unlabeled or unidentified container is found at a worksite, it should be handled according to existing regulations for the handling of hazardous materials and wastes until the substance is identified.
- Well-planned worksites should be controlled so that **movement of containerized chemicals and substances** is minimal or non-existent.
- Employees occupied in the on-site movement of containers and drums must be **trained** in the potential hazards and informed of the potential for health effects.

Moving and Storing Compressed Gas Cylinders

Compressed gas cylinders are dangerous when dropped, exposed to ignition sources, or corroded. Precautions are to be taken when moving and storing cylinders of compressed gas:

- **Valve protection caps** should be placed over all valves. Valve protection caps are not "handles" and should not be used for lifting and moving cylinders.
- Before hoisting cylinders aloft, compressed gas cylinders must be **stabilized** on a cradle, slingboard, or pallet. Magnets or choker slings may not be used for moving gas cylinders.
- When moving cylinders, they should not be **dropped** or permitted to crash into each other violently. The recommended method is to tilt them and roll them on their bottom edges.

Precautions are to be taken when moving cylinders of compressed gas by motor vehicle.

- It is important to recognize that cylinders of **oxygen** must be separated from **gas** cylinders or other combustible materials when moving them by truck.
- If moved and then stored, the oxygen cylinders must be set apart at a distance of **20 feet or more**. If this cannot be done, then a **noncombustible barrier** must be erected to a height of at least 5 feet. The barrier must have a fire-resistance rating of at least $\frac{1}{2}$ hour.
- Cylinders moved by motor vehicle must be loaded in a **vertical position**. Unless a special carrier is used, gas regulators must be removed and replaced with valve protection caps before the cylinders are moved.
- Unless transported by a dedicated cylinder truck with racks, a **chain or stabilizing device** should be used to maintain cylinders in an upright position.

Preventive Safety Measures for Workers Using a Conveyor Mechanism to Move Materials

There are specific dangers associated with conveyor mechanisms used to move materials from one area to another:

- Hands, gloves, or clothing may be caught in **rollers or pinch points** as materials are moving. Loose flowing clothing should not be worn when working near conveyors.
- **Emergency buttons or pull cords** set up to halt the conveyor must be conveniently located and accessible to prevent minor accidents from becoming major ones.
- Before a stopped conveyor is restarted, a complete **inspection** of the conveyor should be conducted by safety personnel to determine the origin of the stoppage.
- **Safety nets or barricades** should be constructed to prevent materials from falling from conveyors and striking nearby workers.
- Conveyors are intended to move materials, not **people**. Employees should not sit on or attempt to ride conveyors.

Common Hazards and Controls Associated with Material Storage

Controls for Storing Materials Safely

To store materials safely, the following actions should be taken:

- Use good **housekeeping procedures** so that everything has a place where it belongs.
- Don't **stack** materials too high. Stacks that are too high take little force to be knocked over. In addition, if materials are stacked too high, lower boxes can be crushed and the stack could then tip over.
- Use **crossties** for lumber and bagged materials.
- Use **stepping back procedures** when materials are stacked in several rows.
- Use **retaining walls** to restrain bulk materials.
- Use **racks** to keep items like drums, rolls, and piping from falling.
- Use **protective barriers** to keep industrial trucks from running into racks.
- Design storage areas with **wide enough aisles** for industrial trucks if they will later be used to move the stored materials.
- Use **netting** to catch items that may fall from overhead storage areas.

Storing Flammable or Combustible Materials Within a Construction Area

The use of proper storage containers for flammable or combustible materials cannot be overlooked in the interest of worker safety.

- If combustible or flammable liquids must be stored in an amount more than one gallon, only approved **metal safety cans** may be used.
- Flammable and combustible materials must not be stored on **stairwells, hallways, or any passages** where workers routinely travel.
- If **more than 25 gallons** of flammable or combustible liquids must be stored, the method of containment should be constructed solidly and according to the following specifications:
  - Metal containers are preferable. If wooden cabinets are used, they must be constructed of exterior plywood no less than one inch thick. Joints must be rebated and fastened with wood screws to prevent separation. Hinges must be strong and secure and the cabinets should be clearly labeled as "flammable" or "combustible" and painted with a fire-resistant paint.

Storage and Handling of Flammable Materials

Flammable liquids must be stored in approved flammable containers; for example, flammable liquids will generally be supplied in glass bottles or metal drums, not in plastic containers. If a secondary container is to be used (e.g., to transport a small quantity of gasoline), the container must be approved to hold flammable liquids. Flammable liquids must be stored in approved flammable liquid storage cabinets, or in segregated rooms that have separate ventilation and whose walls meet fire resistance ratings. Aisles must be maintained at 3 feet in width, and egress routes must be kept clear in areas where flammable liquids are stored.

Storing Flammable and Combustible Materials Indoors

When storing flammable or combustible materials indoors, the following actions should be taken: Store only small amounts in occupied areas or buildings; larger amounts need to be stored in separate facilities. Follow **National Fire Protection Association (NFPA) standards** when designing storerooms for flammable and combustible materials, including standards for ventilation, static electricity grounding systems, explosion-proof light switches and fixtures, self-closing doors with raised sills, signage, and floor contours. Follow NFPA standards for storage cabinets designed to hold small quantities of flammable and combustible items. When transferring flammables from one container to another, ensure that the containers are touching each other or are connected to a grounding rod or line. Store flammable liquids in closed containers. When dispensing flammable liquids from drums, use gravity or suction pumps rather than pressurizing the drum. Use safety cans to move flammable liquids from their storage area to their point of use. Use plunger cans if you need to wet cleaning cloths with a flammable liquid. Store cloths that have been contaminated by flammable liquids in a small self-closing container that you empty regularly.

Requirements Applying to Specifications for Concrete Mixing, Storage, and Moving

OSHA regulations regarding concrete storage, mixing, and moving operations are designed to protect workers from injuries stemming from uncontrolled flows of concrete:

- Containers and bulk storage bins for concrete should be equipped with **tapered bottoms** to control the volume being poured.
- Concrete silos used to store bulk concrete must be equipped with **mechanical or pneumatic drives** to start and control the flow of material.
- Concrete silos and bulk storage bins should have **prohibited entry systems** which do not permit employees to access them unless the operating systems have been shut down, tagged, or equipped with lockouts which prevent operation.
- Concrete buckets must be equipped with mechanically operated valves which have **safety blocks** to prevent accidental dumping of materials.

Storage Stacking Requirements for Masonry Blocks, Bricks, and Concrete

OSHA regulations in **1926.250** offer the provisions for stacking and storing various types of material: bagged materials, bricks, masonry blocks, and both used and new lumber. The provisions of 1926.250 state:

- **Bagged materials** like concrete, lime, fertilizers, and feed should be stored and stacked by stepping back the layers and cross-keying. Cross-keying refers to the practice of changing the direction of placing the bagged materials and interlocking them with layers below.

- **Bricks** must not be stacked more than seven feet in height. For heights of more than 4 feet, the brick should be tapered back and away at a slope of 2 inches per foot of height.
- **Masonry blocks** stacked higher than six feet must be tapered back a length of one-half block for every layer above the six-foot level.

Control and Storage Measures for Chemicals

Chemicals should be stored separately according to key physical and/or chemical properties. They may be stored alphabetically, but only within a designated group. For example, oxidizing agents should NEVER be stored near organic solvents. Additionally, some chemicals are limited in the amounts that can be stored on-site at any one time. Chemicals must be stored in appropriate containers. For example, some organic solvents will dissolve some types of plastic. Alkaline solutions often etch glass containers. The containers must be appropriately sealed to minimize escape of vapors. Such chemicals are often stored in a cabinet that is vented into the exhaust system. Bottles of chemicals must be scrupulously labeled with the name as provided on the SDS sheet, potential hazards, and the contact information of the manufacturer. Caution signs should be posted at entrances to chemical storage rooms that warn of the particular hazard(s) present. Finally, chemicals must be in a secure location at the proper temperature and pressure needed to ensure stability.

Control and Storage Measures for Corrosive Materials

Corrosive materials are those that can dissolve metal and severely damage human tissue upon contact. They are usually of either a very high pH (greater than 12) or low pH (less than 2) but can also be other types of chemicals. Because they can dissolve metal, they should be stored in plastic containers. Corrosive materials must be stored away from flammable materials and incompatible metals. Acids and bases should be stored separately from one another due to the risk of exothermic (heat-producing) reactions if they are accidentally mixed. If possible, they should be stored in an enclosed cabinet or room with ventilation that does not go into the work area.

Incompatibility

Incompatibility refers to one or more chemicals that when mixed together, create a hazardous reaction. A hazardous reaction can refer to generation of heat (to the point of starting a fire), creation of a hazardous gas (e.g., hydrogen gas, which is flammable), or creation of a toxic gas. Common incompatibilities are strong acids and strong bases (heat and hydrogen gas), ammonia and bleach (toxic fumes), and strong acids with oxidizers. Incompatible chemicals must be stored in well-ventilated areas well apart from one another to ensure they do not mix accidentally.

Hazardous Materials Containment

Hazardous materials containment refers to properly containing hazardous materials in the event of an unexpected spill. Considerations should be made when designing containment as follows:

- Containment capacity: What should the containment capacity be? The design criteria are to contain 110 percent of the largest container or 10 percent of the aggregate volume stored. In addition, if the area is subject to storm water, containment capacity must be provided to contain the storm water also.
- Compatibility: Ensure that the materials of construction of the containment device are compatible with the hazardous materials being contained.

- Inspection frequency: Plans should be made to inspect the containment for spilled material on a regular basis. A corrective action process should be put in place to address any deficiencies discovered.
- Emergency planning: What will the procedure be if material is spilled in the containment? Who will be notified, and how will the material be cleaned up?

If the spill is small and the hazards are known, the spill can generally be cleaned up by personnel using appropriate personal protection gear and methods specific to the chemical involved. Exceptions include mercury and radioactive substances. Spills can typically be handled using specifically designed adsorption spill kits or via neutralization. Powdered neutralizers should be applied to the spill from the outside to the inside of the spill in a circular manner. Depending on the chemical, the neutralized material should be placed in a carefully labeled container for appropriate disposal. All sources of ignition must be controlled if the substance is flammable. At least two people should be present in case the person cleaning the spill needs emergency assistance. If large amounts of chemicals have spilled or there is a possibility the spill can result in an uncontrolled release of a hazardous substance, the designated HAZMAT team must be summoned.

Controls for Preventing or Reducing Severity of Explosions

The exact controls used to prevent an explosion or reduce the severity of an explosion will change depending on the materials and use environment. However, some general guidelines include the following:

- Limit the amount of explosive material stored in any one area.
- If large amounts of explosive materials must be stored, they should be in a remote area.
- Regularly clean areas where explosives are stored so dust will not accumulate.
- Eliminate sources of ignition such as lighters, moving belts, and electrical equipment.
- Store fuels and oxidizers in separate locations.
- Install extinguishing and suppression systems to put out fires before an explosion can occur.
- Use vents in any containers where explosive mixtures could lead to explosions.
- Use distance and barriers to separate explosive materials from each other and from populated areas.
- Train anyone handling, using, and distributing explosives.

Common Hazards and Controls Associated with Housekeeping

Housecleaning, Housekeeping, and Sanitation

Housecleaning refers to the process of cleaning an area: sweeping, wiping surfaces, throwing away trash, etc. **Housekeeping** means putting things away where they belong. Every tool, piece of equipment, and material should have a designated storage area. Hazardous materials should have special storage areas designed specifically for them. Sanitation means ensuring the facility is clean and germ-free.

It includes such issues as the following:

- Safe drinking water
- Clean working toilets
- Clean areas to prepare and eat food
- Insect and rodent control

Housecleaning, housekeeping, and sanitation can all reduce hazards. For example, putting equipment where it belongs means that no one is going to trip over that item and the item cannot fall on anyone. Providing a sanitary environment also reduces disease transmission and can lessen exposure to hazardous substances.

Effective housekeeping is important from both environmental and safety standpoints. Poor housekeeping can increase the volume of waste generated by causing spills and by accumulating out-of-date materials that must be discarded. From a safety standpoint, poor housekeeping can create slip, trip, and fall hazards; can block emergency exits; and can block emergency equipment such as fire extinguishers. Setting **housekeeping policies** and enforcing them, along with employee training and an inspection program, are the key elements of an effective housekeeping program. Effective housekeeping begins with only necessary materials in the work area, with everything having a designated location, and with clear communication of these guidelines (all elements of a 5S program). Employees should be trained to clean up as they go along throughout their shift and should set aside time at the end of shift to clean up completely. All tools should be replaced into storage lockers and containers returned to storage locations. Surfaces should be kept clean. Effective housekeeping promotes a safe workplace and increases job satisfaction because it is more pleasant to work in an orderly environment.

Controlling Hazardous Chemicals

Part of housekeeping is removing dust and cleaning up spills. **Hazardous dust** needs to be regularly vacuumed from surfaces so that it will not become airborne. A vacuum that traps the contaminants must be used. Materials can become airborne when they are loaded, unloaded, or transferred to other containers. Transferring within a closed transfer or exhaust system can protect workers from being exposed to airborne dust and vapors. For **liquids**, it is also helpful to use drip pans or containers to collect overfill spills and leaks. Leak detection programs can include both automatic sensors and regular visual inspections of valves and pipes. The sensors can trigger alarms or even shut down a process. Repairing leaks quickly minimizes any potential exposures. Workers and supervisors who use hazardous chemicals need to receive training on what hazards they face and how to protect themselves. This training will help them stay safe and is also required by OSHA standards and by law in some states.

Safety Hazards Associated with Poor Housekeeping

There are many safety hazards associated with poor housekeeping. There is a potential for slips, trips, and falls if there are objects lying on the floor such as boxes, tools, cords, and equipment. Material spilled on the floor also increases the chance of slipping, whether the material is oil, water, or a solid substance. Poor housekeeping and management of oily rags and other flammable materials is a fire hazard if the materials are not properly stored or disposed of. Materials that are precariously stacked overhead are a hazard to employees, who can be struck by falling objects. Improper storage and handling of hazardous materials can result in exposure to harmful organic vapors. Improperly managed containers also pose a spill hazard. Good housekeeping practices contribute to a safe work environment, are more pleasant to work in, and increase employee morale. They also contribute to increased productivity because employees are able to find objects they need when they need them.

Importance of Housekeeping for a Safe Workplace

Proper housekeeping and cleanliness are key contributors to a safe workplace. Good housekeeping diminishes slip, trip, and fall hazards by ensuring that tools and work items are put away when not in use. Good housekeeping practices clean up spilled materials promptly, getting them out of the aisles so that employees won't slip on them. Strict adherence to the practices will help to minimize waste and assist in proper dust control. Proper waste management and storage practices can also minimize fire hazards. Good housekeeping can also help manage tools appropriately and ensure they are not damaged by improper storage practices.

Materials Given Special Consideration Regarding Fire Hazards

Housekeeping is important in controlling fire hazards, especially storage and handling of rags and wipes soaked in flammable materials. If one accumulates rags and wipes soaked with solvents or similar materials, they should be stored in metal canisters rated for flammable rag storage and emptied daily into suitable containers. If rags and wipes soaked with flammable solvents are stored in piles outside of the metal canister, they can build up heat and create conditions favorable to spontaneous combustion. It is also important to store flammable liquids in an orderly manner in cabinets or rooms rated for flammable chemicals.

Contribution of Poor Housekeeping to Fire Hazards

Poor housekeeping can increase fire hazards in a number of ways. Poor housekeeping can lead to accumulation of oily or solvent-laden rags and wipes, which are a combustible hazard. Poor housekeeping will also contribute to a higher level of combustible debris lying around, such as boxes, pallets, and packaging material. Poor housekeeping can lead to blocking of paths of egress and exits, which contributes to fire and life safety hazards. In addition, poor housekeeping can lead to accumulation of dust on surfaces and in electrical panels, which can be combustible or cause a short in an electrical panel that leads to a fire.

Requirements for Maintaining Clearance Around Electrical Panels and Fire Extinguishers

Occupational Safety and Health Administration (OSHA) regulations do require that electrical panels have a minimum of 30 inches of clearance around them for voltages between 120 to 250 volts. Higher voltage panels require greater clearances (29 CFR 1910.303). In most work areas, lines are painted on the floor to indicate the clearance required as a means of reminding employees not to store materials in these areas. In contrast, there is no specific requirement for a clearance around fire extinguishers; however, OSHA regulations do state that fire extinguishers must remain accessible at all times and that employees must not have to travel more than 75 feet to reach a fire extinguisher if they are working in an area with potentially combustible materials (29 CFR 1910.157).

Common Hazards and Controls Associated with Powder Actuated Tools

A powder-actuated tool is a forceful **nail gun** that uses a small amount of a **chemical propellant** to force the nail or other fastener into the substrate. The mechanism is somewhat similar to the operation of a firearm. Powder-actuated tools use a cartridge with the fastener loaded in. They can be of either high or low velocity; however, the difference in velocity does not imply a difference in the hazard posed by this equipment. Various types of fasteners are available for use with these tools (such as nails or threaded studs), and these fasteners can be driven into steel or concrete.

Powder-actuated tools deliver sharp fasteners such as nails at high velocity and force. Therefore, they pose an extreme **puncture hazard**. The force and mode of operation also pose a **noise exposure hazard**. There is also a risk of **flying objects** if the fastener is not properly delivered into the substrate. They pose a **vibrational hazard** from the vibration of the tool. For these reasons, eye and hearing protection must be worn when using these tools, along with foot protection. Specific training in the use of these tools must be completed and documented.

Guards and Controls for Powder-Actuated Tools

Powder-actuated tools should be **inspected** prior to each use. The manufacturer's literature should be reviewed to determine the engineered safety features of the tool. Ensure that the tool is equipped with an **ignition safety feature** so that it doesn't fire when pressed against a surface. Ensure that it is equipped with a **drop safety feature** so that it will not fire if dropped. Inspect the fastener cartridges before use to ensure they are intact and not compromised. There are no particular guards required, but warning signs stating "Power-Actuated Tool in Use" must be on the tool case, on the tool itself, and posted in the work area.

Materials to Avoid When Using Powder-Actuated Tools

Powder-actuated tools drive sharp objects at high force to fasten two pieces together. As such, any material that is **brittle** or will **shatter** should not be fastened with a powder-actuated tool. Examples of these types of material are cast iron, marble, glazed ceramic tiles, glass blocks, slate and other types of natural stone, and clay bricks. If powder-actuated tools will be used to fasten into **concrete**, the concrete must have appropriate **compressive strength** to ensure proper fastening. Concrete of compressive strength greater than 8,500 psi is not suitable for powder-actuated fasteners.

Common Hazards and Controls Associated with Hand and Power Tools

Hazards and Injuries Associated with Tools and Machines

Hazards associated with tools and machines include the following:

- Being struck by a tool, machine, or machine part
- Being struck by flying debris from materials the tool or machine is acting on, such as concrete chips
- Getting caught in a machine or tool

Less direct hazards also exist. For example, if a machine requires repeated motion, cumulative stress disorders can result. For powered tools, electrical hazards are also present. Tools and machines are a major source of injury, including the following:

- Cuts
- Abrasions
- Puncture wounds
- Tissue tears
- Crushing injuries
- Fractures
- Carpal tunnel syndrome
- Bursitis
- Tendonitis

As many as 8% of injuries involving lost time are related to hand tools.

Hand Tool Inspections

The term **hand tools** refers to handheld tools that are not powered. This includes screwdrivers, hammers, axes, scissors, box cutters, and pliers. All tools should be inspected before every use to ensure they are in good operational order. One should examine hand tools to ensure the metal is not cracked or thinning to the point that it may break during use. One should also check the handles to ensure they are firmly seated and that any grip coverings are in place and not loose. Tools that are comprised of several parts should be checked to make sure that the parts are securely attached.

Power Tool Inspections

Power tools refer to handheld tools powered by electricity, batteries, compressed air, or internal combustion engines. These tools include power saws, drills, nail guns, grinders, cutting tools, sanders, and riveters. Some inspection elements are common to other handheld tools, such as checking that the metal parts are not cracked or worn and that all parts are firmly seated and not loose. The electrical cord should be inspected to make sure it is not worn or the insulation frayed. The electrical prongs should be inspected to make sure they are not bent or missing (especially the round grounding prong). If there are cutting blades in the tool, they should be inspected to make sure they are not too worn or in need of sharpening. Dull blades are unsafe due to the extra force required to use them that may cause one's grip to slip.

Pneumatic Power Tools

The development of pneumatic hoses to power construction tools is a relatively recent innovation in the construction industry and has presented a new potential for safety problems. OSHA guidelines in 1926.302 recommend:

- Pneumatic power tools should be locked and secured in place to prevent the tool from being forcefully ejected from the power source. Safety clips should be used for this purpose.
- Pneumatic power nailers and fasteners operating at more than 100 psi should have built-in safety mechanisms which prevent triggering unless the muzzle is pressed against the desired surface.
- Compressed air at more than 30 psi. cannot be used for cleaning purposes.
- Pneumatic hoses more than $\frac{1}{2}$ inch inside diameter must be equipped with safety devices which reduce pressure if hose lines are ruptured.

Management System to Ensure Tools Are Inspected Regularly

An effective management system to ensure tools are inspected regularly must first identify what types of tools are on hand and in which locations and departments. A schedule of inspection and a responsibility matrix (by job title) must be developed. Finally, the system must include inspection checklists to document findings and follow up corrective actions. Calendar systems can be used to schedule inspections and ensure deficiencies are corrected in a timely fashion. Coupled with routine inspections, random inspections of tools in use on the floor are important as cross-checks to ensure the inspection program is operating properly.

Common Hazards and Controls Associated with Asbestos Exposure

Asbestos in the Workplace

Asbestos is the common name for a group of minerals like chrysotile, amosite, crocidolite, tremolite, and other mineral compounds used and found in the construction workplace:

- Surfacing materials contain asbestos substances. These may be sprayed, brushed, or troweled onto surfaces. Wallboard, floor tiles, and mastics are all commonly found to contain asbestos materials.
- Asbestos products are found in thermal insulation piping, tanks, ducts, boilers, and other heating and cooling equipment structures.

- There are four classes of asbestos-related construction activity listed according to the degree of hazard and contact. These categories range from Class I to Class IV.
- Each exposure class has a specific individual protocol for handling ACM (asbestos containing material). OSHA requires that the process of handling ACM be managed by "competent persons." In this case, competent persons are those who have been certified in asbestos handling procedures.

Employers must control worker exposure to asbestos to the extent possible, monitoring air quality, providing protective equipment, and medical testing for workers who are subject to Class I, II, or III work activities for 30 days or more.

- Employers must retain training records up to one year extending from the last day of a worker's employment.
- Employers must retain medical records during the period the worker is employed and for 30 years afterward.
- Employers must establish hygienic decontamination areas for workers.
- Employers must establish limiting control zones which permit access only to authorized persons. Eating, drinking, smoking, and applying cosmetics shall not be permitted in restricted areas to prevent ingestion or other contamination.

Approaches for Dealing with Asbestos

The first decision to make regarding **asbestos in the workplace** is whether it should be left alone or abated. If the asbestos is not friable and is not in an area where it will be disturbed, it may be safer to leave it alone. However, if the asbestos is friable or is, for example, in an area scheduled for renovation, then the **asbestos-containing material (ACM)** must be removed, enclosed, or encapsulated. If the decision is to leave the asbestos alone, an asbestos-monitoring program must be put in place to track whether the condition of the asbestos changes. Any work done with asbestos needs to be performed by trained and licensed personnel wearing high efficiency respirators and disposable personal protective equipment that covers the entire body. *Removing* ACM is the most expensive choice, but once the ACM is removed, no further work or monitoring is needed. When ACM is removed, the area where it is located must first be enclosed in tough plastic walls. That area then must be ventilated by HEPA filtered negative air machines. The ACM is covered by a liquid solution that keeps asbestos fibers in place and is then put in leak proof containers for disposal. If ACM can be encapsulated, it is sprayed with a sealant that binds the asbestos fibers together. The sealant hardens into a tough skin. ACM can also be *enclosed.* This involves placing airtight walls around the ACM. Warning signs need to be placed on the walls and their location needs to be marked on the building plans. If ACM is enclosed or *encapsulated*, an asbestos-monitoring program needs to be set up to track whether the condition of the asbestos changes.

Common Hazards and Controls Associated with Lead Exposure

Lead Contamination in Construction Sites

Though the use of lead-based paint in housing construction was banned in the 1970s, there are still a surprising number of construction processes through which a worker may be exposed. Workers at the highest risk of lead exposure are those involved in:

- Abrasive blasting, which has the capacity to put lead in the air
- Welding, cutting, and burning on steel structures, which create lead particles
- Careless or accidental burning of materials containing lead

- Using lead-containing mortar
- Power tool cleaning without dust collection systems
- Rivet busting
- Cleanup activities where dry expendable abrasives are used
- Movement and removal of abrasive blasting enclosures
- Manual dry scraping and sanding
- Manual demolition of structures
- Heat-gun applications
- Power tool cleaning with dust collection systems

OSHA PEL Standards Which Apply to Worker Exposure to Lead in the Atmosphere

The PEL or Permissible Exposure Limit for airborne lead particles is 50 micrograms of lead per cubic meter of air expressed a 50 $\mu g/m^3$ per eight-hour period.

- The standard PEL for lead is based on a TWA or time-weighted average.
- A daily exposure may exceed 50 $\mu g/m^3$ if the worker hours are averaged so as not to exceed the PELs for an eight-hour period.
- If worker exposure to lead will be maintained for a ten-hour workday, the PELs are reduced to 40 $\mu g/m^3$ according to a formula.

The purpose of the revised formulas for workdays exceeding eight hours is to recognize true workplace demands.

Increasing Protections Against Lead in the Workplace

Hygiene and access to washing facilities are an important part of an employer's lead monitoring and control program. The objective of any program is to minimize the amount of lead uptake through ingestion, inhalation, or through skin contact.

- The employer must maintain separate areas for the consumption of food or beverages. The same applies for areas where tobacco products are used. Cosmetics should not be applied in an area where lead is present since these absorb particles of lead which then come in contact with skin.
- When airborne lead exposure exceeds the PEL, the employer must provide clothing changing areas and showers. Disposable protective clothing must be isolated and disposed of appropriately. It is important to prevent cross-contamination between PPE and clothing.

It is both vital and mandatory to monitor, train, and educate workers as to the dangers of lead exposure in the workplace. The following measures should be taken to protect workers and their families:

- Remove protective and disposable work clothing at the end of the work shift. Avoid cross-contamination with street clothing.
- Wash hands, face, and any other exposed body parts before changing into street clothing. A shower is the best means of washing lead particles from the body if such facilities are available.
- Wash hands, face, arms, and other exposed body parts prior to eating, drinking, smoking, or applying cosmetics.
- Store street clothes in an area which is safe from contamination from lead dust.

Common Hazards and Controls Associated with Noise Exposure

The primary hazard of noise is **hearing loss**. Noise-induced hearing loss is related to the amount of time a person is exposed to the noise, the decibel level, the frequency, and whether the noise is continuous or intermittent. Types of noise-induced hearing loss include the following:

- **Temporary threshold shift** is caused by a short exposure to loud noise.
- **Permanent threshold shift** is caused by continuous exposure to noise.
- **Temporary or permanent acoustic trauma** is caused by a loud noise, as from an explosion.

In addition to hearing loss, noise can also interfere with communication. Noise can make it difficult to hear warnings and sirens and even to communicate normally. It also interferes with learning, causes a startle response and other physiological problems such as high blood pressure and ulcers, and makes people irritable and frustrated.

A **baseline audiogram** is an audiogram (hearing test) done after a quiet period and used as a comparison for future audiograms to see if hearing thresholds have changed. The **decibel** (dB, often adjusted to dBA) is a unit that defines the intensity of sound. **Hazardous noise** is any sound that can cause permanent hearing loss in a specified population. OSHA has established allowable daily amounts of noise that workers can safely be exposed to. The **noise dose** is the percentage of this daily exposure that a particular sound meets. A **noise-induced hearing loss** is any sensorineural hearing loss that can be linked to noise and for which no other cause can be identified. The **threshold of hearing** is 1 dBA. This is the weakest sound that a healthy human can hear in a quiet setting. The **threshold of pain** is 140 dBA. This is the maximum level of sound that a human can hear without pain.

Continuous and Impact Noise

Since decibel levels of noise and duration of exposure are primary factors that indicate the type and degree of noise level protection required for construction site workers, it is important to know the difference between "impact" and "continuous" noises as OSHA regulations describe them. When sound levels are measured at a construction site, permissible worker exposures are clearly described in OSHA tables.

- For a full 8-hour workday, a worker must not be exposed to noise greater than an 85-decibel level. As the level of sound begins to exceed that level, the length of permitted exposure time begins to decline. If continuous sound levels rise to 115 decibels, a worker is permitted no more than 15 minutes of exposure, according to the OSHA table.
- Impact noise levels are short-duration sounds emitting from such equipment as nail guns, air hammers, or punch press equipment. A worker must not be exposed to decibel levels that exceed 140 dB of impact noise.

Measuring Noise Levels in a Work Area

There are two basic types of noise metering instruments available for on-site noise measurement to ensure compliance with OSHA noise regulations:

- **General area noise meters** are designed to measure the noise level in the areas where they have been placed. These instruments may be moved to other areas but have the disadvantage of providing only generalized data. Workers are often moving through a construction site, where the noise levels vary according to the types of activity and machinery being used. A generalized metering scheme provides little data about the individual exposure record.
- **Personal noise meter devices** are designed to be worn by working personnel. Personal noise meters obtain individual readings. This type of noise metering has a practical advantage, since it is common for workers to move through various work areas with different levels of noise production.

Whatever type of device is used, it is important to follow proper handling procedures:

- Check batteries prior to use. Batteries should be removed from any sound meter that will be stored for more than 5 days.
- Use a windscreen to protect the microphone in areas where dust and airborne debris may be lifted into the air by excavating machinery or equipment.
- Never cover the microphone pickup with plastic or other material that will distort reception and invalidate the readings.
- Calibrate the meter periodically using the manufacturer-recommended calibration device.

Determining Adequate Level of Hearing Protection

The OSHA hearing conservation standard establishes a permissible exposure limit to noise at 85 decibels (dBA). Exposures above 85 dBA require employees to wear hearing protection in the form of muffs or earplugs. Hearing protection devices have a Noise Reduction Rating provided by the manufacturer. Theoretically, wearing the device reduces the noise exposure by that number of decibels. In practice, one applies a safety factor. This calculation requires the "A-weighted" sound-level readings from a noise meter (the A-level readings are expressed as "dBA" and represent the exposure as heard by the human ear, with very high and low frequencies screened out). The calculated employee exposure wearing the hearing protection is as follows:

$$Estimated\ Exposure\ (dBA) = TWA\ (dBA) - (NRR - 7)$$

If C-level noise readings are available, the Noise Reduction Rating can be subtracted directly from the dBC reading to determine estimated employee exposure.

Hearing Protection for Workers Exposed to Noise Levels Exceeding 90 dB

All employees exposed to an 8-hour time-weighted average noise of 85 dB or greater should be provided with appropriate hearing protection at no cost to them.

- Employers must ensure that hearing protection is being worn by workers subject to high-decibel sound levels (85 dB) during an 8-hour time-weighted period.
- Hearing protection is required of any employee or worker who has experienced a threshold shift in hearing level.

- Employees must be allowed to select the type of hearing protectors that they need from a variety of approved protectors.
- It is the employer's responsibility to reevaluate and inspect hearing protection effectiveness when changes in work activity expose the worker to greater levels of noise.

Control Methods for Exposure to Excessive Levels of Noise

Engineering controls are the first consideration for control of excessive exposure to noise. Sound-dampening foam products can be used to line enclosures and dampen noise from machinery. Where noise can't be avoided, its effects can be reduced by grouping and enclosing noisy processes in a soundproof area so that people working in other areas are not bothered. Design features that can help reduce noise include the following:

- Controlling the direction of the source
- Reducing flow rates
- Reducing driving forces
- Controlling vibrating surfaces
- Using barriers and shields
- Building with sound-absorbing materials

In addition to controlling the noise source itself, you can also protect the workers by requiring **protection** such as earplugs or earmuffs.

Earplugs are soft foam plugs that expand to block noise from entering the ear canal. Earmuffs surround the entire ear and prevent noise from entering. Either may be used to achieve the proper level of noise reduction. Devices are assigned a Noise Reduction Rating (NRR) by the manufacturer after testing to determine how effectively they block sound.

Common Hazards and Controls Associated with Radiation Exposure

There are three types of radiation: alpha, beta, and gamma. Proper shielding is the best protection from the harmful effects of radiation.

Alpha radiation is made up of small, positively charged particles. Due to an alpha particle's size and characteristics, it cannot travel great distances and is easily stopped by clothing, gloves, or a piece of paper. Care must be taken not to breathe alpha particles into the lungs, as they will damage the lungs and potentially cause cancer.

Beta radiation has a higher energy level than alpha radiation and therefore a greater ability to penetrate surfaces. However, it can be easily blocked by a layer of aluminum foil or similar material.

Gamma radiation has the highest energy level and has the most ability to penetrate the human body. Lead is the most common shielding used for gamma radiation; several centimeters is usually sufficient to shield from gamma radiation. Water can also be used for shielding; nuclear reactors and power plants use water shielding. Several feet of water are needed to shield from gamma radiation.

Non-Ionizing Radiation

OSHA defines **non-ionizing radiation** as a series of energy waves from electric and magnetic fields. As with other radiation sources, these types of radiation travel at the speed of light. There are several sources of non-ionizing radiation:

- Power lines, electrical equipment, and household electric wiring produce extremely low frequency (ELF) radiation.
- Radio frequency and microwave radiation may be absorbed through the skin and body. At high levels, this type of radiation can cause tissue damage due to the heating of the body. Radio transmitters and cell phones are examples.
- Infrared radiation is emitted from such sources as heat lamps and lasers that operate from this wavelength.
- Visible light radiation can damage eyes and skin if it is too intense. The visible light spectrum is what we normally see with our eyes.
- UV or ultraviolet radiation sources come from welding arcs and UV lasers.

Ionizing Radiation

Ionizing radiation is radiation that can produce ions when it interacts with atoms and molecules. Types of ionizing radiation include x-rays, alpha particles, beta particles, gamma radiation, and neutrons. Ionizing radiation can come from natural sources, such as cosmic radiation and radioactive soils, and from artificial sources, such as television sets, diagnostic x-rays, and nuclear fuels. Exposure to ionizing radiation damages human cells, especially rapidly developing cells. It is especially dangerous for infants and children, who have the most rapidly developing cells. Exposure to high doses of ionizing radiation causes radiation sickness, characterized by weakness, sleepiness, stupor, tremors, convulsions, and eventually death. Low doses may cause more delayed effects, such as genetic effects, cancers, cataracts, and shortened life span.

Preventing Damage Caused by Radiation

The damage caused by ionizing radiation depends on the type and dose of the radiation, the tissue and organs exposed, and the age of the person being exposed. The best way to control potential damage is to limit the amount of radiation people are exposed to by limiting the amount of source material. It is also important to limit the amount of time people are exposed to radiation. Other ways to reduce exposure to ionizing radiation include the following:

- Increasing the distance between people and sources of ionizing radiation.
- Using shielding such as air, hydrogen, and water to protect people from sources of radiation. The material used as a shield depends on the type of radiation.
- Using barriers such as walls and fences to keep people away from sources of radiation.
- Using liners and protective materials to keep contaminated waste from leaching into groundwater.

A key concept in radiation protection is to keep the radiation dose "as low as reasonably achievable" (ALARA). Regardless of whether the radiation is alpha, beta, or gamma particles, both time and distance can be used to reduce the amount of exposure. Work schedules should be arranged so that exposure time is limited. Reducing the time exposed to radiation reduces the absorbed dose in a directly proportional manner. Distance away from the radiation source can also be used as a means of reducing absorbed radiation. As one moves away from the radioactive source, the exposure decreases according to the inverse square of the distance, as illustrated by the following formula: intensity is proportional to $\frac{1}{x^2}$ where x is the distance from the radiation source.

Safety Controls for Radiation

Warnings need to mark any areas where ionizing radiation is located, as well as equipment that uses ionizing radiation. In addition to signs on these areas, flashing lights and audio signals can serve as additional warnings.

Evacuation is a tool used to remove people from an area where a significant amount of ionizing radiation has been released.

Security procedures need to be in place to keep sources of ionizing radiation from getting into the wrong hands. Procedures can include physical monitoring, controlled entry and exit, and manifest systems.

Dosimetry measures people's exposure to ionizing radiation. It is necessary because we cannot see or feel this radiation. People who work with or near ionizing radiation also need training about its hazards and how to protect themselves and others.

System design and analysis can help prevent dangerous exposure to ionizing radiation by anticipating and preventing possible sources of failure.

Radiation Counters

There are several different methods and devices for detecting and determining exposure levels of ionizing radiation. Common devices used to determine exposure to individuals are dosimeters or film badges that are worn by the person. When exposed to ionizing radiation, elements within the film react and, when developed, result in a permanent radiation exposure record. **Dosimeters** similarly react to the ionizing radiation and register a measurement of the exposure, and—in the case of a pocket dosimeter, which uses an ionization chamber—they can be read immediately. **Ionization chambers** measure radiation exposure by the conductivity created by interaction of radiation with the enclosed gas. These chambers are used by themselves, in dosimeters, and are modified for use in Geiger-Mueller counters. **Scintillation instruments**, which use phosphors or crystals to produce light in reaction to radiation, have been found to be extremely sensitive and useful when detecting low levels of exposure.

Common Hazards and Controls Associated with Silica Exposure

Jobs or Tasks That May Be Associated with Silica Exposure

Silica (chemical formula SiO_2) is a naturally-occurring mineral that is the main ingredient in **glass**. Silica can exist in both crystalline and amorphous forms. **Crystalline silica** is a health hazard because the dust produced from the material forms sharp shards that can be inhaled. They can lodge deep in the lungs and cause **silicosis**, which is a debilitating and incurable disease. In the construction industry, crystalline silica may be encountered in cutting and fabricating granite and other naturally-occurring rocks, in concrete, drywall, bricks, grout, sand, and tile. Any activity using these materials that creates dust creates a potential exposure to crystalline silica that needs to be evaluated.

Elements of a Written Silica Exposure Control Plan

The purpose of an exposure control plan is to identify potential materials and tasks that could expose workers to **crystalline silica dust** and present the methods that will be used to prevent and control that exposure. The written plan must formally assign responsibility to a qualified, competent person to carry out the elements of the exposure control plan. The plan must specify procedures that will be used to control exposures, such as the types of saws that will be used in cutting silica-containing materials to minimize dust. The plan must detail the housekeeping procedures that will be used to clean up dust that potentially contains silica while preventing re-entrainment into the atmosphere (for example, wet sweeping or HEPA filter vacuums). The plan must also account for the medical monitoring program offered to employees to provide chest x-rays and spirometry. The plan must also outline the employee training with regard to silica exposure and controls, and must outline the records that will be maintained in support of implementing the control plan.

Engineering Controls for Silica

Engineering controls for any safety hazard are always preferable to control with personal protective equipment alone. The **OSHA silica exposure rule** mandates engineering controls when using saws to cut material that potentially produces an exposure to silica. The saw must be fitted with an **integrated water delivery system** that wets the material during the cutting process to reduce the quantity of dust produced. The use of engineering controls must be augmented by use of **respiratory protection** if the sawing task will be done for more than four hours per day.

Housekeeping Controls for Silica

Housekeeping is important to minimize exposure to silica. Best practices for housekeeping are designed to prevent build up, disturbance, and re-entrainment of any dust that may contain silica. Dust should be cleaned up regularly using wet sweeping or vacuuming with a HEPA-filter vacuum. **Wet sweeping** involves misting the area to be cleaned with a light mist of water (for example, from a garden sprayer) and then sweeping up the material. The swept-up material should be stored in a covered container. Vacuums with **HEPA filters** are necessary to ensure that the dust isn't emitted into the atmosphere in the exhaust of the vacuum. Using compressed air to blow dust from surfaces should be expressly forbidden.

Recordkeeping Requirements of the OSHA Silica Rule

The OSHA silica rule requires employers to keep records to demonstrate compliance. **Exposure assessments** for silica exposure must be kept, showing that the exposure was assessed to determine if it exceeded the action level of 25 $\mu g/m^3$ (averaged over an eight-hour day). Records of medical monitoring assessments of workers must also be maintained. In the context of the silica exposure rule, **medical monitoring** must be offered, including chest x-rays and lung function tests (spirometry). **Spirometry records** are also required if employees will be required to wear respirators. Records of respirator fit tests and respirator training must be maintained in the event employees are required to wear respiratory protection.

Common Hazards and Controls Associated with Chemical Exposure

CHST Responsibilities in the Area of Hazardous Chemicals

The CHST has responsibility for worker protection from chemical hazards in the workplace. Accident preventive strategies pertaining to hazardous chemicals may include the following:

- A spill prevention plan should be formulated for each type of chemical containing product used in the workplace. The safety data sheets (SDS) for that particular product should be reviewed and considered in devising the specific plan for that product. A spill-kit should be placed close to where the chemical will be used in advance.

- Workers in the vicinity of hazardous chemicals and materials should be informed of proper handling techniques and protective measures. This information can be provided through oral instruction and reinforced by appropriate signage. Language and instructions should be clear and understandable.
- In addition to verbal directives, workers should be encouraged to read the SDS themselves and to take the recommended personal protection measures. The CHST should emphasize the importance of following SDS requirements for safe storage and disposal of hazardous materials.

Permissible Exposure Limits (PEL)

The major distinction between Permissible Exposure Limits and TLVs is that TLVs are guidelines set by an organization, while PELs are regulations set by OSHA. They are both valuable in ensuring a safe workplace, but PELs have the power of legal enforcement behind them. The PEL is measured and reported as a time-weighted average (TWA). The OSH Act (General Duty Clause) was enacted in 1970 and requires employers to provide a safe working environment, which includes ensuring that toxins are maintained below the PEL determined for the chemical (or other guideline if OSHA has not determined a PEL for the substance).

Threshold Limit Values (TLV)

The American Conference of Governmental Industrial Hygienists, Inc (ACGIH) has developed guidelines, called Threshold Limit Values, some of which have been adopted by OSHA, that estimate the limit that a worker may be exposed to a substance during a standard 40-hour workweek without experiencing undesirable effects. The guidelines are based on inhalation data obtained from scientific journal articles. The two most common units are $\frac{\text{mg}}{\text{m}^3}$ and parts per million (ppm). The formula for converting between the two units is:

$$\text{ppm} = \frac{\frac{\text{mg}}{\text{m}^3} \times 24.45}{MW}$$

In this formula, MW references the molecular weight of the substance.

$$MW = \sum[(atomic\ mass) \times (subscript)]$$

The units of MW are $\frac{\text{grams}}{\text{mole}}$. Care needs to be taken in interpreting reported TLV values. Some values are averaged over a typical 40-hour work week (TWA), some are taken as 15 min (STEL) snapshots, four times a day. The "ceiling" TLV is the maximum exposure at any time for any duration.

Important Terms Related to Hazard Exposure

- **Dose threshold** is the minimum dose of a substance needed to produce a measurable effect. Thresholds are determined, in part, by observing changes in body tissues, growth rates, food intake, and organ weight.
- **Lethal dose** is the amount of a substance that is likely to cause death. The lethal dose of a substance is determined by testing on animals.
- **Lethal concentration** is the amount of an inhaled substance that is likely to cause death.
- **Latency period** is the amount of time between exposure to a chemical and observable effects from that chemical. The effects may be immediate, as with a chemical burn, or delayed, as with cancer.

- **Acute exposure** refers to disease or effects that occur after only one exposure to a chemical while **chronic exposure** refers to disease or effects that occur only after repeated exposure to a chemical.
- **Local effects** are effects from a chemical that injure the skin, eyes, or respiratory system.
- **Systemic effects** are effects from a chemical that damage organs or biological functioning.
- **Carcinogens** are substances that produce cancer in animals or humans.
- **Mutagens** are substances that change the genetic structure of an animal or human, affecting the health of future generations.
- **Teratogens** are substances that cause a fetus to be malformed.
- **Irritants**, including **nuisance dusts**, are substances that irritate the skin, eyes, and the inner linings of the nose, throat, mouth, and upper respiratory tract. These substances do not do any irreversible damage.
- **Asphyxiants** are materials that displace oxygen, interfering with breathing and oxygen transport in the blood.
- **Narcotics** and **anesthetics** keep the central nervous system from operating correctly. Used at the right dose, these substances cause no serious or irreversible effects. However, if too much of the substance is used, unconsciousness or death can result.

ENGINEERING CONTROLS FOR PROTECTING PEOPLE FROM EXPOSURE TO HAZARDOUS CHEMICALS

Engineering controls include substitution, isolation, and ventilation. **Substitution** means substituting a non-hazardous or less hazardous material for a hazardous material. For the substitution to be effective, the new material needs to work as effectively as the old material. **Isolation** means creating a barrier between workers and the source of contamination. This barrier could be a glove box that encloses the hazardous material or a separate enclosure that workers can only access remotely. Isolation can also mean separating hazardous and non-hazardous processes. **Ventilation** protects workers from airborne contaminants. With general ventilation, fresh air replaces or dilutes contaminated air. Local exhaust ventilation captures contaminants before they reach people and moves the contaminants to a safe area for treatment.

CHEMICAL SUBSTITUTION

Substitution of safer chemicals for more toxic or dangerous chemicals should always be explored as part of harm reduction. A safer chemical is one that is less toxic to workers or poses fewer physical hazards such as flammability. A safer chemical can also result in a product that is safer for use by the end user or public. Using safer chemicals can save a company money on personal protective equipment (not required if employees are not exposed). It can also greatly decrease associated hazardous waste disposal costs if toxic waste does not have to be disposed of.

VENTILATION

Ventilation refers to the movement of air and exchange of fresh air for trapped air. Ventilation is important in all indoor environments but is especially important if toxic or harmful vapors are in the work area. For example, in a laboratory setting, ventilation hoods are installed, and workers are instructed to use all volatile solvents inside the ventilation hood. The ventilation hood is equipped with a fan to pull air out of the hood and exhaust it outside. This keeps the vapors out of the work area and away from the breathing zone of the worker. In industrial environments, ventilation hoods operate similarly and evacuate vapors to the exterior of the building, either with capture technology or without depending on the contaminant. The American Conference of Governmental Industrial Hygienists (ACGIH) publishes "Industrial Ventilation: A Manual of Recommended

Practice for Design, which outlines ventilation recommendations for types of chemical atmospheres and work environments and is the standard reference text.

Administrative Controls for Protecting People from Exposure to Hazardous Chemicals

One work modification that can help protect workers from **chemical hazards** involves reducing an individual worker's *exposure time* to hazardous chemicals during any one work shift. For example, workers can share an activity or task so that each worker remains below the exposure limit. When it comes to chemical hazards, one aspect of *personal hygiene* involves cleansing any contaminated skin whether the skin was contaminated during regular work tasks or because of an accident or spill. Procedures should be in place to specify which soaps and cleaners to use for which chemicals. There may also be specific washing stations and showers for different purposes. Hygiene may also include washing eyes that become contaminated as well as safe areas for changing clothes and for eating and drinking. *Personal protective equipment* provides a final level of protection against chemical hazards. Such equipment may include protective clothing, eyewear, creams and lotions, and respiratory equipment.

Chemical Process Safety Management

Chemical process safety management is an OSHA regulation designed to provide safe operation of highly hazardous chemical processes, such as in refineries and large chemical manufacturing plants. The regulation may be found in 29 CFR 1910.119. The overall required elements must document the process, the chemicals used, and the reactor vessel construction and must conduct a hazard analysis that takes into account potential failure paths and how catastrophic releases and explosions can be prevented. In the case of pressure relief systems, consideration must be made to install and document the type of pressure relief system, and it must be ensured that the valve materials are compatible with the chemicals in the process. Pressure relief systems may be **spring-loaded**, which provide simple overpressure protection for a tank or piping system. A second type of pressure relief system is a **balanced bellows** or **balanced piston system**, which is appropriate to use when superimposed backpressure is variable. Chemical compatibility of tanks, piping, valves, and tools with the reactant and product chemicals must be investigated and documented.

In chemical process safety, detailed process flow diagrams of the entire process must be produced. The process chemistry must be thoroughly documented, and maximum inventories of chemical reactants, intermediates, and products must be accounted for. Safe upper and lower limits of temperature, pressure, and flows must be calculated and documented. Once this is completed, a robust management of change to the process must be implemented that includes a full assessment of potential impacts of process changes. For each change, a documented technical basis for the change must be completed along with an analysis of its impact on employee safety, a list of operational procedures that will need to be altered due to the change, a timeline for the change, and a consideration of authorization needed to implement the change. The hazard assessment portion of chemical process safety can be any accepted method that considers potential failures of the process and how to prevent a resulting catastrophe. Examples of acceptable hazard assessment methods include failure mode analysis, fault tree analysis, and a hazard and operability study.

Common Hazards and Controls Associated with Working in Extreme Temperatures

Wet Bulb Globe Thermometer Index

WBGT or Wet Bulb Globe Thermometer index readings are taken with thermometers which measure wet-bulb temperature, dry-bulb temperature, and globe thermometer temperature. The readings obtained are then fed into a formula or index. The formula or index stated below is an index of outdoor WBGT readings:

$$WBGT = 0.7 \times NWB + 0.2 \times GT + 0.1 \times DB$$

- $WBGT$ refers to the wet bulb globe temperature index.
- NWB refers to natural wet-bulb temperature.
- DB refers to dry-bulb temperature ("regular" air temperature)
- GT refers to globe thermometer temperature.

Thermometers used for these measurements should be placed upon a stand which permits free air-flow. Placement should be representative of real worksite conditions.

Heat is measured in calorie units. The impact of heat upon the human body can depend on a variety of factors like humidity, direct sunlight, and even the metabolic heat caused by work activity. The WBGT measures:

- Globe temperature: This refers to the temperature inside a blackened, hollow globe. It measures simple radiant heat.
- Natural wet bulb (NWB) temperature: This measures the effects of evaporation and convection. In other words, the amount of air movement or convection over a human body contributes to its stress protections. NWB measurements employ a "wet sensor," essentially a wet cotton wick.
- Dry bulb (DB) temperature is measured by a thermal sensor, such as an ordinary mercury-in-glass thermometer, that is shielded from direct radiant energy sources.

Health Hazards Associated with Hot Weather Work

Heat exhaustion and **heat prostration** are different names for the same illness. They are caused by a victim failing to drink enough water to replace fluids lost to sweat when working in a hot environment. Symptoms include the following: cold, clammy skin; fatigue; nausea; headache; giddiness; and low-volume, highly concentrated urine output. Treatment involves moving the victim to a cool area for rest and replacing fluids. **Heat cramps** are muscle cramps during or after work in a hot environment. They occur because of excess body salts lost during sweating. Treatment involves replacing body salts by drinking fluids such as sports drinks. **Heat fatigue** occurs in people who aren't used to working in a hot environment. Symptoms include reduced performance at tasks requiring vigilance or mental acuity. Victims need time to acclimate to the hot environment and training on ways to work safely in a hot environment.

A **heat illness** is any illness primarily caused by prolonged exposure to heat. **Heatstroke** occurs when a person's thermal regulatory system fails. Symptoms include lack of sweating, hot and dry skin, fever, and mental confusion. Victims need to be cooled immediately or loss of consciousness, convulsion, coma, or even death can result. **Sunstroke** is a type of heatstroke caused by too much sun exposure. **Heat hyperpyrexia** is a mild form of heatstroke with lesser symptoms. **Heat syncope** affects individuals who aren't used to a hot environment and who have been standing for a long time. The victim faints because blood flows more to the arms and legs and less to the brain. The victim needs to lie down in a cool area. **Heat rash** is also called prickly heat. It occurs when sweat glands become plugged, leading to inflammation and prickly blisters on the skin. Treatment can include cold compresses, cool showers, cooling lotion, steroid creams, and ointments containing hydrocortisone. During treatment, victims must keep their skin dry and avoid heat.

HEATSTROKE AND APPROPRIATE FIRST-AID MEASURES

Heatstroke is an affliction often occurring in construction zones. Workers must be able to recognize its symptoms and to deliver an appropriate first-aid response. The human body afflicted with heatstroke loses its ability to adapt to heat stresses and to regulate itself appropriately.

Heatstroke is characterized by:

- A markedly elevated body temperature, generally greater than 104 °F
- A loss of mental alertness and disorientation are likely to occur
- Hot and dry skin in some instances, though sweating may be profuse
- Headache, rapid heartbeat, rapid and shallow breathing, and blood pressure vacillation

Immediate treatment for heatstroke:

- Move the person out of the sun and into a shady area.
- Direct air onto the person with a fan or newspaper. It is necessary to cool the person and lower the body temperature. Cool, damp cloths or sheets may be used, or cool water sprays.
- Attend the heatstroke victim until professional help arrives. Do not leave the heatstroke victim unattended. Unless instructed by a physician, a heatstroke victim should never be sent home alone.

CONDUCTING A WORKSITE INVESTIGATION OF HEAT STRESS

Safety officials conducting an investigation into a worksite where heat stress injuries are reported should take the following steps:

- Review the on-site OSHA Log or any other existing reports of heat stress injury.
- Interview employees to determine whether the employer has taken action to protect workers from heat stress. Are water supplies adequate? Are cooling areas available? Does the employer provide training in heat stress indicators?
- Make a visual inspection of the site by walking around it to assess potential problems. Determine the locations of heat sources like machinery, furnaces, or boilers.
- Take heat measurements at various locations. Use the heat stress index to determine the wet bulb globe temperature (WBGT).

MEASURING FOR HEAT STRESS RESPONSE

Individual heat stress monitoring is recommended when the worker is under heavy loads (burning 500 kcal/hr) and the temperature is at least 69.8 °F. Oral temperature and water loss must be measured, along with heart rate monitoring.

- Oral temperature must be taken before the employee drinks water. Drinking water before taking the temperature will have the effect of "cheating" the thermometer. The work cycle should be shortened for any worker whose oral temperature exceeds 37.6 °C or 99.68 °F.

- Body water loss measurements are taken by weighing the worker at the beginning and end of each day. Body water weight loss should not be more than 1.5 percent of total body weight per day. Fluid intake should be increased if loss approaches that amount.

Controls for Reducing and Eliminating Heat Stress and Thermal Injuries

The keys to reducing heat stress and thermal injuries are:

- Control the source by keeping heat sources away from occupied areas.
- Modify the environment through ventilation, shielding, barriers, and air conditioning.
- Adjust activities by making the work easier, limiting time spent in hot environments, and requiring periodic rest breaks.
- Provide protective equipment such as water-cooled and air-cooled clothing, reflective clothing, protective eyewear, gloves, and insulated materials.
- Incorporate physiological and medical examinations and monitoring to identify high-risk individuals.
- Develop a training program to help workers acclimatize to hot environments and learn safe work habits.

Controls for Reducing and Eliminating Cold Stress and Thermal Injuries

A cold environment can be measured according to the air temperature, humidity, mean radiant temperature of surrounding surfaces, air speed, and core body temperature of people in extremely cold temperatures. The keys to preventing injury from **cold environments** are as follows:

- Modifying the environment by providing heat sources and using screens or enclosures to reduce wind speed
- Adjusting activities to minimize time in cold areas and requiring regular breaks in a warm area
- Providing protective clothing with insulated layers that both wick moisture away from the body and provide a windscreen
- Providing gloves, hats, wicking socks, and insulated boots to protect vulnerable extremities
- Allowing employees time to become acclimated to the cold environment
- Training employees on practices and procedures for staying safe in a cold environment

Health Hazards Associated with Cold Weather Work

Trenchfoot occurs when a person spends an extended time inactive with moist skin, at temperatures that are cold but not freezing. Bloods vessels in the feet and legs constrict, causing numbness, a pale appearance, swelling, and pain. Treatment involves soaking the feet in warm water. However, the numbness can last for several weeks even after the feet are warmed. **Chilblains** are an itching and reddening of the skin caused by exposure to the cold. Fingers, toes, and ears are the most susceptible. Gentle warming and treatment with calamine lotion or witch hazel can lessen chilblains. Itchy red hives can occur in some people when their bodies develop an allergic reaction to the cold. The hives may be accompanied by vomiting, rapid heart rate, and swollen nasal passages. Cold compresses, cool showers, and antihistamines can help relieve the symptoms.

Frostbite and hypothermia are the most dangerous cold hazards. **Frostbite** occurs when the temperature of body tissue goes below the freezing point. It leads to tissue damage. The amount of damage depends on how deeply the tissue is frozen. Severe frostbite can lead to the victim losing a damaged finger or toe. Frostbitten skin is usually white or gray and the victim may or may not feel pain. To treat frostbite, the damaged body part must be submerged in room-temperature water so it can warm up slowly. Hypothermia occurs when a victim's body temperature drops below normal. Symptoms include shivering, numbness, disorientation, amnesia, and poor judgment. Eventually, unconsciousness, muscular rigidity, heart failure, and even death can result. Warm liquids and moderate movement can help warm a victim who is still conscious. An unconscious victim needs to be wrapped warmly and taken for medical treatment.

Common Hazards and Controls Associated with Vibration and Impact Exposures

Exposure to vibration can be either whole-body or isolated to a body part, most commonly **hand and arm vibration** caused by use of power tools. Vibration jostles nerves and muscles in unnatural ways and, over time, can cause adverse health effects. Repeated exposure over long periods of time results in **musculoskeletal disorders** such as tenosynovitis and carpal tunnel syndrome. These disorders are the result of inflammation of the tendon and its sheath due to repeated use or vibrational exposure. Whole body vibration exposure can cause more generalized symptoms such as fatigue, headache, and loss of balance. Over time, the effects can be debilitating, yet difficult to diagnose and ascribe to only one task or activity.

Measurement of Vibration

Vibration is measured in meters per second squared (m/s^2) as a measure of acceleration. The higher the value, the greater the exposure to vibration. The harm caused is a **dose-response relationship** in that people exposed to greater vibration for longer amounts of time potentially suffer more effects. Instruments can measure this acceleration for **impact** (a jackhammer, for example) or **non-impact tools** (a saw or other power tool). Manufacturers of power tools publish data on the vibration magnitude of their tools. There are no currently established exposure limits for vibration, so it is important to evaluate vibration exposure as part of a **comprehensive risk assessmen**t, since actions can be taken to control exposures before harm is caused.

Design of Tools to Reduce Vibration Hazards

The best solution to a safety hazard is always to **engineer** the hazard out of the job. In the case of a job that exposes a worker to vibration through hand tools, first consider how the job could be done with fixed tools. When this is not possible, aspects of tools can be **redesigned** to reduce vibrational hazard. For example, a different tool with lower vibrational risk could be used. Perhaps wheels or disks on grinding tools could be replaced or better balanced to reduce vibration. If the tool is powered by compressed air, perhaps fewer pounds per square inch could be used. This will reduce the vibrational force of the tool. Also check for different hand tools with a lower intrinsic vibration. Inquire with manufacturer representatives to find a better solution.

Vibration Hazard Training

Workers need to be aware of the risks posed by vibrational hazards. They need to be trained in the proper **grip** to use with a hand or power tool. They need to know how to keep their tools in top **operating shape** so less force will be needed for use (for example, making sure blades are sharpened and grinding wheels regularly replaced). Workers should **dress** warmly enough to ensure circulation to their hands or other body parts affected by vibration. Workers should **grip** tools just as hard as necessary. Gripping too hard will result in more vibrational energy being transferred to the worker's hands and arms. Workers should wear the recommended **gloves** for the job; some gloves can reduce the impact of vibrational exposure.

Recommended Exposure Limits for Vibration Exposure

There are no currently established OSHA exposure limits for vibration. However, the American Conference of Governmental Industrial Hygienists (ACGIH) has developed a **Threshold Limit Value (TLV)** for hand-arm vibration that recommends a maximum eight-hour day vibration exposure total value of $\mathbf{4\ m/s^2}$. Maintaining this level will keep most workers from developing symptoms associated with vibration exposure. As a point of comparison, jackhammers are in the $8 - 15\ \mathrm{m/s^2}$ range, a chainsaw is typically in the $6\ \mathrm{m/s^2}$ range, and a random orbital sander is in the $7 - 10\ \mathrm{m/s^2}$ range. In order to use these tools without causing harm to the workers, job rotation and exposure limitation must be used to reduce exposure to stay below recommendations.

Common Biological Hazards and Controls

Pathogens

A **pathogen** is a biological organism that causes a disease in humans. Pathogens require a host for nutrients and protection from the environment. The pathogen utilizes the host body's resources to thrive and replicate. A pathogen may come from the environment (such as a fungal spore), an animal (rabies), or another person (hepatitis A).

The primary mode of causing harm to employees is through infections. Viruses, bacteria, and fungi can cause infections, particularly in immunocompromised individuals. Bacteria, once established, can release toxic byproducts that can harm the host. Parasites can divert nutrients from the host, eventually causing starvation and death. Insects can spread viral and bacterial infections, which can lead to sickness and, in some cases, death.

Classes of Pathogens

Pathogens fall into one of four classes: viruses, bacteria, fungi, and parasites.

- **Viruses** consist of a piece of genetic code protected by a protein coating. They require a living host in order to replicate. The virus attaches itself to a cell and hijacks the replication process, forcing the cell to replicate the virus. The cell eventually bursts, releasing viruses to invade nearby cells. Viruses cause infections and can be transmitted from one person to another.
- **Bacteria** are single-cell organisms that have a nucleus, but do not have a cell membrane. Bacteria come in a variety of forms, with some being beneficial, if not necessary, for human survival. Bacteria replicate independent of the host, and pathogenic bacteria can release toxic agents into the host organism or cause infections.
- **Fungi** are single-cell organisms with a thick membrane that cause disease in humans. Fungal infections can be mild, such as athlete's foot, or severe, such as aspergillosis. Fungi are prevalent throughout the world, with only about 300 being pathogenic.
- **Parasites** rely on the host for nourishment and, if left unchecked, may kill the host by diverting nutrients. Human pathogenic parasites fall into three main groups:
  - **Protozoa**: single-celled organisms that replicate within the body
  - **Helminths**: worms that can live inside or on the body
  - **Ectoparasites**: parasites that live on the outside of the body, including ticks and mosquitoes

Protection from Pathogens

The method of protecting an employee from biological pathogens depends on the pathogens that the employee may encounter.

Vaccines may protect employees from certain pathogens, primarily viruses. The use of insect repellent can deter disease-carrying mosquitoes and flies. Proper handwashing and sanitation practices can protect against bacterial and parasitic infections. Fungal spores can be addressed by using respiratory protection, such as filtering facepieces or an air-purifying respirator with a HEPA filter attachment.

The Globally Harmonized System of Classification and Labeling of Chemicals (GHS)

Hazard communication refers to the Occupational Safety and Health Administration (OSHA) Hazard Communication Standard found in 29 CFR 1910.1200. The Hazard Communication Standard governs the requirements to notify workers of chemical hazards faced at work and to provide information on protection from hazards. GHS refers to an international standard developed by the United Nations to guide hazardous chemical labeling, warning systems, and safety data sheets. The GHS was developed to standardize hazard warning terminology, pictograms, and safety data sheets worldwide so that international commerce could be improved and language barriers overcome. The OSHA Hazard Communication Standard includes the requirements of the GHS. As of 2015, all workplaces in the United States are required to have safety data sheets available on site that conform to the GHS system and to use these in their notification and training programs.

SDS (Safety Data Sheets)

Safety Data Sheets (formerly Material Safety Data Sheets) provide information on the physical and chemical properties of a substance as well as potential health and environmental concerns. OSHA requires that all chemicals be labeled appropriately and that SDS be readily available in the workplace. The hazard communication standard also requires employees to be trained, and for the employer to maintain records of the training given. The format for SDS includes sixteen sections. The required sections are as follows:

I—Identification
II—Hazard Identification
III—Composition/Information on Ingredients
IV—First Aid Measures
V—Firefighting Measures
VI—Accidental Release Measures
VII—Handling/Storage Requirements
VIII— Exposure Controls/Personal Protection
IX—Physical/Chemical Properties
X—Stability/Reactivity
XI—Toxicological Information
XII—Ecological Information
XIII—Disposal Considerations
XIV—Transportation Information
XV—Regulatory Information
XVI—Other Information

SDS provide a number of indicators for possible health threats of a particular chemical. They are required to provide all known information regarding carcinogenicity of a substance (known or potential cancer-causing risks). Carcinogenic risks are published in the National Toxicology Program report (NTP), the International Agency for Research on Cancer (IACR), and the Occupation Safety and Health Administration (OSHA). Toxicity levels are indicated by numbers called the LD_{50}

and the LC_{50}. LD_{50} refers to the dose at which 50% of the test subjects were killed. LC_{50} is the lethal concentration at which 50% of test subjects were killed. Dosages are typically normalized to include the mass of the possible toxin divided by the mass of the test subject. LD_{50} values may also include descriptors that indicate the mode of administration of the dose (intravenously or orally) and the timeframe for death after administration. Limits for exposure to a particular chemical are also provided. These can be measured as the OSHA permissible exposure limit (PEL) and/or the Threshold Limit Values (TLV), which are published by the American Conference of Governmental Industrial Hygienists (ACGIH).

SDS often recommend the usage of chemical protective clothing (CPC). Protective eye goggles with splash guards and air vents should be used when handling chemicals. Face shields should be used when working with large quantities of a substance and are most effective when used in conjunction with safety goggles. If the mode of possible hazard is through contact and/or absorption on skin, appropriate gloves should be worn. Gloves are chosen based upon their permeability to and reactivity with the chemical in use. Personal respiratory equipment may be indicated if fume hoods do not provide adequate ventilation of fumes or airborne particulates. Body protection depends on the level of protection needed and ranges from rubberized aprons to full suits that are evaluated for their permeability and leak protection. Closed-toed protective shoes should always be used when working with chemicals.

Requirements for Labels

The term "label" under the GHS of Classification and Labeling of Chemicals refers to the label on the container. Under GHS, it's required to contain certain elements; these requirements apply whether the label is affixed by the manufacturer or whether the chemical is placed into a smaller, secondary container in the workplace. The label must include the identification of the chemical, the manufacturer's name and US contact information, the applicable GHS pictograms, the applicable signal words (either "danger" or "warning," as applicable), and precautionary statements (measures to reduce risk from exposure to the chemical).

Pictograms

The pictograms used in the GHS system are simple pictures used to convey hazards posed by the chemical. They are meant to be universally understandable by people with diverse language and reading fluencies. They are as follows:

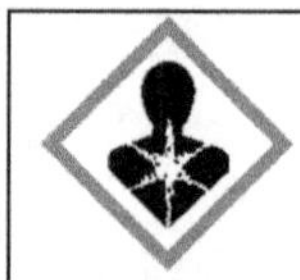 *Health Hazard*	 *Flammable*	 *Sensitizer/ Irritant*
 *Gas Under Pressure*	 *Oxidizer*	 *Corrosive*
 *Reactive*	 *Environmental Hazard*	 *Poison*

Signal Words

Under the GHS hazard communication and safety data sheet system, the term "signal word" is used to describe one word that summarizes the degree of danger posed by the substances. There are only two signal words: "danger" and "warning." The word "danger" is used for more hazardous substances that present immediate hazards such as flammability, reactivity, poison, and so on. The word "warning" is used for lesser hazards such as irritants, environmental hazards, and less toxic substances. The signal word is used on the label to provide a quick and easily understandable indication of the degree of hazard posed by the substance.

Basic Safety Through Design

The hierarchy of controls refers to the preferred methods of controlling health and safety hazards. In order of most preferred first, these include the following:

1. Elimination is completely eliminating the hazard through process changes.
2. Substitution is substituting a lesser hazard, for example, changing from an organic solvent parts washer to an aqueous parts washer.
3. Engineering controls are physical modifications to a work station that serve to reduce or eliminate hazards. An example of engineering controls is to install duct work and exhaust ventilation to remove fumes from the breathing zone of the worker.
4. Administrative control refers to worker management as a means of controlling the hazards. An example of administrative controls is job rotation to limit an employee's exposure to repetitive motion.
5. PPE is garments or auxiliary equipment used to protect workers. Examples of PPE include respirators, gloves, Tyvek suits, welding hoods, and steel-toed boots. This is the last item in the hierarchy because PPE is often uncomfortable to wear, and its effectiveness relies on employee compliance, whereas the other methods of hazard elimination do not.

Elimination of Hazards and Substitution to Mitigate Hazards

Elimination of hazards refers to making process changes that completely eliminate the hazard rather than developing a work-around or protective device. An example would be changing a process to eliminate solvent use. Another example of elimination of hazards would be to change the way parts are delivered to an assembly line that eliminates the need to lifting heavy boxes. **Substitution** is another mechanism to mitigate hazards. Examples of substitution include substituting chemicals in a process that are of lower toxicity than the original chemical or substituting a tool that requires awkward, forceful grasping with one that uses a more ergonomically favorable grip.

Engineering Controls

Engineering controls to mitigate occupational hazards are changes in the way the process is designed or the physical controls on the process that make it unlikely or impossible to be injured by the hazard. An example of engineering controls is installing ventilation systems to remove hazardous vapors from the worker's breathing zone, eliminating chemical or hazardous dust exposure. Another example is to install permanent access platforms with proper guardrails to provide elevated machinery access that will eliminate the hazard of using an aerial lift or ladder to access at heights. Another example of an engineering control is to install a switch on an electrical testing device or press brake that requires two hands to activate it; this eliminates the possibility of inserting one's hands into the point of operation while it is operational.

Administrative Controls

Work practice controls are administrative in nature and seek to control risk by policies and procedures rather than physical barriers. Examples of work practice controls are as follows: implementing a policy that any lifting of objects more than fifty pounds requires two people or instructing an employee operating a grinder to always wear a face shield during the grinding. Clearly, work practice controls are potentially not as effective as physical controls because they rely on employees following instructions and policies, whereas physical controls do not present the option to circumvent the control and be exposed to the risk.

Risks Associated with Multiple Trades Working Simultaneously in Work Area

Project Planning Tools

Project planning involves creating a written plan of the steps to complete a project with a timeline. There are several software packages and visual representation tools available for project planning. These tools can be used to lay out a **project timeline** step by step, allowing potential hazards and risks posed by simultaneous work to be assessed. For example, a **Gantt chart** represents various tasks across time with bars showing how long each task lasts, enabling visualization of tasks that occur at the same time. Then one can list hazards posed by each activity during a given time period, linking to the areas the work activities will be occurring.

Bystander Hazards

When multiple tradespeople are working on a site, communication is essential to avoid bystander hazards. When a person not directly involved in the work activity is inadvertently injured by the work activity, this is a **bystander hazard**. For example, a bystander who is unaware that heavy equipment is working in the area could walk into the equipment path and be struck. A bystander could also be struck by flying debris during a demolition activity or by falling items from overhead work. Measures to prevent bystander hazards include holding a central **site safety and communication meeting** each day to ensure that all workers are aware of the day's planned work, discussing potential hazards, and planning controls.

Principles of Ergonomics as Applied to Construction Practices and Material Handling

Ergonomics is the study of the use of the human body in the work environment. It refers to how the body is configured, the motions it is required to take, and the forces applied to muscles and bones by the work. Ergonomics seeks to adapt the work demands to make them as easy as possible for the body to endure to prevent injury. Anyone who has participated in physical activities and been sore the next day understands the effects of overexertion. Overexertion at work over long periods of time can lead to injuries that can be costly to treat. Even in the absence of an injury situation, improved ergonomics can improve productivity and comfort in the workplace.

Good **ergonomics** improves output and performance and reduces error and accidents by making the environment comfortable and user-friendly. The four general principles of ergonomics that apply to safety engineering are as follows:

- **People versus machines**. This principle states that people are better at some jobs, such as reasoning inductively and handling unexpected occurrences, while machines are better at other jobs, such as repetitive operations and deductive reasoning.
- **Change the job, not the person**. People have limits, and not recognizing those limits can cause errors, hazards, and accidents. It is better to change a job, equipment, or environment to fit the person rather than trying to change the person to fit the job.
- **Work smart**. Productivity can be improved and errors reduce by finding new and better ways to do a job.
- **People are different**. People differ in their age, height, weight, reaction time, strength, coordination, attitudes, etc. Designers and managers need to adjust jobs accordingly.

Organizational, Behavioral, and Psychological Influences on Ergonomic Injuries

Organizational, behavioral, and psychological factors can influence the likelihood and severity of ergonomic injuries. **Organizational factors** include how the work processes are designed and laid out, the time demands that are placed upon the workers, and how long the cycle times are. Workstation design is the most important organizational factor. **Behavioral factors** that impact ergonomic injuries include observing proper lifting behavior, observing the buddy system for lifts over 25 pounds, and observing proper neutral postures. Poor postures may cause the employee to feel pain, which results in uneven muscle strains. The most important **psychological contributor** to ergonomic injuries is stress. Stress and perceived lack of control lead to poor job satisfaction and promote a feeling of monotony at work. These stress factors can increase perceptions of pain, decrease blood flow, create muscle tension, and therefore increase the likelihood of an ergonomic injury.

The Four Major Ergonomic Stressors

When considering the likelihood of an ergonomic injury, tasks must be evaluated with regard to four major stressors.

- **Repetition**—How often does a certain movement have to be repeated during the work day? Higher numbers of repetitions create more stress on bones, muscles, and tendons, thus increasing likelihood of injury.
- **Force**—How much force must be applied when performing the task? Force also includes how much weight must be lifted. Any lifting greater than 25 pounds is considered heavy. The heavier the lifting (or force), the greater the likelihood of injury.
- **Posture**—This refers to the orientation of the body or body part. Neutral postures are the least stressful. The farther from neutral the body is required to be, the greater the likelihood of injury.
- **Vibration**—This refers to the amount of vibratory movement required. Vibration is generally caused by holding power tools (e.g., a drill or jackhammer). Exposure to vibration for long periods of time can damage nerves.

All of these **risk factors** combine to produce an overall risk of injury. The common theme among all of the risk factors is the degree to which the body is required to *deviate* from its natural, neutral state. The greater the deviation or demand on the body, the greater the likelihood of injury. An additional contributing factor is the amount of rest allowed between exposure to risk factors. Over time, with insufficient rest, tendons and ligaments can become inflamed and painful and can lead to an ergonomic injury. Ergonomic injuries sometimes require surgery to correct. This is a cost to both the employer and the employee.

Proper Lifting Technique

Use of proper lifting techniques can prevent back strains in the workplace. The goal of proper lifting is to put as little stress on the spine as possible. The spine is meant to be straight, and lifting with a bent, curved-over spine puts extreme stress on it, even if the weight lifted is relatively small. To properly lift, place the feet hip-distance apart, and face the object to be lifted. Squat to pick up the object, then pull the object close to the body at the waist to move it. Stand up, using the strong thigh muscles to accomplish the work of lifting. Do not twist the spine when lifting or moving the object. Reverse the process to put down the object. Peer assessments should be conducted regularly to reinforce proper lifting techniques as people tend to get lazy and revert to poor lifting techniques.

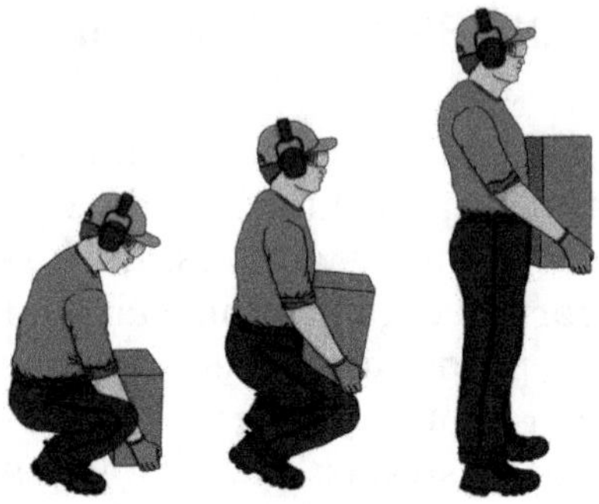

Neutral Standing Posture

A neutral standing posture can help prevent fatigue and muscle strain. The feet should be kept shoulder-width apart for optimal support. Shifting one's weight from one foot to the other is a sign of fatigue that indicates inattention to neutral standing posture. Hips should be in line with the shoulders and the knees. The arms should hang comfortably at one's side, and the shoulders should not be hunched or tensed. The neck should stay straight (head not bent over), and the spine should stay straight, not twisted.

Repetitive Stress Injuries

A repetitive stress injury is caused by using the same muscles, tendons, and bones repetitively several times over the course of a work day. Repetitive stress injuries occur because of strains on muscles, tendons, and ligaments. If caught early, and the stress is eliminated early, they are almost always reversible. However, repetitive stress injuries can also lead to employees having to have surgery and spend weeks away from work, followed by months of light duty. A common example of a repetitive stress injury is carpal tunnel syndrome, which is a disorder of the tendons and nerves in the wrists and hands experienced by some workers who must do repetitive work with their hands (such as data entry operators, typists, or hand assembly workers).

Techniques to Decrease Likelihood of Injury

Repetitive stress injuries are caused by using the same motion (and thus, the same muscle groups and bones) over and over again. Another contributing factor is the force applied to the motion. To reduce the likelihood, it is important to quantify the loads and the numbers of repetitions. Configure the work areas to ensure that items used most often are within close reach and that long reaches or limb extensions are only required infrequently. Break up loads into smaller weights. It is also important to evaluate whether employees' work stations are configured to ensure neutral postures are used most of the time. In addition, job rotation can also be important to changing the motions employees have to make over the course of the work day.

Minimizing or Eliminating Lifting Hazards

Minimizing or eliminating lifting hazards should be pursued as an effective injury prevention tool. Tools to do this include ensuring that all loads lifted with regularity are less than 25 pounds, and heavier loads that must be lifted must use a buddy system (two-person lift) or assistive device (forklift, drum lifter, or vacuum-powered lifter). Pallet jacks that can be height adjusted to the work level can minimize or eliminate the need to carry a package from floor height to work table height, which decreases the risks posed. Carts can be used to transport items that must be moved from one location to another rather than carrying them.

NIOSH Lifting Equation

The NIOSH lifting equation uses several factors to calculate a **recommended weight limit (RWL)** for a certain task. This RWL is the maximum weight that almost any worker can lift over an eight-hour shift without increased risk of musculoskeletal injury. The equation is as follows:

$$\text{RWL} = (51\text{ lbs.}) \times (\text{Horizontal Multiplier}) \times (\text{Vertical Multiplier}) \times (\text{Distance Moved}) \\ \times (\text{Asymmetry Angle}) \times (\text{Frequency Multiplier}) \times (\text{Coupling Multiplier})$$

The horizontal and vertical multipliers are how far the load must be moved in the horizontal and vertical directions. The asymmetry angle is the number of degrees from vertical the body must be moved. The frequency multiplier ranges from 0.2 to 15 times per hour. The coupling multiplier is a measure of how well the person's grip is on the load to be moved, ranging from 1 (good) to 3 (poor).

Rapid Upper Limb Assessment

A Rapid Upper Limb Assessment (**RULA**) is an ergonomic evaluation method that can be used to screen and quantify the degree of ergonomic risk from a given task or work flow. The tool requires observation and evaluation of a work task with regard to body posture, required application of force, and degree of repetition in the tasks. Body postures receive higher scores the more they deviate from neutral. Force or lifting is evaluated from 1 kilogram to 10 kilograms; any forces or lifting greater than 10 kilograms is assigned the highest score. Repetition is scored from intermittent to frequent. Each body part evaluated is assigned a number from 1 to 7. Scores of 1 to 2 are considered acceptable ergonomically, 3 to 4 may require change, 5 to 6 require change in the near time frame, and scores of 7 require immediate evaluation and change. The RULA is most appropriate for seated or sedentary tasks that do not involve the entire body.

Rapid Whole-Body Assessment

A Rapid Whole-Body Assessment (**RWBA**, also referred to as Rapid Entire Body Assessment or REBA) is an ergonomic evaluation tool used to assess degree of ergonomic strain or risk to the entire body from a given work task sequence. It is used as a screening tool to provide a relative risk ranking that can be used as a guide to determine which work tasks to redesign first. Assessment is similar to that of the RULA, and is expanded to the entire body—neck, arms, legs, trunk, and wrists. Scores obtained in the RWBA range from 1 to 15 and are evaluated as follows:

- 1–3: low risk
- 4–7: medium risk, action may be necessary in the future

- 8–10: high risk, action is necessary in the short term
- 11–15: very high risk, action is necessary immediately to redesign the work task

The RWBA is best applied to tasks that involve the *entire body* rather than just the upper body.

Requirements, Usage, and Limitations of Personal Protective Equipment

Personal protective equipment is any specialized clothing or equipment worn or used to protect a person from hazards. Personal protective equipment can include HAZMAT suits, goggles, gloves, respiration equipment, hard hats, and more. Personal protective equipment can form an essential part of a safety plan, but it should never be a primary means for controlling hazards. Personal protective equipment forms a barrier between the user and a hazard, but it does not remove the hazard. It is far better to remove the hazard, if possible. Personal protective equipment has limited success because the user must have the right equipment and know how to use it properly, often in an emergency situation. In addition, the personal protective equipment must fit properly and must be well-maintained.

In order to be effective, a program for personal protective equipment must include detailed, written procedures that address how to select, manage, use, and maintain personal protective equipment. These procedures must be enforced and supported by management and should include standards and rules for the following:

- Wearing and using personal protective equipment.
- Inspecting and testing personal protective equipment to ensure it is in good condition and working properly.
- Maintaining, repairing, cleaning, and replacing personal protective equipment.

Another important element of a personal protective program is to ensure that users understand and accept the importance of personal protective equipment. Allowing users to participate in selecting the personal protective equipment they will wear can help them "buy into" the program and be more likely to use their personal protective equipment when it is needed.

Four areas where personal protective equipment is a major factor in accident prevention are in the areas of head protection, face and eye protection, feet protection, and hand protection.

- **Hand protection**: Gloves should be appropriate to the task. To prevent concrete poisoning, non-porous rubber gloves must be used. Specific glove designs should be used for welding, whereas another non-conductive type is used for electrical work.
- **Head protection**: Hard hats are the rule and should be regularly inspected for damage or wear.
- **Eye and face protection**: A combination of safety glasses and facial shields may be used for construction site activities like electrical work, welding, cutting, grinding, and nailing.
- **Foot protection**: Steel-toed shoes and boots are the rule when working around forklifts or around other heavy and rolling objects. Safety plated shoes can prevent sharp objects from penetrating the soles, and non-slip soles provide better grip on walking surfaces.

PPE for Hands, Fingers, Arms, Feet, and Legs

Hands, feet, arms, fingers, and legs need protection from heat and cold, sharp objects, falling objects, chemicals, radiation, and electricity. Gloves and mittens protect the hands and fingers and can even extend up the wrist and arm. They can be made of different materials according to the protection needed: for example, lead is used for radiation protection and leather for protection from sparks. Gloves can be fingerless to protect just the hands; finger guards are used to protect just the fingers. Creams and lotions also protect hands and fingers from water, solvents, and irritants. Feet can be protected by the appropriate type of shoes: safety shoes with steel toes and insoles; metatarsal or instep guards; insulated shoes; and slip-resistant, conductive, or non-conductive soles. Rubber or plastic boots provide protection against water, mud, and chemicals while non-sparking and non-conductive shoes are useful for people working around electricity or where there is a danger of explosion. Shin guards and leggings protect legs from falling and moving objects and from cuts from saws and other equipment.

Work Gloves

Work gloves fall into four basic categories based on the type of materials used in making the glove and the type of materials they are designed to handle.

- Gloves made of **leather**, **canvas**, or **metal meshes** are designed for general-purpose material handling. Most will protect from splinters or from sharp edges. Metal mesh gloves can protect from puncture but should never be used near electrical lines.
- **Coated fabric gloves** are often used for work like brick-laying. They may offer some protection from slippage and from chemicals but are generally considered "light duty" and not protective in extreme situations. Coated gloves are made from fabric and an applied coating of rubber or vinyl.
- **Insulating rubber gloves** are specifically designed to protect against electric shock. Most insulating gloves have the limitation of not being able to protect against other hazards like punctures. In most cases, rubber insulating gloves are covered with a second pair of durable leather gloves which can protect from pinching and punctures or similar work problems
- **Chemical and liquid-resistant gloves** are made according to the materials they are designed to handle. Design of these gloves may vary. Some may cover the hands, wrists, and beyond the elbows.

PPE for the Head and Face

Personal protective equipment for the head can protect wearers from being hit by falling or flying objects, from bumping their heads, and from having their hair caught in a machine or set on fire. Helmets, bump caps, and hard hats are examples of this type of head protection. Hoods and soft caps are another type of head protection that also protects the face and neck. Hoods may include hardhat sections as well as air supply lines, visors, and other protective features. They provide protection from heat, sparks, flames, chemicals, molten metals, dust, and chemicals. Head protection can also aid sanitation by keeping hair and skin particles from contaminating the work. This is especially important in processes involving food and clean room work. Hairnets and caps offer this type of protection. Face shields and welding helmets protect the face from sparks, molten metal, and liquid splashes. They should always be used in conjunction with eye protection such as goggles or safety spectacles.

Hard Hats and Helmets

There are two helmet types specifically designed to protect the wearer from electric shock: Class E and Class G. A third type, Class C, may be worn by utility workers but offers no protection against electrical shock; its primary purpose is to protect from head injury due to bumping or dropped objects.

- **Class E**: This type of helmet offers the greatest protection against electrical shock from high voltage conductors. Class E (think E, for electrical) helmets are proof tested at 20,000 volts. They are also tested for the ability to sustain an amount of force, the ability to repel electrical shock without leakage, and the capacity to function without burn-through.
- **Class G**: This type of helmet will protect against electrical shock but not of a voltage exceeding 2,200 volts, applied for 60 seconds. Class G helmets were formerly known as Class A helmets until the more recent change. Think "G", for general use.
- **Class C**: These helmets afford very little or no electrical shock protection. They protect the head only from bumps, jolts, and dropped objects.

OSHA standards do not legally require the use of hard hats in areas where workers are not exposed to hazards from falling or flying objects or electrical shocks. However, employers may mandate the use of hard hats, though such mandates may require negotiation with union representatives.

PPE for the Eyes

The CHST should be familiar with the various types of eye protection which may be used to protect against specific hazards peculiar to the type of task performed. Eyes need to be protected from flying particles and objects, splashing liquids, excessive light, and radiation. Types of eye protection include spectacles, goggles, and welding helmets.

- Safety spectacles: These simple protective eyeglasses are constructed of metal or plastic with impact-resistant lenses. Side shields are available on some models.
- Goggles: These are tight-fitting eye protection devices which are designed to completely cover the eyes. Goggles provide protection from flying objects, dust, and splashed chemicals. Goggles may be fitted over personal eyeglasses of the employee.
- Full-face welding shields: Vulcanized fiber or fiberglass face shields are fitted with a filtered lens. Designed to protect against intense radiant light, they also protect from flying debris produced during the welding, soldering, or brazing process.

Respiratory Protection

Respiratory protection programs are required of employers within any site where airborne materials (including gases) may pose a threat to worker health. When a worker enters the respiratory program of an employer, the worker is subject to a medical evaluation (by a licensed health care practitioner or LHCP) of his fitness to use and wear respirators. Respirators are designed for different purposes:

- **An atmosphere-supplying respirator** refers to a means of supplying the user with an independent air supply usually contained in a tank or storage device. These types are designed for areas when it is impractical or dangerous to use ambient air.
- **An air-purifying respirator** is one that uses ambient air and filters it through a canister or cartridge to remove toxic elements. This type of respirator is useful when air contamination is not of such volume and extent as to overwhelm the filtering cartridge.

The chief concern in choosing the appropriate respirator is the material hazards presented by the worksite. Air purifying respirators cannot protect against gases, in other words. Other concerns may also impact the CHST's decision in choosing a respirator:

- The physical configuration of the jobsite is a factor in choosing the correct respirator. Tight areas and close quarters may not permit the use of **SCBA** (self-contained breathing apparatus) due to their size and bulk. Similarly, a respirator with a long hose is not suitable for working around machinery or other obstacles.
- The worker's medical condition is another factor determining choice of respirator. Negative pressure respirators wouldn't be appropriate for workers who have asthma, for example.

Worker comfort should be considered also. A comfortable worker is a more productive worker. Masks and helmets should meet fit requirements.

Body Protection

Personal protective equipment for the body provides protection from hazardous materials, biohazards, heat, fire, sparks, molten metal, and dangerous liquids. This clothing is often combined with other types of personal protective equipment, such as respiratory equipment and eye protection. Types of personal protective equipment for the body include the following:

- Coats and smocks to protect clothing from spills
- Coveralls, which may include hoods and boots
- Aprons, which protect the front of a person from spills and splatters
- Full body suits for working with substances that present a danger to life or health (may include cooling units to help lower the wearer's body temperature)
- Fire entry suits
- Rainwear
- High-visibility clothing for people working on road construction or in traffic
- Personal flotation devices for people working around or on water
- Puncture-resistant or cut-resistant clothing for protection from ballistic objects, power saws and other cutting equipment

Some personal protective equipment is designed to be disposable, especially clothing contaminated with hazardous materials.

Storage and Maintenance

Proper storage and maintenance of personal protective equipment is important for personal hygiene and for proper operation of the equipment. The following covers storage and maintenance procedures for respirators, hearing protection, and gloves:

Respirators—Respirators must be stored in a sealed plastic bag to protect them from being contaminated by dust. Best practice dictates that respirators are only stored after they are thoroughly cleaned and dry. Maintenance of respirators requires that all inhalation and exhalation valve flaps be in place and replaced if worn, and that filter cartridges be replaced according to a pre-established schedule or whenever breakthrough is detected.

Hearing Protection—Ear plug hearing protection should only be inserted using clean hands. If ear plugs are soiled or fall on the ground, they should be replaced. They should be stored in a plastic bag away from potential dust contamination. Replacement of disposable earplugs should be done weekly if they are worn daily, or if they seem to be losing elasticity. Ear muff hearing protection needs only to be cleaned off periodically. They should be replaced if they are cracked or worn and do not provide a good seal around the ears.

Gloves—Gloves should be stored in a locker or other area away from dust. They must be examined before each use to ensure there are no holes or tears. Gloves should be replaced when worn out.

Inspection

Once the proper personal protective equipment is selected in accordance with a job hazard analysis, it must be properly inspected and maintained. Whenever possible, employees should be assigned their own PPE devices and should not have to share with another employee. Certain types of PPE must be inspected for proper configuration before each use, such as fall protection equipment and respirators. High-value and high-risk PPE should be inventoried and documented inspections conducted (for example, fall protection equipment and SCBA respirators). Equipment must be assigned to a particular person so that any deviations from inspection procedures or misplaced equipment can be identified and the offending employee counseled. All PPE used for emergency purposes must be inspected at least monthly, and these inspections should be documented to show they have been conducted.

Basic Testing and Monitoring Equipment

Industrial Hygiene Sample and Indicator Media

There are three basic types of sampling methods used to determine concentration levels of gases and vapors: a spot-check or snapshot sample, a long-term sample, and a passive sample. A **spot-check** is a quick sample of the air in a specific location, which is good for determining exposure risk at the moment of the sample. A **long-term sample** uses a device to sample the air over a period of eight hours or more to determine the contamination level for an average work day or the length of the sample. This provides a more thorough evaluation of the air environment for long-term exposure. The **passive sample method** is where a worker wears a sample device that simply warns when contamination levels are too high. Although the passive sample seems like the simplest method for keeping workers out of harm's way, these devices provide no measurement capability for preventive measures.

Sampling Equipment

Below are several hand-held and portable instruments that are used in industrial hygiene applications:

- **Noise meters and dosimeters** are used to conduct sound level exposure surveys and are worn for an entire shift to determine whether the OSHA Permissible Exposure Limit of 85 decibels was exceeded. Results and reports can be downloaded to a computer for storage and documentation.
- **Gas meters** are used for real-time analysis of various gas levels, such as oxygen, carbon monoxide, and hydrogen sulfide. These are mandatory for permit-required confined space entry.
- **Gas detection tubes** are glass tubes containing a chemical-specific reactive substance used to measure concentrations of chemicals in the atmosphere. They are available for a variety of chemicals used in industry and provide an accurate assessment of exposure. Some are designed to be used with a hand-held pump and others are worn by the employee and sample the air by passive diffusion.
- **Photoionization detectors** are portable hand-held instruments that measure the concentration of volatile organic compounds (VOCs) in the air. They are not chemical-

specific but measure all VOCs in the surrounding air. Photoionization detectors provide a way to assess exposure levels and establish the level of respiratory protection needed in a given environment.

Colorimetric Indicator Tubes

Colorimetric indicator tubes are a popular method for monitoring for airborne contaminants since the tubes provide a quick response and are easy to use. The tubes work by drawing a certain amount of air through the tube, passing through a layer of cotton and a conditioning filter, then interacting with an indicator ruler which changes color according to the concentration, and then back out through another cotton plug. The color changing indicator line is read against a direct reading scale which displays the concentration, usually in parts per million. The colorimetric indicator tubes provide an excellent warning signal for certain hazardous contaminants and can be used to monitor for several different chemicals. Inaccurate readings are at a higher risk factor, however, because only the manufacture can calibrate the tubes. Once used, these tubes must be disposed of as hazardous waste.

pH Strips

The pH scale goes from 1–14 and is a measure of acidity (free hydrogen ions in solution) or alkalinity (free hydroxyl ions in solution). pH strips are plastic or paper strips coated with a substance that changes color based on the amount of hydrogen or hydroxyl ions in a solution. Pure water is considered neutral with a pH of 7. The acidity or alkalinity of a substance can be quickly determined with pH strips. If the pH is below 2 or above 12 then it is classified as a corrosive hazardous substance. A pH analysis has a fifteen-minute holding time, which is not long enough to get a sample to the laboratory, making the test strip a useful alternative. For regulatory compliance, pH determinations should ultimately be confirmed by use of a calibrated pH meter. To maintain accuracy, pH strips must be protected from humidity and other gases.

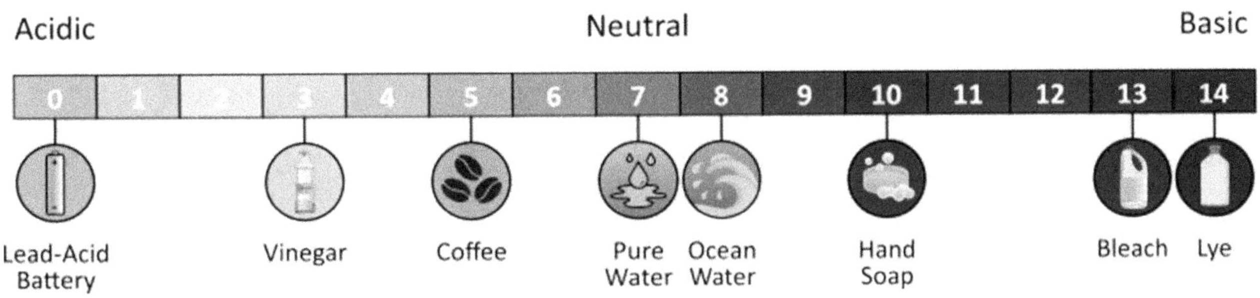

Light Meters

Light is measured in units of lux, which is one lumen per square meter. A lumen is the SI measurement of the intensity of light in a cone-shaped area. Light levels can be measured with hand-held instruments. Appropriate light levels are important for various tasks in order to reduce eye strain and avoid injury. To get a sense of the scale of lux values, a bright summer day is about 100,000 lux, an overcast summer day is about 1,000 lux, and twilight is about 10 lux. In occupational settings, a normal office environment should be at 250–500 lux. A supermarket or mechanical workshop should be approximately 750 lux. Detailed drawing work requires 1,500–2,000 lux for comfort. Consideration should also be given to the type of light used, as different types have different qualities.

Constant-Flow Sampling Pumps

Direct reading instruments take snapshots of the air quality and provide almost instant readings of the concentration levels of specified contaminants in the air. These readings are reliable enough for detection and early warning of possible hazardous exposure levels of the contaminants, but for complete accuracy and regulatory compliance verification, the constant flow sample pump is the measurement device used by OSHA. The constant flow air pump draws a measured amount of air through a filter medium for a given period of time to measure exposure levels for up to eight hours in a single day. The contaminants captured by the filter medium are analyzed by a laboratory to measure the concentration and determine the time-weighted exposure levels. This most accurate measurement of the contaminant level is used to determine whether or not a facility is in compliance with OSHA permissible exposure level limits regulated by OSHA.

Chapter Quiz

Ready to see how well you retained what you just read? Scan the QR code to go directly to the chapter quiz interface for this study guide. If you're using a computer, simply visit the online resources page at **mometrix.com/resources719/chst** and click the Chapter Quizzes link.

Emergency Preparedness, Incident Investigation, and Response

Transform passive reading into active learning! After immersing yourself in this chapter, put your comprehension to the test by taking a quiz. The insights you gained will stay with you longer this way. Scan the QR code to go directly to the chapter quiz interface for this study guide. If you're using a computer, simply visit the online resources page at **mometrix.com/resources719/chst** and click the Chapter Quizzes link.

Proper Fire Protection and Prevention Methods

Classes of Fires

The United States fire classification system identifies the first three primary classifications of fire types as A, B, and C.

- **Class A**: This is the most common type of fire and refers to the ignition of combustible materials like wood, cloth, and paper. Class A fires can be extinguished with water but the use of Class A rated fire extinguishers works as well.
- **Class B**: This type of fire is the result of ignition of flammable liquids. Water will not extinguish Class B fires. A jet of water may have the opposite effect, forcing the fire to spread or splatter. Class B fires should be put out with foam, CO_2, or dry chemicals.
- **Class C**: Class C fires are electrical. One of the first responses should be to cut off the power source, reducing the fire to Class A type of event.
- **Class D**: Metals fires can originate with flammable metals like titanium and magnesium. NFPA recommends dry powder to extinguish them. **Dry powder** is different in composition from **dry chemicals**, and the two must not be confused. Each of these compounds will have a different effect upon a fire.
- **Class K**: This refers to fires from cooking oils. Do not apply jets of water as this will cause the fire to spread and splatter. Class K fires may be put out with dry chemicals from suitable fire extinguishers.

Fire Extinguisher Inspection Criteria

Fire extinguishers must be fully **charged** and must have been charged with fire suppression agent within the last year to ensure they are effective. To inspect a fire extinguisher, first note that it is in the correct **location**. Best practice is to mark the locations of fire extinguishers with red arrows that point to the extinguisher. The signs should be mounted high enough on the wall to be spotted from a distance in an emergency. Check that the **pin** is intact and that the **seal** has not been broken. Check the **gauge** that indicates whether the extinguisher is fully charged. Check that there is an **inspection tag** in place, and that the recharge date is within the past year. The recharge date will usually be indicated by punching the month and year of the last recharge. Finally, check the back of the inspection tag to confirm that each **monthly inspection** has been completed and add initials and date for the current inspection.

Fire Sprinkler Inspection

The National Fire Protection Association publishes **NFPA 25**, the *Standard for the Inspection, Testing, and Maintenance of Water-Based Fire Protection Systems*. There are separate inspection elements for quarterly, annual, and five-year inspections. Inspections must be performed by a trained and competent person. The quarterly inspection must include operation of the water flow alarm devices, the valve supervisor signal devices, and the integrity of the fire department connections. The sprinkler heads must be inspected for proper installation, evidence of damage or leaks, and proper clearance (1 inch).

Components of Emergency Action Plans

Emergency action plans should be tailored to a specific work site in the construction industry. Plans should include the location of the nearest **hospital** or clinic that is available to treat injured workers, locations of **emergency equipment** such as fire extinguishers and spill kits, and the location of the **assembly area**. Although these are elements common to all emergency plans, it is important to realize that this information will change for each site and should be updated accordingly. In addition, emergency planning should take into account that multiple activities may occur on a site, requiring coordination and communication. Special attention should also be paid to the location of **utilities** and any disruption in utilities that may be caused by the construction activities.

Emergency Evacuation Maps

An emergency evacuation map should be created for each work site and should be posted at visible and accessible areas near exits. An emergency evacuation map should show the basic **outline** of the structure, including the interior walls. It should show all interior and exterior **doors**, and the direction that they open. It should highlight available **exit routes** from each occupied workspace. The map should show the location of the **Assembly Area**. The location of **emergency equipment**, such as fire extinguishers, eye washes, emergency spill kits, and safety showers, should also be marked.

Site Risk Assessments

Risk assessment is an essential element of an emergency action plan. Risk assessment is necessary to anticipate what types of emergencies may occur and should be planned for. For example, depending on the geographical location, different **weather emergencies** must be planned for. In some areas of the country, plans must be made for extreme weather events or tornadoes, whereas in other parts, planning for wildfires and earthquakes is necessary. Additionally, the risk assessment will help to plan for potential **health and safety emergencies**. For example, if trenching will be conducted, possible emergencies and responses associated with trenching must be assessed. Overhead hazards must be anticipated and planned for. If hazardous chemicals are in use, it is important to assess the potential size of a hazardous materials spill and to have the proper response equipment on site. Many other risk assessment considerations can be made as well. An emergency action plan attempts to answer the question of what to do if something goes wrong.

Common Elements of Response Plans for Environmental Hazards

An environmental emergency response plan should include information on how to prepare for and respond to a **chemical spill**. The plan must include a description of where chemicals are stored, what types of chemicals are on site, and where the spill response equipment is located. The contents of each **spill kit** should be included and these must be regularly inspected to ensure the supplies are still intact. The response plan should describe how to clean up simple spills and the notification requirements for spills that may be too large for on-site personnel to clean up. The emergency response plan should be used as a training tool and a reference document but **training and drills** must be conducted to prepare employee responders in advance of an emergency.

When to Notify Authorities

If a hazardous substance or chemical is released off-site into a waterway or could potentially contaminate a waterway or sewer system, regulatory authorities must always be notified. Both the US EPA and US DOT define a "**Reportable Quantity**" as the amount of a hazardous substance that, if released into the environment, requires reporting to the National Response Center. Reportable quantities may be found in the **US DOT Hazardous Materials Table, Appendix A** (49 CFR §172.101). A release to the environment means chemicals released on land, to the air, or into a waterway. Pure chemicals, chemical mixtures, and EPA hazardous wastes each have specific reportable quantities.

Agency Contact Information to Include in Environmental Response Plan

An environmental emergency response plan must include the contact information for the **company emergency coordinator** and **alternative emergency contacts** if that person is unavailable. To act as emergency coordinator, the person must be fully aware of all hazardous materials at the site and have complete site access, including keys to access all locked areas and codes to disarm any alarms in case of emergency. The emergency response plan should also include **agency contact information** for agencies that may need to be notified of a chemical emergency or spill. This includes the sewer agency, the water quality control board for notifications of stormwater pollution, the US EPA National Response Center for notifications of chemical spills, the Department of Fish and Game if wildlife is harmed, and non-emergency contact information for the local fire department.

Emergency Response System

Incident Command System (ICS)

An ICS is a standardized, multi-agency, management system designed to apply to all types of hazards from small to catastrophic. The task was mandated by Homeland Security Presidential Directive 5 and is now regulated by the National Incident Management System (NIMS). OSHA rule 1910.120 mandated the implementation of an ICS for all companies that are involved with hazardous materials. FEMA (https://training.fema.gov) provides extensive training materials to help in the development of an ICS. Deployment of an ICS is required when a hazard or natural disaster has a high potential of resulting in harm to humans, facilities, or the environment.

Utilization of an ICS helps ensure safety of people and cost-effective use of resources in order to effectively manage a hazard response program.

The Incident Command System is designed to manage emergency situations effectively and efficiently. It is a standardized approach to command, control, and coordination of all types of emergency responses. The structure and format of the ICS is part of the National Response System codified in the National Contingency Plan; and is widely used by government agencies when responding to disasters. The ICS is under the direction of an On-Scene Coordinator. This individual acts to coordinate the involvement of various agencies and serves as a clearinghouse for information related to the incident so that those involved coordinate effectively. The ICS is organized with divisions responsible for operations, planning, logistics and finance, with additional support for safety and information gathering and dissemination.

Crisis Management

The concept of crisis management refers to an organization or company being prepared to handle all aspects of an unexpected event in order to ensure that the situation is properly **managed** and that employees, interested parties, and the public are properly **informed and cared for**. Response to **emergencies** is one type of crisis management that an organization should be prepared for. It is important to anticipate potential events in planning for a response. Consideration must be given to **communication** inside and outside of the organization. Consideration should also be given to the **short-term response** to a crisis (what to do during the emergency) and to the **long-term and follow-up responses** to a crisis (how to rebuild and who to notify).

Business continuity planning can result in a smoother response to crisis management. Although it may not be possible to consider every possible emergency, consideration can be given to fire, earthquakes, tornadoes and other weather events, chemical spills, and disruption in the workforce due to pandemic illness or a labor strike. Envisioning **potential scenarios** is a good first step to take. Consideration should be given to handling disruptions in internal and external communication, and conducting business if information technology or computer systems are disrupted. **Drills or tabletop exercises** can be used to simulate a crisis so participants can practice their responses. This type of planning can also help to reveal shortcomings in the planning process and areas of potential improvement.

Emergency First Aid Equipment and Procedures

Eyewashes and Safety Showers

OSHA requirements for emergency eyewashes and showers are found in 29 CFR 1910.151. Requirements for eyewashes and showers are similar. They are required in areas where an employee might be exposed to corrosive or hazardous chemicals, such as strong acids or bases. Eyewashes and showers are required in work areas with anhydrous ammonia and forklift battery charging stations. The path to the eyewash or shower must be unobstructed, and require no more than 10 seconds of travel time from the hazard. Both eyewashes and showers must deliver tepid water. The flow must be able to be sustained for a minimum of fifteen minutes, and the valve that delivers the water must stay open on its own and not shut off unless someone physically shuts it off.

Fire Extinguishers

OSHA regulations require a written fire prevention plan that is communicated to all employees (29 CFR 1910.39). Employers with fewer than ten employees may verbally tell their employees the plan. The plan must list the fire hazards present in the workplace, and procedures to control accumulation of flammable and combustible materials. There must be written procedures for maintenance of heat-producing equipment to prevent accidental build-up of heat and ignition of combustible materials. Part of the fire prevention plan must describe where fire extinguishers are located, and what type of fire extinguisher is appropriate for the fire hazard. Fire extinguishers should be located near areas with fire hazards. Employees must be trained in how to use fire extinguishers and taught that they are only to be used for small fires. A back up extinguisher should be available as well. Employees should be instructed to never let a fire back them into a corner; they should always have an exit route when using a fire extinguisher.

AED

An automated external defibrillator (AED) is a device that can be used in an emergency situation when a person's heart has stopped beating (cardiac arrest). It works by delivering an electrical impulse (shock) that restarts the stopped heart. These should only be used by those that have been trained to use them in an approved first aid course. The device consists of two adhesive pads that are wired to the defibrillator. The device can assess the person's heart rhythm and recommend whether defibrillation is indicated. These pads are placed on either side of the heart on a dry chest, generally the patient's top right and lower left torso, per the equipment's directions. CPR should continue until the pads are in place. Hold CPR and press "analyze" to allow the device to analyze the rhythm and identify the indication for defibrillation. If defibrillation is indicated (in the case of ventricular fibrillation or pulseless ventricular tachycardia), the machine will announce that the pads are being charged and then instruct its user to administer the shock. Prior to administering the shock, all individuals should confirm that they are "clear" (not contacting the patient). Immediately after the shock is delivered, CPR should resume. The AED will continue to provide instructions for further rhythm analysis and shocks. Continue CPR and follow the AED's instructions until emergency responders arrive.

Responding to Media Inquiries

As part of emergency response planning and business continuity planning, it is important to plan a **response to media inquiries**. The most important part of the planning is to define who is authorized to speak to the media. Most often, this will be the President or CEO, but could also be an assigned **public relations officer** in a large organization. All other personnel should be instructed to **refer** inquiries to this person. When responding to the media, it is important to be factually accurate without being alarmist. Communication must also be maintained with agency first responders to coordinate responses to media.

Potential First Aid or Medical Needs

Availability of First Aid Kits

First aid supplies must be on hand for the trained first aid responders to use when providing first aid. There are no specific lists of supplies that are required by OSHA. The reference for the **minimum contents of a first aid kit** may be found in **ANSI standard Z308.1**, *Minimum Requirements for Workplace First Aid Kits.* This ANSI standard requires adhesive bandages, adhesive tape, topical antibiotic treatment, antiseptic wipes, burn dressing, burn treatment supplies, a breathing barrier to provide CPR, a cold pack, eye covering and eye wash, exam gloves, sterile gauze pads, hand sanitizer, and a triangular bandage that can be used as a sling. There may be specific hazards at a given workplace that require more supplies, or more extensive supplies. For example, if workers are working with glass or ceramics and there is greater risk of cuts, it may be necessary to stock more extensive first aid supplies designed to stop bleeding.

Availability of Trained First Aid Responders

The OSHA regulations regarding trained first aid responders may be found in **29 CFR §1910.151**. This standard requires that **trained first aid providers** be available if there is no clinic or hospital in near proximity, as defined by OSHA (within 3 to 4 minutes). Unless the work site is actually at a hospital or clinic, it is unlikely that a medical facility is near enough to provide the proper first aid when needed. In addition to requiring a person or persons trained in administering first aid, several OSHA regulations require trained personnel in **cardio-pulmonary resuscitation** (CPR). Also be aware that if workers may be providing first aid as part of their job duties, they must be trained in the hazards of **bloodborne pathogens** and be offered Hepatitis B vaccination as is required by the bloodborne pathogen standard.

CPR

CPR should only be performed by trained and qualified people. A well-stocked first aid kit will include a **CPR face shield** that is used to perform CPR without direct physical contact between the CPR first responder and the victim. In addition, the CPR kit will have **latex gloves** and **antiseptic wipes** to clean off the victim and the face shield. Recent guidelines for first response to cardiac events indicates that rapid chest compressions can be just as effective as chest compressions combined with the rescue breaths. However, a lack of CPR equipment should not prevent a first responder from providing life-saving CPR.

Universal Precautions

"Universal precautions" is a term used in the control of occupational exposure to blood-borne pathogens. The premise is that any and all blood and bodily fluids are potentially contaminated with blood-borne pathogens such as HIV or hepatitis and should, therefore, be treated as such. The universal precautions to be employed are to wear proper protective equipment when handling these items (latex gloves and eye protection) and to ensure that all surfaces contaminated with blood are properly cleaned with a bleach solution to kill any viruses that might be present. Solid waste contaminated with blood must be contained in a plastic bag to prevent any exposure to those who may be handling it.

Blood-Borne Pathogens

Blood-borne pathogens are disease-causing agents that can be transmitted from one person to another by contact with the blood or bodily fluids of a person who is infected with them. Examples of blood-borne pathogens include hepatitis, human immunodeficiency virus (HIV), and Ebola virus. Employees may be at risk of infection by blood-borne pathogens while administering first aid or during an injury situation. They are a concern because the diseases that are transmitted by contact with blood or bodily fluid are serious, with severe and sometimes fatal consequences. In addition, they do not have successful cures available, so if one becomes infected, it will impact the person for the rest of his or her life.

Disposing of First Aid Materials Contaminated with Blood

During the course of administering first aid, an employer will likely generate waste materials contaminated with blood. The Bloodborne Pathogen Standard defines "regulated waste" as that which is contaminated with blood that is liquid or free-flowing. This regulated waste must be closed and contained to prevent potential contamination and exposure and should be disposed of through a licensed medical waste company. However, if there are small amounts of blood on first aid materials such as gauze pads, Band-Aids, or wound dressings that do not contain free-flowing blood, these items can be placed in a plastic bag and then placed in the regular trash and disposed of with the regular solid waste.

Chapter Quiz

Ready to see how well you retained what you just read? Scan the QR code to go directly to the chapter quiz interface for this study guide. If you're using a computer, simply visit the online resources page at **mometrix.com/resources719/chst** and click the Chapter Quizzes link.

Safety Program Development, Implementation, and Sustainment

Transform passive reading into active learning! After immersing yourself in this chapter, put your comprehension to the test by taking a quiz. The insights you gained will stay with you longer this way. Scan the QR code to go directly to the chapter quiz interface for this study guide. If you're using a computer, simply visit the online resources page at **mometrix.com/resources719/chst** and click the Chapter Quizzes link.

Applicable Health and Safety Standards and Best Practices

Safety Training for New Hires

All newly hired employees should go through **safety orientation** on their first day. At a minimum, this training should include an introduction to workplace hazards. It should also review emergency egress routes, the location of the emergency assembly area, and emergency action procedures. There should be a review of the required **personal protective equipment** for the job, and employees should learn how to obtain new PPE supplies. If an employee will be required to wear a respirator, a **fit test** should be conducted. If complete safety training cannot be conducted on the first day of employment, the employee must not conduct unsupervised work until training is completed.

Management Practices for Handling Chemicals

Best management practices for handling chemicals include always wearing the proper **PPE** (safety glasses or goggles, chemical-resistant gloves, a respirator if needed). Read the **Safety Data Sheet** and any manufacturer's literature prior to using a chemical to understand its proper use and the risks it poses. If chemicals are placed into another container, be sure to label the container. Always purchase the smallest amount of chemical needed for a job to minimize waste. Use chemicals in well-ventilated areas. Clean up spills immediately and properly dispose of the waste. Never eat, drink, or smoke when using chemicals. Always wash hands and face after using chemicals.

Waste Minimization Techniques for Construction

Waste minimization is the practice of creating as little waste as possible to minimize disposal. **Waste minimization practices** for a construction setting are based upon proper inventory management and estimation of quantities of materials necessary for a job. For example, one can calculate the amount of paint needed to perform a job using either past experience or an online calculator. This will help to minimize waste by avoiding purchasing too much paint. Another waste minimization technique is substitution of less toxic materials that can be discarded as solid waste instead of hazardous waste. For example, use water-based cleaners rather than solvent-based cleaners that would have to be managed as hazardous waste.

Management Practices for Fall Protection Programs

A fall protection program should include a written plan that outlines the steps that will be taken to prevent falls. The plan should include an assessment of the types of work that will be conducted at heights, the type of fall protection equipment that will be used, and the type of training that will educate employees in equipment use. The employees must learn how to inspect the fall protection harness and how to tie the lanyard to a fixed object that will provide the fall arrest. In a manufacturing environment with routine tasks that require fall protection, tie-off points should be pre-identified and clearly marked so that employees are consistently using the correct points.

Management Practices for Confined Space Safety

A confined space is defined as one that is not designed for continuous occupancy but can still be entered by a body part or entire body. Confined spaces must be entered from time to time for maintenance or repair activities. The first step in **confined space safety** is to identify whether confined spaces exist, and if so, whether they are permit-required or not. A **permit-required confined space** is one that may have a hazardous atmosphere, either from low oxygen concentration or a flammable or toxic vapor present. Once the confined spaces are identified, each one must have a **sign** posted at the entrance indicating that it is a confined space and stating whether a permit is required. Anytime someone enters a confined space, a **sentry** must be posted at the entrance to call for assistance if an employee is injured in the confined space.

Common Components of Site-Specific Safety Plans

For work that is performed at different job sites, it is important to create a **risk-based site safety plan** for each job site. The risk assessment should consider the overall process that will be conducted, as well as the site-specific risks. For example, is the area rural or urban? How close are emergency personnel in case of a fire or injury? Is emergency equipment available? What are the risks posed by use of heavy machinery? Does the site work pose risks of chemical exposure? How does the terrain or topography influence the site risks? Do any weather-related risks need to be accounted for?

Each site-specific site safety plan should list the nearest **medical facilities**, including urgent care clinics for minor injuries and hospitals for major injuries. There should be an assessment of the **personal protective equipment** needed at the job site. There should be an assessment of what other work will be conducted at the same time, and whether this will pose additional hazards. The location of power, potable water, and washing facilities should be investigated and accounted for. It should be determined if there are any extraordinary communication challenges posed by the site (for example, is it out of mobile phone signal range?).

Worksite Assessment or Audit Processes

When conducting an audit of a construction work site for compliance with health and safety regulations, examine records and work site physical conditions. **Records** include training records to show that employees have been trained in hazard communication, respiratory protection (if applicable), lockout/tagout procedures (if applicable), emergency evacuation procedures, PPE requirements, proper tool safety, and ladder safety. **Physical conditions** include evidence that safe electrical practices are followed, proper PPE is worn, and emergency equipment such as fire extinguishers is available. Routine **safety tailgate meetings** should be conducted and records of these should be maintained.

Electrical Safety Inspections

A routine general electrical safety inspection should be conducted at a construction site. A more in-depth inspection should be held for activities conducted by electricians. The **general electrical safety inspection** should examine all power tools to ensure the cords are intact and no wires are exposed. Power tools should be properly grounded and double insulated. Extension cords should not be used in daisy chains. There should be no exposure to live electricity. A lockout/tagout program should be in place and employees trained according to their role as *authorized* or *affected* employees. Any overhead work that could be impacted by overhead power lines must be planned and carried out to avoid touching the wires. Sources of water near electricity should be eliminated.

Roles, Responsibilities, and Lines of Authority as They Relate to Safety Management

Responsibility of Supervisors with Regard to Employee Safety

The role of the supervisor is very important in employee safety. The supervisor is directly responsible for **communicating work instructions** and for providing employees with **on-the-job training** in how to safely conduct the work. The supervisor enforces **policies** on safe work practices and wearing the required personal protective equipment. The supervisor is responsible for providing **disciplinary action** if the employee refuses to follow safe work practice requirements. The supervisor is also important as a **communicator with management** regarding implementation of safe procedures and recommendations for continuous improvement.

Responsibility of Employees with Regard to Employee Safety

An employee is responsible for his own and his coworkers' safety. An employee has the responsibility to follow all established **safe work procedures** and to wear all required **personal protective equipment**. For example, the employee must wear safety glasses if required, even if it is not preferred. The employee is responsible for communicating to the supervisor and to management if an **unsafe condition** is present and should participate in developing a solution. The employee must not take shortcuts that will endanger his and others' safety. The employee must also report to work well rested and sober so that he can give proper attention to the job and perform it safely.

Responsibilities of the Employer with Regard to Employee Safety

The employer has the ultimate responsibility for creating a safe work environment. The employer must design the company's **work procedures and operations** with safety in mind. Proper **job hazard analyses** should be conducted for each operational process, and hazards should be considered so they can be minimized. The preferred method is to engineer out hazards. The employer is responsible for documenting the **safe work procedures** and for communicating this information to the employees. The employer must properly **train** the employees and provide them with the proper **personal protective equipment** if needed. The employer must listen to and consider any **employee complaints** received about safe work practices, and must not retaliate against any complainants.

Recommended Equipment Inspection Records or Logs

Job Site Safety Inspection Sheets

A general job site safety inspection sheet should include an **inspection** for general safety best practices. For example, are employees wearing proper PPE for the jobs assigned? Are the tools in good repair? Are there slip/trip/fall hazards present in the workplace? Are any ergonomic risk factors present that can be engineered out of the work site? Are emergency egress routes established, and do employees know what to do in case of an emergency? Are fire extinguishers and emergency spill kits available? Are hazardous and/or flammable materials stored properly? If ladders are in use, are they inspected regularly?

POWER TOOL INSPECTION SHEETS

Power tools should be inspected prior to each use to ensure they are safe to operate. They should be examined to ensure they are free of **grease and oil**. One should ensure their **power source** is intact (for example, battery not cracked and power cord not frayed or wires exposed). The **guards** or fail-safe mechanisms must be present and operational. If the power tool has a **blade or drill bit**, it should be examined to ensure it is securely affixed to the tool and is not cracked or damaged. The tool should be inspected to be sure that it is **grounded and double-insulated**. If it has an electrical cord, the **third prong** should be intact. Also, ensure that the proper **PPE** is available for use with the tool (for example, if use of the tool will create flying particles, a face shield is recommended in addition to safety glasses).

INSPECTION OF FALL PROTECTION EQUIPMENT

Fall protection equipment consists of a **harness** designed to arrest a fall before it can cause serious or fatal injury. Regular inspections are important to ensure the equipment will work when needed. The **D-rings and buckles** should be inspected to ensure they are present, are not damaged with nicks or abrasions that might weaken the metal, and are not corroded. Any **stitching** on the harness must be intact and not frayed or unraveled. The **webbing material** the harness is constructed of should similarly be inspected for fraying. **Lanyards** used in conjunction with harnesses must be inspected for the same elements. Be aware that if a fall protection assembly has been used to arrest a fall it must be **retired and replaced**.

Basic Risk Management Concepts

The risk management process typically involves five steps to complete when addressing issues:

1. Identify and analyze loss exposures — The process of **risk identification** helps determine those risks that create loss exposure to the organization. The process of **risk analysis** attempts to determine the potential severity of loss to the organization should the identified risk occur.
2. Consider alternative risk techniques — Risk techniques involve two strategies:
    a. **Risk control** — preventing or mitigating the impact of losses
    b. **Risk financing** — paying for unavoidable losses
3. Select those risk management techniques that are best suited to manage the specific risk.
4. Implement the selected risk management technique.
5. Evaluate and monitor the effectiveness of the chosen technique in managing the identified risk.

JOB SAFETY ANALYSIS

Job safety analysis (JSA) is an approach to safety that determines specific behaviors within a defined operation that are appropriate. JSA evaluates an operation by looking at each activity performed by employees and then identifying all risks that correspond to those activities. JSA provides specific instructions and procedures for completing each of the identified actions appropriately to minimize risks and maximize safety. This analysis also considers a variety of different job components including motion, efficiency based on time, specific movements, and pace. It looks at both normal job conditions as well as examining job activities related to possible emergencies or non-routine conditions, which is when accidents most commonly occur. JSA is often incorporated into safety training programs and operations guides.

WHAT-IF/CHECKLIST ANALYSIS

A **what-if/checklist analysis** combines a "what-if" analysis with a checklist process to produce a failure evaluation tool that is more robust than either of the two would be on their own.

What-if is a brainstorming technique that is used to identify events and their potential outcomes that could impact a process. A **team leader** guides the group through the process to keep the team within the confines of the process being evaluated. The phrase "what if..." is used to identify potential negative events. Once events have been identified, the team assesses each one for the likely source of errors and whether the result has an acceptable level of risk. For those events deemed unacceptable, **corrective actions** are identified and implemented before the event can occur. This process is limited in that only events that are considered can be evaluated and has a high level of subjectivity.

A **checklist analysis** relies on published standards, codes, or industry practices to determine whether a process meets all the criteria. Responses can be "yes," "no," or "not applicable." This method relies on a walk-through method to verify the existence of necessary controls or processes.

The **what-if/checklist** combination uses the brainstorming process to identify controls necessary to address the failures anticipated by the group.

METHODS OF MANAGING RISK

Risk transfer is when the degree of potential loss is known or unknown and is no longer managed by the primary entity. Risk can be transferred by hiring subcontractors to conduct the work, purchasing insurance, or including contractual language such as indemnification.

Risk control is when the degree of potential loss is known, and measures are put in place to keep the risk at an established threshold. Control methods can include contractual language such as establishing parameters for a project, or environmental, physical, or chemical hazard control methods such as engineering, administrative, and PPE methods.

Risk acceptance is when the degree of potential loss is known and is below the predetermined acceptable threshold established by management. Risk can never be completely eliminated—the goal is to reduce it to the lowest level possible while still considering economic, personnel, and industry standards.

BUILDER'S RISK INSURANCE

Builder's risk insurance is a policy that covers the structures under construction while they are being built. While each policy differs in its exact coverage, it is designed to protect the builder from risk of loss to the structure due to an **external event**. For example, builder's risk insurance may cover against a loss to the structure due to vandalism or an extreme natural event such as a tornado or fire. Or it may protect the contractor against a loss due to weather (for example, if the supplies already purchased by the contractor are damaged in a weather event before they can be used). This type of insurance is distinct from general business liability insurance.

GENERAL LIABILITY INSURANCE

General liability insurance for the construction trade covers losses attributed to the **contractor** in the areas of property damage, faulty workmanship, and bodily injury. For example, if a worker is driving a forklift at a job site and accidentally damages a parked car that belongs to the facility owner, a general liability insurance policy can offer coverage for the damages. Or, if a visitor is touring the job site and trips over a tool that a worker left on the floor, general liability insurance will cover that. If a client alleges faulty workmanship, a claim can be lodged against your general liability insurance policy. In addition, contractors who subcontract work are required to maintain general liability coverage to insure against losses caused by these subcontractors.

Data Gathering Techniques and Procedures Used in Incident Investigations

The primary purpose for investigating accidents is to prevent future accidents from happening. **Investigations** can also identify causes of accidents and injuries, provide evidence for legal claims and lawsuits, and help assess the amount of loss and damage. After an accident, an accident investigation should begin as soon as all emergency steps have been taken to care for the injured parties and to bring the emergency situation under control. Beginning the accident investigation quickly offers several benefits:

- Immediate investigations produce **more accurate results** because witnesses' memories are fresh and untainted.
- Immediate investigations allow the investigator to study the **accident scene** itself before it is changed.
- Immediate investigations send a message that the company **cares** about employees' safety.
- Immediate investigation demonstrates the company's commitment to discovering the cause of the accident and thus **preventing future accidents**.

Accident investigations should begin as soon as possible. The first tool an accident investigator needs is **rope** or **security tape**. Stretching rope or tape around the accident scene will help keep people out of the area, keeping the scene secure, undamaged, and unchanged. Even with rope or tape, though, evidence at accident scenes can dissipate, so the investigator should take **photos** or a **video** of the site as soon as possible. Tape measures can be used to mark where items are located and ID tags can be used for marking evidence. Investigators may also need tape recorders to interview **witnesses** about where they were and what they saw. Particular types of equipment may be needed depending on the type of accident. For example, Geiger counters are needed for radiation releases while colorimeters, sampling equipment, and clean specimen jars are needed for chemical spills.

Accident investigation can be expensive, so it is not always possible to investigate every accident. When determining whether to investigate an accident, managers need to consider the following:

- The **cost and severity** of the accident. Accidents with high losses, whether in life, injury, or property damage, need to be investigated.
- The **frequency** of the accident. If similar accidents occur frequently, they need to be investigated.
- **Public interest** in the accident. If the accident affects the community or is otherwise of special interest to the public, it needs to be investigated to provide factual information and protect the company image.
- The **potential losses** caused by the accident. If the accident may have large losses in life or property damage, it should be investigated.

Recordkeeping

Recordkeeping is one of the most important aspects of occupational health and safety because it provides a basis for independent verification that compliance was achieved. It is also critical for having a properly documented record in the event of an injury. Records must be kept of initial and annual OSHA training classes to show that employees have been properly instructed regarding potential hazards of their jobs and how to protect themselves. Medical clearance records must be maintained. Industrial hygiene testing records must be retained to show that exposure levels are within acceptable limits. Routine safety inspection records must be kept to show that equipment is maintained properly and that emergency equipment is available (i.e. fire extinguishers). Employee injury documentation must include accident investigations, work status reports from the physician, and records of restricted duty offered and accepted.

Ethical Practices in Recordkeeping

Organizations will maintain a variety of records, including personnel files, medical records, legal documents, and financial records. Thus, record keeping for an organization must be done ethically.

The **integrity** of these files is paramount in maintaining the truthfulness of the information contained within. Records must be as accurate and complete as possible for their intended use. Those responsible for records must maintain the security of the records to prevent tampering or manipulation of the information they contain. Records should be secure from unauthorized access or disclosure and should only be released to those legally allowed to have them. Records should only be maintained for the requisite time period and be properly destroyed when no longer needed.

Techniques for Determining the Root Cause of Accidents or Incidents

Root cause analysis methodology is used to determine the most fundamental reason for a system failure or mistake that has led to an injury or equipment failure. The goal of a **root cause analysis** is to identify the central cause; this will allow the central cause to be fixed, thereby preventing a recurrence of that particular event. An effective methodology to conduct a root cause analysis is the **Five Why method**. This method uses around five *why* questions to determine the root of the problem. An example of a Five Why analysis is as follows:

- Q: Why did the valve leak oil? A: The valve had not been replaced.
- Q: Why wasn't the valve replaced? A: The workers didn't know it needed to be replaced.
- Q: Why didn't the workers know it needed to be replaced? A: No one told them.
- Q: Why didn't anyone tell them? A: There is no written replacement schedule available to them.
- Q: Why isn't there a written replacement schedule? A: No one has written one.

This process results in an identifiable task that can be undertaken to prevent a recurrence of this problem.

Post-Incident/Accident Reporting and Follow-Up Procedures

Incident Reporting System

An effective **incident reporting system** must have participation from the right people in the organization, which are those people with firsthand knowledge of the incidents being reported. In the case of injuries and near misses, diligent participation of the direct supervisors is most important. They must be properly informed of the type of incidents that need to be reported, exactly what information to report, and how to report it. In addition, the reporters must be given guidance on how the severity of the incident impacts the reporting of the incident. For example, minor incidents such as issuing a bandage for a small cut may need to be only recorded on a first aid log, whereas an injury requiring medical attention must be recorded but also requires notification of the manager on duty so that the insurance reporting and incident investigation process can begin.

A robust incident reporting system that **records** each incident without fail, and that **reports** the required information about each incident, can provide valuable information to improve future incident response. Over time, the incident log can be examined for patterns that lead to improved prevention of incidents. For example, temporal investigation may reveal that incidents occur disproportionately on the swing shift or just after break time. Incidents may be more likely to occur in one particular area; discovering these patterns can allow the organization to focus resources on prevention and response in particular areas. Analysis of incident data on leaking equipment, for example, may reveal that a certain piece of machinery develops leaks more frequently than others in the work center; the appropriate response that can be developed from this information is to put that piece of machinery on a more frequent preventative maintenance schedule. The point of tracking incidents and collecting data is to use that data as a continuous improvement tool that provides feedback on performance and guides active response.

Eyewitness Interviews

The purpose of interviewing eyewitnesses to accidents is to obtain information about what happened from real-time observers. Each observer witnesses and remembers different aspects of an incident. This information can be used to assist in **root cause analysis** and planning of **corrective action**. It is important to ask open-ended questions that allow the eyewitness to tell the story of what happened rather than to ask leading questions that make it seem as if the answer is already known. For example, ask "What happened after that?" to get a sequential idea of what has happened. Ask follow-up questions if a detail is unclear. In addition, the eyewitness should be asked to sign the statement provided as evidence of its veracity.

Documentation Requirements of Occupational Injuries and Illnesses

Recordable Injuries

A recordable injury is one that must be recorded on an employer's **OSHA 300 log**. A recordable injury is a **work-related** injury or illness that results in a fatality, loss of consciousness, days away from work, days of restricted duty work, or requires medical treatment beyond first aid. Recordable illnesses also include work-related cancer, hearing loss, cracked bones or teeth, or puncture injuries. The number of recordable injuries, benchmarked according to industry and total employee hours worked, is used to set **worker's compensation insurance rates** and is an important **safety metric** that needs to be tracked.

"First Aid" Injuries

OSHA has a strict definition of what types of injuries can be considered "first aid." It is in the best interest of an employer to classify as many injuries as possible as first aid, because this reduces insurance costs. OSHA considers first aid to be a medical treatment that is **one-time** (with one follow up visit allowed) and is generally a **minor** scratch, cut, burn, splinter, or debris in the eye. Injuries that result in the doctor prescribing medication, or in days away from work or days of restricted duty, cannot be considered first aid cases.

OSHA 300 Logs

The OSHA 300 log is a required OSHA recordkeeping element. It serves as the legal record of **recordable injuries** at a given work site. The information to be recorded includes the injured worker's name, the date of the injury, the job title, the work area the employee was in when injured, and a brief description of the injury (for example, employee experienced a lower back strain when moving heavy boxes). The log also requires the injury to be **categorized** as a case that involves death, days away from work, job transfer or restricted duty, or as an "other recordable case." If the worker experiences days away from work and/or restricted duty, the **number of days** must be tracked and recorded. If the recordable case is not an injury but an illness, one must also record the **type of illness** (skin disorder, respiratory condition, poisoning, hearing loss, or other). Be aware that **temporary agency workers** must be recorded on the OSHA log of their current location/employer when the injury occurred.

The OSHA 300A log is the **summary sheet** that accompanies the log of recordable injuries at a work site (the OSHA 300 log). On the 300A summary sheet, the **total number of recordable cases** is presented, along with the number of cases that resulted in **days away from work**, and the total number of cases that resulted in **restricted work duty**. In addition, the total number of **occupational illnesses** is recorded. The company **Standard Industrial Code** is noted, along with the total number of employee hours worked in the year. Note that this figure should not include vacation or holiday time and that it should include all workers at the site, including office workers. The OSHA 300A is an official summary record for the year of occupational injuries and illnesses and must be signed by a company official. It must be posted in an area accessible to the workforce from **February 1 through April 30** of each year.

Electronic Submittal of OSHA 300 Records

OSHA recently published a rule requiring electronic submittal of OSHA 300 records to OSHA by certain industries and businesses. The rule may be found in 29 CFR §1904.41, **Recording and Reporting Occupational Injuries and Illness**. Businesses with more than 250 employees at any time during the previous calendar year that are required to keep OSHA 300 and 300A records must electronically submit the **Form 300A information** by March 2 of the next year (starting in 2019; in 2017 and 2018 the reporting deadlines were different). Businesses in specified industries with 20 - 250 employees must also report. The specified industries are diverse industries such as manufacturing, agriculture, utilities, construction, various types of stores, trucking, warehousing, nursing care, etc. These industries are listed in Appendix A to 29 CFR §1904 Subpart E.

Beginning in 2024, businesses in designated industries with more than 100 employees are also required to electronically submit their forms 300 and 301 once a year.

Occupational Illnesses

An occupational illness (as opposed to an injury) is an illness or condition that can be attributed to an **occupational exposure in a workplace** through a physical, chemical, or biological cause. Examples of occupational illnesses are poisoning caused by exposure to a toxic chemical (such as exposure to lead dust or fumes), or hearing loss caused by overexposure to noise. Other examples that could occur in a construction setting are occupational lung diseases such as asbestosis (from exposure to asbestos) or silicosis (from exposure to silica dust). Contact dermatitis (skin irritation or inflammation) can be caused by exposure to a chemical or to latex in gloves. Carpal tunnel syndrome or other repetitive use disorders are also occupational illnesses.

DART

The DART statistic is the number of injury and illness cases with "**D**ays **A**way and/or **R**estricted Duty **T**ime" in a given year. The statistic is normalized by the number of hours worked and refers to the number of **recordable incidents** per 100 full-time workers. The calculation is as follows:

$$DART\ Rate = ((Total\ Number\ of\ DART\ Incidents) \times (200{,}000)) \div (Total\ Hours\ Worked)$$

The 200,000 figure normalizes the statistic to 100 full-time workers and allows the DART rate to be compared across industries and workplaces of different numbers of employees. A general rule is that the **DART rate** should be less than the **Total Case Incident Rate**, which indicates that only a portion of recordable cases was severe enough to result in restricted duty or lost work time.

TCR

The TCIR is the **Total Case Incident Rate** (sometime called the IR, or Incident Rate), and represents the total number of recordable injuries and illnesses per 100 employees. The calculation is as follows:

$$TCIR\ Rate = (Total\ Number\ of\ Recordable\ Cases) \times (200{,}000) \div (Total\ Hours\ Worked)$$

The TCIR is tracked by the **Bureau of Labor Statistics** for a wide variety of industries and published by **Standard Industrial Classification code**. This is a useful tool in benchmarking a given company's safety performance against others in the same industry.

Documenting Occupational Injuries

After ensuring that an injured worker has the proper medical treatment, it is important to follow a protocol to properly **document** the injury, both for legal and insurance purposes. A **statement** should be taken from both the injured worker and any eyewitnesses regarding what happened. A **root cause analysis** should be done to determine how processes or procedures can be altered to prevent recurrences. Records of all **medical visits** should be maintained. If a worker is on **restricted duty**, a formal record should be kept, stating whether restricted duty is offered to the worker, and the conditions of acceptance. If the injury requires reassessment of a **job hazard analysis**, this record should also be maintained.

Reporting Severe Injuries

A severe injury is defined as any occupational injury that results in an **amputation**, **loss of an eye**, **inpatient hospitalization**, or a **fatality**. Work-related fatalities must be reported to OSHA within **eight hours** and other types of severe injuries must be reported within **24 hours**. An amputation is the loss of any limb or appendage, even the tip of the finger. The company must report the name of the affected employee, the location the injury occurred, the company name, the type of event, a brief description of the incident, and the contact person who can be reached for further information. OSHA concentrates investigation efforts on workplaces with severe or serious injuries, and generally follows up with a **site visit** to verify information and compliance status related to the severe injury event.

Determining Whether an Injury Is Work-Related

The regulation that states the definition of "work-related" is found in 29 CFR §1904.5(a), which states that an injury or illness is work-related if "an event or exposure in the work environment either caused or contributed to the resulting condition." Sometimes this is obvious; for example, if an employee is using a saw and is cut by the saw, this is obviously work-related. In the case of musculoskeletal strains, it is not always so clear. However, if a worker claims that an injury is work-related, it is **assumed to be so** unless proven otherwise. In other words, the worker is given the benefit of the doubt. Even if the worker's off-hours activities may have contributed to the injury (for example, participation in a sports league), an injury will be designated work-related if it could have been caused by the work.

Maintaining OSHA Logs for Multiple Job Sites

OSHA regulations regarding recordkeeping are found in 29 CFR Part 1904. The rules for recording injuries that may occur at **remote job sites** are relevant in the construction industry; for example, an electrical contractor may have employees performing work at several job sites simultaneously. For employees that work at a given job site for **less than one year**, the employer may maintain one OSHA 300 log and record all recordable injuries on the one log. If an employee is working at a given location for **longer than one year**, individual logs must be maintained for each site. The log may be maintained at a central location or corporate headquarters as long as it can be provided to regulators upon request in a reasonable length of time.

Chapter Quiz

Ready to see how well you retained what you just read? Scan the QR code to go directly to the chapter quiz interface for this study guide. If you're using a computer, simply visit the online resources page at **mometrix.com/resources719/chst** and click the Chapter Quizzes link.

Leadership, Communication, and Training

Transform passive reading into active learning! After immersing yourself in this chapter, put your comprehension to the test by taking a quiz. The insights you gained will stay with you longer this way. Scan the QR code to go directly to the chapter quiz interface for this study guide. If you're using a computer, simply visit the online resources page at **mometrix.com/resources719/chst** and click the Chapter Quizzes link.

Available Training Delivery Methods and Instructional Materials

Classroom Training

Classroom training is often used for safety training, but it has its pros and cons. On the positive side, it offers a chance for workers to dedicate their attention to the task without distractions. It allows for questions to be asked to clarify concepts. Some types of information are best presented in a classroom; for example, hazard communication and toxicology of chemical exposures are best presented in a classroom setting. On the negative side, it can be difficult to make the training class lively and engaging, and the participants may get sleepy. It may not be interactive enough. Some people do their best learning by doing and not by listening to another person give a presentation.

On-the-Job Training

On-the-job training is training given outside of the classroom setting in the work area. It is usually used for teaching the actual mechanics of a production job; however, it is also a great opportunity to present **risk and hazard information** in a real-world setting. A positive reason to do on-the-job training is to provide an opportunity for **interaction and hands-on training**. This can help solidify the concepts and make the training more relevant. A negative aspect of doing on-the-job training is that it is not well-suited to providing information about academic concepts such as chemical exposures or hearing conservation. It is also not a good method for large groups or noisy work areas.

Online Training

Online, web-based training classes are useful because they are available anytime and do not require scheduling the physical space and the trainer. They can be very cost-effective and can be tailored to the specific topic to be delivered. Programs purchased from a training program vendor can also keep track of the training records. They can also have built-in **competency assessments**. On the negative side, they provide little if any interaction and may be difficult for some people to pay attention to. They do not provide an opportunity for the student to ask questions and get clarification on concepts that aren't understood. Some people may not be familiar with how to use a computer, depending on literacy levels and background.

Training Delivery Mediums and Technologies

Safety training can be difficult and tedious, but it is critical for worker health. Classroom training is beneficial because content can be tailored to the audience and work environment. It also allows for personal interaction and hands-on training activities that enhance learning. The disadvantages include the difficulty of scheduling a class that can accommodate everybody who needs to take the class. Usually multiple times and locations must be offered. Online training classes are an attractive option because they can be completed whenever the individual has the time. However, they tend to be generic and not tailored to the audience or specific work environment. Sitting in front of a computer terminal is often boring and it is more difficult for employees to understand and retain information delivered this way.

Choosing Appropriate Training

A variety of training techniques should be used to teach and reinforce concepts. The best training programs use a combination of all three methods. Traditional classroom lecture-style training is useful when the topic is academic in nature. For example, it can be helpful to explain lockout/tagout regulations and procedures in a classroom setting. Demonstration training is best when the topic involves physical manipulation of an object. An example of demonstration training is teaching how to use, disassemble, and maintain a respirator. On-the-job training refers to training in specific tasks and occurs in the actual work environment. Written reference material and job aids should be available to reinforce the proper performance of each task. In the example of lockout/tagout above, on-the-job training should be conducted with each piece of equipment to show employees where the lockout points are and to demonstrate proper isolation of the energy source.

Learning Styles

Although there are many theories about learning styles in the educational field, one popular theory that can be applied to safety training classifies learning styles into four types (acronym: VARK):

- **Visual**—This type of learner retains information best by seeing information presented visually, such as reading it for themselves as opposed to listening to a presentation. They can also benefit from information presented in charts or graphs.
- **Auditory**—This type of learner responds to hearing information, such as a lecture or oral presentation.
- **Reading/Writing**—This type of learner responds to reading, followed by note taking or re-writing information. The act of taking notes reinforces the concepts and helps the learner to retain knowledge.
- **Kinesthetic**—A kinesthetic learner learns best by doing an activity related to the concepts. For example, this type of learner would respond to a hands-on exercise of how to disassemble and reassemble a respirator as a way of learning how it works, rather than viewing a slide show presentation on the topic.

The best types of safety training classes integrate aspects that will appeal to all of these learning styles and reinforce concepts using multiple methods.

Appropriate Human Behavior Motivation Methods and Techniques

Behavioral-Based Safety

Behavioral-based safety is a system of improving a safety culture and safety performance that concentrates on recognizing and changing employees' unsafe behaviors. Although it is preferable to improve safety by engineering out hazards, it remains a fact that many on-the-job injuries result from individual employees doing unsafe acts, despite policies and procedures that are put in place to ensure a safe workplace. For example, an employee may lift an object without employing safe lifting procedures or may elect to use a tool without donning the appropriate personal protective equipment (PPE). Implementing a behavior-based safety program aims to identify unsafe acts and uses peer observations and corrective actions to improve safety performance and adherence to safe workplace policies.

<u>KEY ELEMENTS OF SUCCESSFUL IMPLEMENTATION</u>: Behavioral-based safety must first start with a sound assessment of the types of behaviors that are important in a given work environment to reduce the incidence and severity of injuries. Past data on injuries, first aid, and near misses should be evaluated to determine the common unsafe acts that could lead to incidences. These are the behaviors that can be concentrated upon first. Second, teams of safety observers must be recruited for participation in the behavior-based safety program. The line-level workers must be trained in how to make observations of behavior for unsafe acts, and the workers who have performed the unsafe acts need to be notified and corrected on the spot but not in a punitive way. There should also be ongoing feedback to the entire team on the aggregate behavioral observations and how the team is improving over time so that the benefits of the system can be observed.

<u>POTENTIAL PITFALLS</u>: There are potential pitfalls in the execution of a behavioral-based safety program. The successful implementation requires involvement and participation from line-level workers to make safe behavior observations and to provide feedback to their coworkers. If proper training and support are not provided, the observations made will not be useful. In addition, the team has to understand the potential benefits of improving safety behaviors and cannot be made to feel that they are being blamed for injuries that are out of their control. There should be a definite and observable connection between the safe behaviors observed and the types of injuries and incidents being reduced through improved behavioral performance.

Communication Strategies

There are several ways to communicate safety information to workers. One is **written information** in the form of work instructions or safety guidelines. Another is to post reminder signs with **pictures** to communicate the information. A **company newsletter** can be used to publish reminders or other information. **Safety tailgate meetings** can provide valuable time for two-way communication. **Posters** can be used as safety reminders. **Video monitors** can show a slide show presentation on a continuous loop; this is effective to use in a break room or in a situation in which employees are standing in line and are therefore a captive audience.

Clear communication is critical for safety. Job procedures, PPE requirements, and emergency procedures should be communicated in training courses, reviewed in safety meetings, and refreshed and reinforced regularly. To help reinforce these training methods, communication aids should be created and posted in the work areas to serve as a reference and a reminder.

Job procedures—Job procedures should be written in a way that communicates in words the steps needed to complete a given job safely. The procedure should also use pictures and illustrations as much as possible so that communication is visual as well as written. These can be posted at the point of use as a reference and reminder.

Personal Protective Equipment—Required PPE should be listed and posted near the equipment along with a photograph or illustration for clarity.

Emergency Procedures—Emergency exit routes should be clearly indicated on facility evacuation maps that are posted in work areas. It is also helpful to have evacuation routes drawn on the floor in reflective paint. Emergency exits must be indicated with lighted signs. An emergency procedures checklist should be readily available in the work area that lists any equipment that should be de-energized during emergencies. Also, someone must be designated to ensure everyone has evacuated a given area and is accounted for in the assembly area.

When to Consult with Equipment Manufacturers, Suppliers, or Subject Matter Experts

Manufacturers and suppliers can be valuable sources of safety information. They know best how their products work and which product is best for a given hazard. For example, glove manufacturers should be consulted when selecting gloves for particular hazards. They can provide information on chemical and physical resistance of the gloves and suggest alternative products that may be more cost-effective for your application. When you are purchasing a new piece of equipment, consulting the manufacturer in advance to determine potential safety risks of the process is essential. The manufacturer will often be able to provide safety and training materials. Forklift manufacturers provide training materials for their products and usually have personnel that can come on-site to conduct the training. Consulting with manufacturers of safety equipment such as machine guards can provide valuable information on the right solution for your machine guard hazard. Although consulting manufacturers does not always yield a solution, it is an avenue that should be explored.

There are multiple subject matter experts who can be consulted to promote optimum safe practices. For example, one should always consult the maintenance and engineering staff for input on mechanical and machine hazards. They can provide information on the operation of various machines and the potential hazards, what hazardous chemicals may be used in maintenance activities, and use of emergency shut-off mechanisms. Other subject matter experts to consult are industrial hygienists for advice on exposure limits to determine appropriate levels of respiratory protection. Worker's compensation insurance providers may be able to provide invaluable information to help manage worker's compensation claims to minimize costs while providing proper care for employees injured on the job. In order to minimize ergonomic hazards and prevent injuries from ergonomic stresses it is important to consult an ergonomics expert.

Information Confidentiality Requirements

Employee Health Information

Employers are not entitled to all the health and safety information of employees. There are legal privacy protections under the Health Insurance Portability and Accountability Act (HIPAA) that protects employee's health information if it is not connected to a work-related injury, request for sick leave, or provision of health insurance. Health information provided to the employer must be authorized by the employee. Employers are authorized to receive work readiness health information for employees that have work-related injuries. These are not actual medical records but are the physician's assessment of an employee's work restrictions or prognosis. Alcohol and substance abuse records are not considered occupational health records. There are other health-related records employers are obligated to retain. For example, records of employee exposures to toxic materials must be retained for thirty years. This includes industrial hygiene monitoring records and the results of any biological monitoring such as blood lead levels or cadmium in urine levels. Results of audiometry must be similarly retained. The purpose of this requirement is that certain adverse exposure outcomes may take years to manifest (such as hearing loss or cancer) and exposure records need to be available in case there is a possibility that the problem is work-related.

Trade Secrets

A trade secret is business information that the business does not want disclosed to customers or competitors but is not protected by any formal procedure such as a copyright. To be considered a trade secret, the information must not be generally known, it must have commercial value because it is secret, and the organization must have taken steps to keep the information confidential. Steps that organizations must take to keep trade secrets include identifying what the information is and to whom it will be disclosed. Steps must be taken to ensure written documents are stamped or marked confidential and that there is an assured destruction method available to discard written materials (such as shredding). Computer files that contain confidential information must be password protected, and the electronic file storage systems must be designated and access granted only to those who need to know. There should be a documented procedure for granting computer file access. Employees must be given training on the type and nature of trade secret information and the importance of securing this information.

Leadership

Trait Theory of Leadership

According to the trait theory of leadership, all successful leaders have certain personal characteristics in common: initiative, ambition; energy, desire for improvement, self-assurance, competence, and a firm grasp of the strengths and weaknesses of others.

Stogdill's Handbook of Leadership identifies the following traits common to good leaders:

- *Capacity*: knowledge, intelligence, focus, creativity, and judgment
- *Achievement*: accomplishments and education
- *Responsibility*: persistence, reliability, and determination
- *Participation*: cooperation, friendliness, helpfulness, and selflessness
- *Status*: popularity, reputation, and deference from colleagues

One criticism of this leadership theory is that it places all the emphasis on the leader, ignoring the employee, and it fails to differentiate between traits.

Behavioral Theory of Leadership

The behavioral theory of leadership, which began with the early work of Rensis Likert and his colleagues at the University of Michigan who identified job-centered and employee-centered leader behaviors. There are two leadership behaviors at the center of the behavioral theory of leadership: task accomplishment and sensitivity for the feelings of subordinates. A leader who demonstrates both of these behaviors will always be effective. In the work of Likert and his associates, there was special attention to three leadership behaviors:

- *Task performance behavior*: related to specific processes and performance of work tasks
- *Group maintenance behavior*: teambuilding and supportive leadership
- *Decision-making*: may be either autocratic or democratic

An autocratic decision-making style may result in higher performance, but the democratic style seems to result in happier employees. Leaders who make no decisions (that is, leaders who have adopted a laissez-faire style) create low performance and low employee satisfaction. Leaders who pay great attention to maintenance behavior tend to retain their employees far longer.

Situational or Contingent Leadership Theory

According to contingent (also known as situational) leadership theorists, the most effective strategy for leading a group depends on the circumstances. In other words, there is no one set of leadership skills that works in every case. Situational leadership centers around three approaches:

- The first approach emphasizes harmony between manager and situation
- The second approach emphasizes the probability that a manager will have to modify his or her approach in a given situation
- The third approach centers on the way some situations can mitigate or neutralize the skills and style of the leader

At this point, most leadership theorists believe that the second approach is the most consistent with reality.

Transformational Leadership Theory

Transformational leadership theory (Burns, Bass) stresses the impact that leaders can have to facilitate change. Instead of a transactional style of leadership in which change is brought about through a system of rewards and punishments (or withholding of rewards), a transformational leader leads change through showing respect and consideration for individuals, challenging them intellectually, and inspiring and influencing them. Transformational leaders are often personally charismatic and tend to lead through example rather than control. The transformational leader identifies those things that need change along with old methods that are inefficient. They then work with other members of the team to motivate them, help them to find solutions, and develop new and more efficient methods of doing things. The transformational leader provides a sense of direction and empowers others to do their best, all while focusing on collective purpose, pride in work, and achievement.

Ethical Leadership

Ethical leadership is where those in decision-making positions demonstrate, through their own behavior, the conduct that is expected of subordinates. In basic terms, it means leading by example. For a safety professional, this involves adhering to safety requirements at all times while at the worksite—wearing the appropriate personal protective equipment, following applicable procedures, and abiding by all safety regulations.

Ethical leadership is achieved by mutual respect and two-way communication. Safety leadership is not accomplished by threats or strongarm tactics. It is accomplished by humanizing the employees and building relationships. Rules are applied evenly as are expectations—there are no "favorites" and no turning a blind eye. This approach has been shown to be much more effective for creating a positive safety culture within an organization.

Integrity

Integrity can be defined as the adherence to an ethical or moral code. It is the practice of being honest and doing what is right no matter the circumstances. Integrity is difficult to establish but easy to lose.

Integrity is multifaceted. It is demonstrated by following through on promises to locate information, by avoiding gossip, and by being fair and consistent. It also involves putting aside personal gain in favor of the advancement of the organization. Integrity is lost by playing favorites, being dishonest, advancing one's own interests above those of the organization, and by demonstrating to employees that they or their opinions are not valued.

Importance in the Safety Profession

Integrity is a quality of a successful leader. A safety professional is part of the leadership team, even if they are not part of the management team. Their role is to convince employees to follow safe procedures and protocols, to use protective equipment, and to avoid risky behaviors. In some instances, the safety professional must manage the change from a risky culture to a risk-averse culture.

The safety professional must demonstrate their integrity by adhering to health and safety measures at all times to avoid appearing or being hypocritical. Managers must be held to the same safety standards as all others, so that employees can see that the safety professional is fair and consistent, regardless of status. When the safety professional has earned the confidence of a group of employees, the group is more willing to listen and to follow and support safety initiatives. The safety professional must believe in safety and the importance of the worker in order to be successful. The integrity of the safety professional plays a pivotal role in that success.

Chapter Quiz

Ready to see how well you retained what you just read? Scan the QR code to go directly to the chapter quiz interface for this study guide. If you're using a computer, simply visit the online resources page at **mometrix.com/resources719/chst** and click the Chapter Quizzes link.

CHST Practice Test #1

Want to take this practice test in an online interactive format?
Check out the online resources page, which includes interactive practice questions and much more: **mometrix.com/resources719/chst**

1. A workstation is located an equal distance away from three different machines that emit sound pressure levels of 100 dBA, 98 dBA, and 95 dBA. When employees are at this workstation, what sound pressure level are they exposed to?

a. 293 dBA
b. 102.9 dBA
c. 97.7 dBA
d. 98 dBA

2. What is the difference between a safety inspection and a safety audit?

a. Safety inspections can be performed by anyone, whereas audits are generally only performed by upper management.
b. Safety inspections are performed during the actual job tasks, and audits are performed in a non-working environment.
c. Safety inspections focus on safety processes and procedures, and safety audits focus on the actions of people and equipment.
d. Safety inspections focus on physical safety hazards, and safety audits focus on evaluating processes and programs.

3. During an emergency, specific employees can be selected as emergency coordinators or emergency wardens. Which of the following assignments would NOT be automatically assigned to an emergency coordinator or warden?

a. Determining when an emergency exists and activating procedures
b. Coordinating with outside emergency services
c. Assisting in the evacuation of employees
d. Filling the incident commander position

4. OSHA 29 CFR 1910.28 and 1926.501 are fall protection standards for General Industry and Construction Industry, respectively. What is the main difference between these two standards?

a. CFR 1910.28 calls for fall protection practices when an employee is at 4 feet above the walking/working surface, and 1926.501 calls for fall protection at 6 feet above the walking/working surface.
b. General Industry does not require a guardrail system for material handling operations on a dock board when the exposed fall height is less than 10 feet. Construction Industry requires sidewalls to be in place during material handling on a dock board, regardless of height.
c. CFR 1926.501 calls for fall protection at 4 feet above the walking/working surface, and 1910.28 calls for fall protection practices when an employee is at or above 6 feet off the walking/working surface.
d. General Industry and Construction Industry standards are the same in this instance.

5. A small fire breaks out from an electrical outlet inside a portable building being used as a temporary break room and technology server room for a school district. Since the school is under construction, there is a pile of fire extinguishers along the wall. Which type of fire extinguisher should be chosen for this fire?

a. Class A
b. Class B
c. Class ABC
d. Carbon dioxide (CO_2)

6. A safety supervisor writes a violation concerning an air conditioning unit inside a small trailer on the job site that is plugged in using multiple extension cords that lead to a light plant 50 yards away. What is the correct safety recommendation for extension cord use in this situation?

a. Move the light plant within 100 feet of the air conditioner to avoid the use of multiple extension cords.
b. Use a power strip between the light plant and the first extension cord.
c. No violation has actually occurred in this instance.
d. Install an electrical distribution box near the air conditioner so that the unit can be plugged directly into the power source.

7. What leading systematic risk management strategy involves identifying potential deficiencies in design processes and equipment failures and assessing the harmful outcomes to apply actions to safeguard against occurrence?

a. Root cause analysis
b. Job safety analysis
c. Accident investigation
d. Failure Mode and Effects Analysis

8. Which of the following is NOT a cumulative trauma injury?

a. Rheumatoid arthritis
b. Carpal tunnel syndrome
c. Tendonitis
d. Nerve entrapment

9. What is the main objective for designating a spokesperson during a crisis management event?

a. To ensure accurate, consistent, and timely information is delivered to the public
b. To prevent the spread of misinformation
c. To protect the reputation of the company involved in the crisis
d. To intercept negative community feedback

10. The object you have chosen to use to perform testing on personal fall arrest systems (PFAS) is 6 inches wide and 8 inches high. What is the girth of this object?

a. 48 in
b. 1.33 in
c. 28 in
d. 24 in

11. A construction worker is assigned the task of pouring concrete mix into a cement mixer. What would be the PEL when using this material if it contains 20% silica (SiO_2)?

a. 20 mg/m^3
b. 12 mg/m^3
c. 1.2 mg/m^3
d. 0.45 mg/m^3

12. What type of fire extinguishing agent would need to be used in an industrial kitchen fire?

a. Dry chemical
b. Class ABC
c. CO_2
d. Class K

13. What scenario represents the most effective use of universal precautions?

a. A first aid provider immediately disposing of blood-contaminated gloves in the nearest trash receptacle
b. A procedure requiring nurses not to place the plastic cap back on a used needle but rather to place the needle and syringe directly into a biohazard container
c. A company policy requiring health screens after exposure to bloodborne pathogens
d. Cleaning up an injury scene with soap and boiling water

14. In an emergency in which the Incident Command System is utilized, what is the role of the liaison officer?

a. The liaison officer provides services and resources as requested by the incident command.
b. The liaison officer communicates current operations with various agencies involved with the emergency and coordinates media messages with the public relations officer.
c. The liaison officer keeps track of financial expenditures, contracts, and documentation with regards to the emergency.
d. The liaison officer is responsible for protecting the health and safety of the public and the environment.

15. For a job site in a state-owned territory, the state regulatory agency requires that an employee from each contractor on the job site is trained in CPR and first aid. In what document would this new training requirement need to be contained to ensure the information gets passed along to all employees and contractors?

a. Company health, safety, and environmental manual
b. Policies and procedures manual
c. Master service agreement
d. Site-specific safety plan

16. What is the maximum use concentration (MUC) of a chemical with a PEL of 25 ppm using an air-purifying respirator (APR) with an APF of 50?

a. 2 ppm
b. 1,250 ppm
c. 0.5 ppm
d. 10,050 ppm

17. You are evaluating a compressed air gun that construction workers are using for cleaning purposes to determine if it exceeds OSHA's psi limit. You measure the force of the air at 202 lb and the area as 38 cm^2. Does this compressed gas air gun exceed OSHA's psi limits for compressed air used for cleaning purposes?

a. Yes, the psi is over 7,500, so it exceeds OSHA's limit.
b. No, the psi is over 7,500, so it is well under OSHA's limit.
c. No, the psi is around 35, so it is well under OSHA's limit.
d. Yes, the psi is around 35, so it exceeds OSHA's limit.

18. You have identified an employee you believe to be exposed to a significant amount of vibration. In order for controls to be put in place, management has informed you that the amount of vibration must exceed ACGIH's action level ($2.5\ m/s^2$). The employee uses two pieces of equipment (1 and 2) during an 8-hour shift, each for 4 hours. The vibration exposure energy equivalent over an 8-hour shift [A(8)] from equipment 1 is $1.2\ m/s^2$ and from equipment 2 is $1.4\ m/s^2$. Based on the vibration exposure values, is the employee's cumulative overall daily vibration exposure [$A_t(8)$] greater than ACGIH's action level?

a. Yes, the $A_t(8)$ is 2.8 m/s^2, which exceeds ACGIH's action level.
b. No, the $A_t(8)$ is 1.68 m/s^2, which is below ACGIH's action level.
c. No, the $A_t(8)$ is 1.84 m/s^2, which is below ACGIH's action level.
d. Yes, the $A_t(8)$ is 2.6 m/s^2, which exceeds ACGIH's action level.

19. A general construction laborer is operating a piece of equipment used to cut sheet metal. The machine cuts the metal, and a rotating belt slides the sheet metal down to a preset measurement before automatically cutting again. The cuts of sheet metal are slid off the belt into a stacked pile of sheet metal ready for packaging. Each piece of sheet metal is cut five times, and then a new piece of metal needs to be added to the belt. The worker only needs to ensure the metal stays straight on the belt and add new sheets when needed. If the sheet metal shifts or a new piece of metal needs to be added, the worker can use a stop button from a safe distance away to stall the machine until the metal is placed correctly on the belt. During a sheet metal adjustment, a manager walks by the equipment and notices that the cutting equipment is not operating. The manager fails to see the employee kneeling down to adjust the bottom side of the belt. The manager engages the machine, and the belt sucks in the employee's hand, mangling it. According to the domino theory of accident causation, accidents only occur because of careless persons and/or poorly designed and improperly maintained equipment. In this scenario, what example best represents the domino theory?

a. Redesigning the equipment to eliminate the need to access the belt during cutting operations
b. Requiring lockout/tagout procedures to take place when the machinery must be accessed
c. Providing training to all employees on safety measures to take when working on or around the equipment
d. Changing work procedures so that when the equipment needs to be accessed, another employee will stand guard to ensure the equipment cannot go back live

20. According to the Bureau of Labor Statistics, what is the leading cause of work-related deaths?

a. Incidents involving motorized vehicles
b. Incidents involving being struck by another object
c. Incidents involving hazardous gases
d. Incidents involving falling from heights

21. As part of a construction project, several employees need to use a jackhammer to break up some concrete. To protect the employees, you decide to equip each jackhammer with a vacuum dust collection system (VDCS). The job task will require a capture velocity of 2,500 fpm. If each system has a flow rate of 167 cfm, what hose radius must be used in the VDCS?

a. 1.25 in
b. 14.97 in
c. 1.75 in
d. 0.58 in

22. You are supervising a hoisting and rigging task involving a large, 26' square crate. This job requires four anchor points on opposite ends parallel with each other using two crane points, as diagrammed below:

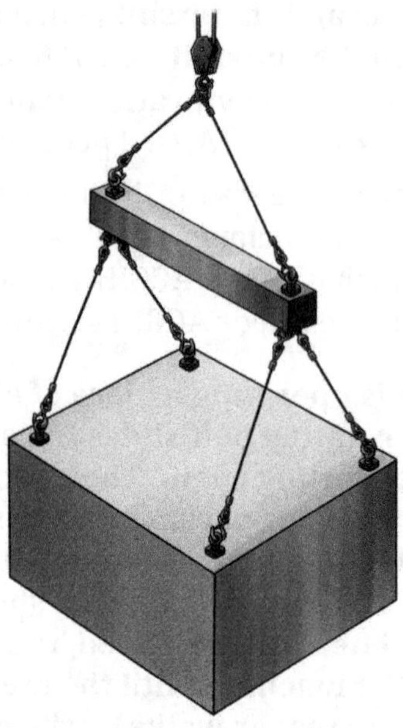

The angle of the ropes to the crate must not be less than 30°, as angles that low drastically increase the amount of tension on the rigging equipment.
It is your job to determine what the angle is. The length of rope from the anchor points on the crate to where the two pieces of rope converge is 15 feet.

a. The angle is barely below 30° (29.9°), so the hoisting and rigging task must not be performed until adjustments are made.
b. The angle is 45°, so the task can be performed.
c. The angle is well below 30° (18°), so the hoisting and rigging task must not be performed until adjustments are made.
d. The angle is 38°, so the task can be performed.

23. During the preparation meeting for a task involving work near a radiological source, it was determined that the employees performing the job could not be exposed to more than 400 mrem for the entire 3-hour task (133 mrem/hr). At a distance of 2 feet from the source, the dose rate is 600 mrem/hr. How close can the workers be to the radiological source while keeping their radiological exposure dose to 400 mrem?

a. 12.54 feet
b. 2.97 feet
c. 8.53 feet
d. 4.25 feet

24. Your company recently installed an eyewash station where the electric forklifts are stored and charged. Following installation, you need to verify that the eyewash station has sufficient flow for 15 minutes according to applicable ANSI standards. To determine the flow rate, you measure the total amount of water discharged from the eyewash station after running for 15 minutes, which equals 5 gallons of water. Does this eyewash station's flow rate meet ANSI's standards?

a. Yes, the eyewash station has a flow rate of 0.33 gallons/min.
b. Yes, the eyewash station has a flow rate of around 5.15 gallons/min.
c. No, the eyewash station has a flow rate of around 0.98 gallons/min.
d. No, the eyewash station has a flow rate of 0.33 gallons/min.

25. During a company fire drill, it comes to the attention of the floor warden that an employee located on the fifth floor is recovering from an injury and is currently in a wheelchair. The injured employee does not have the ability to evacuate the building using the stairs as stated in the emergency action plan. What is the best alternative the company can provide to the employee to ensure her safety?

a. The company can advise the injured employee to shelter in place in a stairwell until the fire department arrives.
b. The company can move the injured employee to a first-floor office to eliminate the need to evacuate from an upper level.
c. The company can assign another employee to assist the injured employee in the event of an emergency.
d. The company can allow the injured employee to use the elevator rather than the stairs in an emergency.

26. You have been asked to determine the daily vibration exposure value [A(8)] when an employee uses a jackhammer with an acceleration of $4.2\ m/s^2$ for 3 hours. What is the A(8) value for this task?

a. The A(8) value is equal to $4.2\ m/s^2$.
b. The A(8) value is equal to $2.6\ m/s^2$.
c. The A(8) value is equal to $1.3\ m/s^2$.
d. The A(8) value is equal to $2.1\ m/s^2$.

27. How often should workers who work at heights daily inspect their fall protection harnesses?

a. Workers should inspect their harnesses before each use, and a competent person must inspect them every month.
b. Workers should inspect their harnesses before each use and at least once every month, and a competent person must inspect them annually.
c. Workers must inspect their harnesses before each use, and a competent person must inspect them annually.
d. Workers must inspect their harnesses before each use, and a competent person must inspect them annually and after an impact.

28. When developing an emergency action plan, what information should also be communicated to local first responders?

a. A checklist with all employee names and emergency contact information
b. The names and contact information of site supervisors for the project
c. The location, amount, and type of flammables or hazardous chemicals stored on site
d. A copy of the emergency action plan to be approved

29. A welder needs to perform work on a construction site within 30 feet of ongoing electrical work and a diesel-powered generator with dual tanks. What type of permit must be obtained before starting welding operations?

a. Critical operations work permit
b. Electrical work permit
c. Hot work permit
d. Energy isolation work permit

30. Your company needs to procure a portable platform to be used on structures in conjunction with overhead line work. The maximum intended load for this equipment is 250 lb. According to 29 CFR 1926, what load rating must the portable platform have?

a. 500 lb
b. 625 lb
c. 750 lb
d. 1,000 lb

31. Workers need to access a distribution panel that is centrally located on a rooftop. The workers will not get closer than 6 feet from the edge of the roof, as the distribution panel can be accessed from a centrally located stairwell. The roof is 63 feet tall. What fall protection method would provide the most protection in this scenario?

a. Safety net
b. Self-retracting lifeline
c. Geronimo line
d. Horizontal lifeline

32. A supervisor notices an employee who has a sudden change in attitude, feelings of paranoia, lack of motivation, slurred speech, and outbursts of anger. What training would be beneficial in helping this supervisor decide what action to take next?

a. Employee termination training
b. Alcohol and substance abuse training
c. Behavior-based safety training
d. Mental health in the workplace training

33. You will be performing personal sampling for respirable silica during the cutting of several cinderblocks by a construction worker. During the pre-calibration of an air sampling pump, you recorded these values:

- Pre-calibration:
    - 2.0154 L/min
    - 1.9843 L/min
    - 2.0004 L/min
- Post-calibration:
    - 2.0034 L/min

What is the average flow rate of the air sampling pump?

a. 2.0017 L/min
b. 2.0009 L/min
c. 2.0034 L/min
d. 2.0000 L/min

34. Protective grounding equipment must have an ampacity at least equal to that of No. 2 AWG copper to safely carry fault currents. If the grounding conductor is labeled XHHW-2, what does this indicate?

a. Thermoplastic insulation, heat-and moisture-resistant, rated for 75°C in wet and dry locations
b. Dual-layer rubber insulation, rated only for dry locations at 90°C
c. Cross-linked polyethylene insulation, heat- and moisture-resistant, rated for 90°C in wet locations
d. Romex bundling insulation, moisture-resistant, rated for 75°C in wet locations

35. When responding to a medical emergency, what should first aid providers always do first?

a. Obtain consent to treat the injured employee.
b. Ask for a witness statement.
c. Contact emergency services.
d. Assess the scene for safety.

36. A metal parts cleaning facility degreases metal parts by submerging them in a solvent bath. The solvent bath consists of acetone (PEL = 1,000 ppm), xylene (PEL = 100 ppm), and n-hexane (PEL = 500 ppm). Using a photoionization detector (PID), you measure the ambient air as containing 475 ppm of acetone, 62 ppm of xylene, and 375 ppm of n-hexane. Based on these measurements, determine the PEL for the organic solvent mixture and whether employees are being overexposed by the solvent bath.

a. TLV = 3.04; employees are not being overexposed.
b. TLV = 1.845; employees are being overexposed.
c. TLV = 0.753; employees are being overexposed.
d. TLV = 0.432; employees are not being overexposed.

37. Which OSHA regulation pertains to emergency response planning, and how often does OSHA recommend emergency response drills occur?

a. OSHA 29 CFR 1926.103; drills should be conducted biannually.
b. OHSA 29 CFR 1910.28; drills should be conducted with each new hire.
c. OSHA 29 CFR 1926.35; drills should be conducted at random and at least annually.
d. OSHA 29 CFR 1910.21; drills should be conducted weekly.

38. According to 40 CFR 264.175(b)(3), a containment system must have sufficient capacity to contain 10% of the volume of its containers or the volume of the largest container, whichever is greater. How many gallons should the secondary containment hold if 10 55-gallon drums are in it?

a. 61 gallons
b. 55 gallons
c. 60.5 gallons
d. 65 gallons

39. A job site has been known to have intermittently low levels of hydrogen sulfide (H_2S) present in the atmosphere. Peak levels range from 10–15 ppm. Knowing this hazard has the potential to exist, what training should the company provide its employees prior to entering the job site?

a. H_2S awareness training should be provided to all employees prior to entering the site.
b. SCBA fit testing and medical review should be conducted for all employees prior to entering the site.
c. Supervisors should undergo H_2S training.
d. No action is necessary at this ppm.

40. A company changes its chemical formula for cleaning equipment. What action should be taken to meet HazCom requirements?

a. Contact the EPA to dispose of the leftover old formula.
b. Document the end date of the old formula and archive the SDS for 30 years.
c. Use the remaining formula before ordering the new chemicals.
d. Report the change to OSHA and mandate specialized training for all workers.

41. You are asked to properly characterize the hazards associated with a chemical solution by determining its pH. The solution contains 0.0001 hydronium [H^+] ions per liter. What is the pH of this solution?

a. 12
b. 2
c. −4
d. 4

42. Prior to a building remediation project, you discover mold growing in one of the building rooms that is 3 m high, 4 m long, and 1.5 m wide. In order to assign the proper PPE, you need to determine how many mold spores are in the ambient air. After using an air sampling pump set to 15 L/min to pull air through a mold trap for 5 minutes, you count 850 mold spores. Based on this, how many mold spores are in the room's air?

a. 204,000 mold spores
b. 52,932 mold spores
c. 850 mold spores
d. 15,350 mold spores

43. In the event of an emergency, what type of device broadcasts the most current weather, evacuation orders, national security messages, and other public safety statements continuously?

a. Cellular phone
b. Internet news
c. NOAA Weather Radio
d. Social media

44. A newly hired safety manager is preparing to conduct site audits and meet employees for the first time. What is the best strategy for the manager to take?

a. The safety manager should inspect the sites and verbally review the findings with the employees in the introductory meeting.
b. The safety manager should inspect the sites and then host a company-wide Zoom meeting to introduce themselves to the employees and include the findings and expectations in the speech.
c. The safety manager should visit the sites in person to speak to the employees prior to conducting any audits or inspections.
d. The safety manager should bring other members of management along for the site audits and have upper management introduce them to the employees.

45. What is the most efficient way for companies to stay up to date with continually changing environmental and safety rules and regulations that affect their industry?

a. Join a professional organization and subscribe to industry alerts.
b. Purchase a copy of the OSHA standards that relate to their industry when there is an available update.
c. During annual program reviews, check the environmental or safety regulations to ensure the program is still relevant.
d. Hire an environmental health and safety officer to monitor potential changes.

46. OSHA 29 CFR 1926.35 requires employers to provide emergency action plans that describe all the following procedures EXCEPT:

a. Procedures that specify employees who are to remain on site to operate critical machinery prior to evacuating
b. Procedures for reporting emergencies to proper authorities
c. Procedures for the proper handling of hazardous chemicals on site
d. Procedures outlining exit routes and emergency evacuation details

47. What is the main goal of an environmental emergency response plan?

a. To contain a spill to prevent discharge into waterways
b. To inform local authorities how to best respond to a facility spill
c. To establish written documentation to provide to the EPA
d. To provide a plan on how to respond, mitigate, and reduce impact in the event of a spill

48. What group of individuals would have the best input when developing new policies and procedures within an organization?

a. Upper management and other industry leaders
b. GA group of middle and front-line managers
c. Safety committee with members from all levels
d. Safety coordinators and regulatory authorities

49. A construction crew is transporting a 16,000-lb load with a 12,000-lb truck. The truck is limited to 5 passengers. What license will the driver need to comply with Department of Transportation regulations?

a. A Class B driver's license with a passenger endorsement
b. A commercial driver's license (CDL)
c. An overweight restricted load permit
d. A Class C driver's license

50. During a tabletop exercise focused on a significant weather event, a construction company discovers that the current written emergency plan provides the address of the company's main corporate office but does not include the specific addresses of the current job sites. The company operates simultaneous jobs throughout a three-state area. What should the company do to ensure emergency services are delivered to the proper locations in an actual emergency?

a. Individualize each emergency action plan to include the address and coordinates of the current job site.
b. Hire a dispatch officer to keep track of job site locations and relay locations to emergency services at the beginning of each job.
c. Change protocol to have all emergency calls made to the corporate office first before the corporate office calls emergency services to provide the specific address.
d. Ensure that each job site supervisor knows the location of the current job and makes any emergency services calls.

51. A construction company has been hired to build a gravel pad for the future site of a drilling rig. The location of the site is heavily wooded and controlled by the Bureau of Land Management (BLM). How could working on BLM land change the job scope?

a. The construction company management would need to ensure that any additional permits and reporting requirements of the BLM were addressed prior to disturbing BLM land.
b. The job scope will not change, as BLM does not have regulatory authority beyond city and state permits.
c. The drilling rig operator would oversee any additional safety and environmental measures required by the BLM.
d. The BLM would only need to be contacted during land reclamation after the rig has moved off-site.

52. You are asked to determine the transmission loss (TL) of a noise control enclosure. The transmission coefficient (τ) for the material is 0.05. What is the TL of the material?

a. 13 dB
b. 0.05 dB
c. 50 dB
d. 5 dB

53. In an area where workers are cutting cement blocks made of Portland cement, you perform an air grab sample to determine the ambient air concentration of hexavalent chromium. The grab sample collected $0.125\ \mu g$ in 20 liters of air. Does this concentration exceed OSHA's action level for hexavalent chromium ($2.5\ \mu g/m^3$) if employees were exposed to this concentration for the entire 8-hour shift?

a. No, the concentration of hexavalent chromium is just under OSHA's action level at $2.3\ \mu g/m^3$.
b. Yes, the concentration of hexavalent chromium is over $50\ \mu g/m^3$, which is well over OSHA's action level.
c. No, the concentration of hexavalent chromium is $0.00625\ \mu g/m^3$, which is well under OSHA's action level.
d. Yes, the concentration of hexavalent chromium is over double the action level at $6.25\ \mu g/m^3$.

54. You are part of a meeting where it is decided that a scissor lift will be used for employees to access HVAC ducting near the ceiling. Your company assigns you the task of determining if the ground bearing pressure of the scissor lift will exceed the floor loading capacity in the area. The floor loading capacity is 200 pounds per square foot (PSF). The scissor lift has a length of 65 in, a width of 30 in, and a gross weight of 2,485 lb. Based on this information, is this scissor lift safe to use in the work area?

a. The scissor lift is safe to use. The ground bearing pressure of the scissor lift is less than 100 PSF, which is less than the floor loading capacity.
b. The scissor lift is not safe to use. The ground bearing pressure of the scissor lift is over 100,000 PSF, which far exceeds the floor loading capacity.
c. The scissor lift is not safe to use. The ground bearing pressure of the scissor lift is over 30,000 PSF, which far exceeds the floor loading capacity.
d. The scissor lift is safe to use. The ground bearing pressure of the scissor lift is 184 PSF, which does not exceed the floor loading capacity.

55. You work for a construction company specializing in asbestos abatement. You are tasked with performing personal air samples to determine worker exposure to asbestos. A particular job will require that the sampled worker be working outside their entire shift. Pre-calibration of the air sampling pump used produced the following values:

- Sampling rate: 2.2 L/min
- Atmospheric pressure: 720 mmHg
- Temperature: 20 °C

At the end of the sampling period, the post-calibration of the air sampling pump was also 2.2 L/min. If the temperature and atmospheric pressure at the sampling site were 0 °C and 750 mmHg, respectively, what would be the corrected actual flow rate?

a. 22 L/min
b. 2.2 L/min
c. 2.08 L/min
d. 4.4 L/min

56. A construction worker drops a tool weighing 8 pounds from a height of 12 feet. All personnel on the ground floor are wearing steel-toe boots rated at 75 foot-pounds. Will their steel-toe boots be sufficient to protect against the falling tool? Why or why not?

a. No, the steel-toe boots are only able to protect against a tool weighing 5 lb from that height.
b. Yes, the steel-toe boots are adequate to protect against the tool even when it is dropped from a height of 15 ft.
c. No, the steel-toe boots are unable to protect against an 8-lb tool from that height.
d. Yes, the steel-toe boots can withstand a 75-lb object dropped from any height under the standard 100 ft used for testing.

57. If an employee responds to a first aid situation and is potentially exposed to infectious material, which of the following is true?

a. The employer must offer a hepatitis B vaccine at no cost to the employee.
b. The employer must provide paid time off to seek medical advice.
c. The employer must provide periodic blood work for 24 months to verify that there is no infection.
d. A potential exposure does not require any additional care from the employer.

58. Construction workers are going to be removing asbestos-containing material (ACM). In order to validate that the controls in place keep their exposure below OSHA's PEL for asbestos, you will be performing personal air sampling on one of the workers. At the end of the sampled employee's 8-hour work shift, you find the following results:

- Pump flow rate: 1 L/min
- Number of identified asbestos fibers in the cassette: 25,000 fibers
- Sampling time: 8 hours

Based on these results, was the employee exposed above the PEL for asbestos? Why or why not?

a. Yes, the employee was exposed to 250 fibers/cc of air, which is above the PEL.
b. No, the employee was exposed to 0.052 fibers/cc of air, which is below the PEL.
c. Yes, the employee was exposed to 52.1 fibers/cc of air, which is above the PEL.
d. No, the employee was exposed to 0.00025 fibers/cc of air, well below the PEL.

59. An employee suddenly slumps to the floor and is now unconscious. Although he is breathing, he must be left alone for emergency services to be contacted. What is the proper procedure for leaving an unconscious victim who does NOT have a suspected spinal injury?

a. Do not attempt to touch or shift the employee.
b. Roll the injured employee into the recovery position.
c. Find a means to take the injured employee with you. Never leave a victim alone.
d. Prop the injured employee up on something so that his airway is clear.

60. Operations meeting the reporting requirements of HAZWOPER must have a written emergency response plan included in the company's health and safety program. Which answer choice correctly lists the components of an emergency response plan?

a. Site security, personnel roles, proof of liability insurance, employee health cards
b. Personnel roles, muster areas, air monitoring, decontamination procedures
c. Emergency response and first aid, PPE and emergency equipment, list of local security contractors, engineering controls
d. Pre-emergency planning, list of PPE vendors, air monitoring, site security, and control

61. You are the health and safety professional for a construction company. Upon arriving at the job site, a worker asks what the flow rate requirements are for his grinding wheel. The wheel is 0.46 feet wide. According to 29 CFR 1926.57 Table D-57.1 (see figure below), what is the minimum exhaust volume (cfm) needed for this wheel?

Table D-57.1 – Grinding and Abrasive Cutting-off Wheels

Wheel diameter, inches (cm)	Wheel width, inches (cm)	Minimum exhaust volume (ft^3/min)
To 9 (22.86)	1.5 (3.81)	220
Over 9 to 16 (22.86 – 40.64)	2 (5.08)	390
Over 16 to 19 (48.26 – 48.26)	3 (7.62)	500
Over 19 to 24 (48.26 – 60.96)	4 (10.16)	610
Over 24 to 30 (60.96 – 76.2)	5 (12.7)	880
Over 30 to 36 (76.2 – 91.44)	6 (15.24)	1,200

a. 250 cfm
b. 680 cfm
c. 980 cfm
d. 1,330 cfm

62. When considering risk reduction, what is the most effective approach for a job that requires remodeling construction work inside a facility with asbestos?

a. Requiring asbestos abatement or remediation prior to the remodeling job
b. Administering fresh PPE at the beginning of each shift and ensuring employees decontaminate at the end of each shift
c. Providing asbestos awareness training to all workers who could be potentially exposed to asbestos
d. Assessing the job tasks and performing as many of the tasks outside the building as possible

63. A crew of welders repairing a petroleum pipeline are working long hours to meet a deadline. One of the crew members is noticeably fatigued and is having difficulty making decisions. What process can be performed during morning safety meetings to ensure employees are alert and ready for work?

a. Pre-job fit-for-duty exam
b. Job safety analysis
c. Medical evaluation
d. Return to work process

64. What strategy is LEAST likely to create an environment in which a work crew or individual would feel pressured to cover up safety incidents?

a. Requiring employees to sign and date the HSE manual stating they will report all incidents
b. Involving employees in corrective actions
c. Offering bonuses related to safety performance
d. Making demotions based on the number of incidents reported

65. The hazard assessment step in the risk management process uses what components to determine overall risk?

a. Impact and financial loss
b. Probability and severity
c. Injury potential and equipment damage
d. Likelihood and damages

66. You are a consultant for a distribution center. In one of their large warehouses that is 100 feet long, 50 feet wide, and 25 feet tall, they need to have at least 3 air changes per hour (ACH). The airflow rate is 7,000 cfm. Is the ACH at least 3 for this warehouse?

a. Yes, the ACH for the warehouse is over 4 ACH.
b. No, the ACH for the warehouse is less than 1 ACH.
c. Yes, the ACH for the warehouse is around 3.4 ACH.
d. No, the ACH for the warehouse is just below 3 ACH.

67. You are observing employees stacking bricks (each brick has a height of 2.25 inches). The employees would like to know how tall the stack of bricks can be. According to the regulations defined in 29 CFR 1926.250(b)(6), what is the maximum number of bricks they are allowed to stack onto each other?

a. 45 bricks
b. 61 bricks
c. 20 bricks
d. 37 bricks

68. OSHA 1926.652 states that employees in an excavation or trench greater than what height shall have protection for cave-ins?

a. 6 feet
b. 4 feet
c. 5 feet
d. 10 feet

69. During a remodeling project on a three-story office, employees will be removing the old masonry and installing brick. The employees will need to access the masonry from a scaffold or aerial lift. According to OSHA 1926.501, what are three fall protection requirements the elevated platform must provide?

a. Toeboards, handrails, and certified tie-off points
b. Handrails, taglines, and canopies
c. Taglines, belly belts, and self-retracting lifelines
d. Handrails, fall restraint system, and a Geronimo line

70. Bloodborne pathogens can potentially be transmitted in all the following bodily fluids EXCEPT:

a. Semen
b. Saliva
c. Synovial fluid
d. Sweat

71. What term is used to describe the maximum number of subordinates or resources that one supervisor can manage effectively during an emergency response incident?

a. Management capabilities
b. Supervisory jurisdiction
c. Span of control
d. Retention rate

72. Multiple sources of electrical power can be present on a construction site. Lockout/tagout is a recommended practice for preventing injuries and property damage due to unexpected energization. What is NOT a component of a lockout/tagout program?

a. Listing sources of potential energy to be isolated
b. Training of all employees affected by the energy isolation
c. Utilizing a multimeter to confirm the de-energization
d. Physically locking equipment that is being de-energized

73. According to OSHA 1926.103, is computer-based training sufficient for training an employee in respiratory protection?

a. No, respiratory protection requires training on the proper use of equipment, which cannot be conducted via computer-based training.
b. No, respiratory protection training must be conducted by a medical professional.
c. Yes, respiratory protection training can be conducted via computer-based training if the employee passes a written exam.
d. Yes, respiratory protection training can be conducted via computer-based training as long as the employee has the option to ask questions during the training.

74. Which of the following demonstrates the use of a crisis management plan?

a. Purchasing extra PPE in case of a catastrophic event within the workplace
b. Providing medical training to employees to enable them to care for injured workers
c. Outlining emergency escape routes and safe gathering areas
d. Designating a spokesperson to interact with the media during the event

75. One of the managers at your company has just approved a few employees to start working 10-hour shifts instead of the usual 8-hour shifts. You plan on performing lead sampling on one of the employees during their 10-hour work shift. Will the OSHA PEL for lead be different than 50 μg/m^3? If yes, what would it be?

a. No, the PEL is never adjusted, regardless of shift length.
b. Yes, the PEL would be adjusted to 10 μg/m^3.
c. Yes, the PEL would be adjusted to 85 μg/m^3.
d. Yes, the PEL would be adjusted to 35 μg/m^3.

76. OSHA recommends that walking surfaces have a static coefficient of friction (μ_s) of 0.5. Your company is installing a pathway as part of the construction of a new business. You are asked to determine the static coefficient of friction of this pathway. An experiment is run where it is determined that an object weighing 3.5 kg has a static frictional force of 20 N between the object and the surface of the pathway. Does this pathway meet OSHA's static coefficient of friction recommendation?

a. Yes, the static coefficient of friction is 0.58, which is above OSHA's recommendation of 0.5.
b. No, the static coefficient of friction is 0.175, which is below OSHA's recommendation of 0.5.
c. Yes, the static coefficient of friction is 5.7, which is well above OSHA's recommendation of 0.5.
d. No, the static coefficient of friction is 0.458, which is just below OSHA's recommendation of 0.5.

77. You work for a company that tests suspension ropes used on adjustable suspension scaffolds. If the maximum intended load of a suspension rope is 750 lb, how much weight must be tested on the suspension rope according to OSHA?

a. 6,000 lb
b. 3,000 lb
c. 1,500 lb
d. 4,500 lb

78. Falls continue to be the leading cause of fatalities in the construction industry, contributing to more than a third of all construction worker deaths. What is the most reported underlying cause of fatal falls?

a. Misuse of portable ladders
b. Unprotected sides, wall openings, and floor holes
c. Lack of proper fall protection or failure to wear fall protection
d. Improper scaffolds

79. How much kinetic energy (K. E.) does a 4,000 kg forklift traveling at 4 m/s have?

a. 16,000 N
b. 64,000 N
c. 32,000 N
d. 35,000 N

80. All of the following should be considered when preparing an emergency contingency plan for a construction location that could potentially experience a significant release of hazardous gases to the atmosphere EXCEPT:

a. Muster areas based on wind direction
b. Evacuation notification and procedures for workers and the community
c. The financial penalty issued by the EPA for hazardous gas release
d. The rescue plan and procedures for employees working in the location

81. You measure the sound pressure to be 1.4 pascals. What is the sound pressure level (L_p)?

a. 104 dB
b. 86 dB
c. 2.9 dB
d. 97 dB

82. A trench is being dug for the underground placement of a fuel storage container that is 52 feet in length, 4 feet in width, and 6 feet in height. Employees need to enter the trench to level the base dirt. How many means of egress would this trench require?

a. 2
b. 4
c. 3
d. 1

83. What proactive safety technique requires an observer to watch an employee while performing his or her job duties, document any unsafe actions or safety violations, and immediately provide feedback on the observation to the employee?

a. Job safety and environmental analysis
b. Behavior-based safety
c. Near-miss reporting
d. Incident reporting

84. Which safety discipline focuses on enhancing safety by designing equipment, products, and systems that operate safely and efficiently in real-world conditions by identifying potential operator errors and modifying the system to eliminate or reduce the chances of those errors occurring?

a. Behavior-based safety
b. Safety engineering
c. Human factors engineering
d. Industrial engineering

85. According to the emergency action plan, employees are to evacuate the fourth floor of an office building through a southeast or northeast stairwell, depending on the proximity of the stairwell to their office space; however, construction is underway, and the southeast stairwell has been closed. How can this evacuation procedure be updated to comply with the means of egress required by the building codes?

a. All employees will be required to exit the floor from the northeast stairwell.
b. Employees will have to be given alternative workspaces on different floors of the building that have two means of egress until the construction is finished.
c. An emergency warden will have to be assigned to the floor to make sure all employees have successfully evacuated in an emergency.
d. A temporary window ladder can be used as a means of egress if it is documented as temporary and included in the emergency action plan.

86. A work site inspection would likely occur after OSHA became aware of any of the following situations. Which of the following would hold the highest priority?

a. A report of an employee whose finger was amputated in a caught-between injury on the job
b. A report of a crane company operating a crane that has failed inspection on a job site
c. An anonymous worker complaint about wrongful termination
d. A random audit on a highly hazardous industry work site, such as a construction project on a high-rise condominium

87. You have been asked to determine the radiation dose that employees are exposed to for their work shifts. Based on the radiation sources, the dose rate and time of exposure for their 8-hour shift is:

- 320 mrem/hr for 3 hours
- 25 mrem/hr for 2.5 hours
- 104 mrem/hr for 2.5 hours

What is the overall radiation dose for these employees?

a. 832,000 mrem
b. 1,282.5 mrem
c. 832.7 mrem
d. 4,985.2 mrem

88. A local exhaust ventilation (LEV) hood is installed to capture respirable silica dust during cinder block cutting operations. If the diameter of the round hood is 14" and the capture velocity (V) must be at least 1,200 fpm, what is the volumetric flow rate (Q) needed?

a. 5,140 cfm
b. 1,400 cfm
c. 1,290 cfm
d. 620 cfm

89. A flammable cabinet in a workroom is used to store less than 30 gallons of flammable liquids. Based on this section of the SDS for sodium, would sodium need to be stored inside the flammable cabinet?

SECTION 5 — FIRE FIGHTING MEASURES	
Flammable solid. Autoignition Temperature: 120-125 °C Severe fire risk. May ignite spontaneously in moist air. Reacts explosively with water producing very dangerous hydrogen gas. In case of fire: Use class D fire extinguisher or dry sand for extinction (P370+P378). Keep away from any possible contact with water, because of violent reaction and possible flash fire (P223).	NFPA Code H-3 F-3 R-2 No Water

a. Yes, this product is explosive and should be stored in the flammable cabinet.
b. No, this product's autoignition temperature is not high enough to need to be stored in the flammable cabinet.
c. Yes, this product produces dangerous gas and should be stored in the flammable cabinet.
d. No, this product is reactive to water and should not be stored in the flammable cabinet.

90. The analytical laboratory has sent you back the results of your sulfur dioxide (MW = 64 g/mol) air samples:

- Sample 1: 2.89 mg/m^3
- Sample 2: 8.65 mg/m^3

What are the results of the two samples in units of ppm?

a. Sample 1: 1.9 ppm; Sample 2: 2.9 ppm
b. Sample 1: 0.8 ppm; Sample 2: 1.5 ppm
c. Sample 1: 1.2 ppm; Sample 2: 1.9 ppm
d. Sample 1: 1.1 ppm; Sample 2: 3.3 ppm

91. OSHA recommends the use of a spotter during forklift operations. What is NOT a primary responsibility of a dedicated spotter?

a. Directing travel when the operator's vision is obscured
b. Assisting the driver with positioning a load that is being lifted
c. Performing daily forklift inspections
d. Following the forklift to help protect the lift and other vehicles and pedestrians

92. Your company is in the process of purchasing a 250 mW laser and has asked you to establish appropriate controls. The maximum permissible exposure (MPE) is 2.55 mW/cm^2. The irradiance (E) of the laser is 650 mW/cm^2. At a wavelength of 500 nm, what is the minimum optical density (OD) required?

a. 40.7
b. 255
c. 1,658
d. 2.4

93. What is the average discharge time of a 5-lb fire extinguisher?

a. 20–25 seconds
b. 10–15 seconds
c. 30–45 seconds
d. 5–10 seconds

94. As part of a procurement order for a new hoist, you review the load tests conducted by the manufacturer. If the rated capacity of the hoist is 1,250 pounds, what is the minimum weight that the manufacturer should use to load test it?

a. 1,562.5 lb
b. 2,500 lb
c. 1,250 lb
d. 1,375 lb

95. You are evaluating a chemical storage room for OSHA compliance. One flammable storage cabinet has three different flammable liquids with the following quantities and physical properties:

- Flammable liquid #1: flash point below 73.4 °F and boiling point above 95 °F (25 gal)
- Flammable liquid #2: flash point above 140 °F and boiling point at or below 199.4 °F (25 gal)
- Flammable liquid #3: flash point below 73.4 °F and boiling point at or below 95 °F (25 gal)

Based on the physical properties and quantities of the three flammable liquids, is this flammable storage cabinet OSHA-compliant? If so, could significantly more of any of the three flammable liquids be added to the flammable storage cabinet and still maintain OSHA compliance? If it is not OSHA-compliant, how much needs to be removed from the cabinet to become OSHA-compliant?

a. The flammable storage cabinet is OSHA-compliant, and significantly more of all three of the flammable liquids could be added to the cabinet.
b. The flammable storage cabinet is not OSHA-compliant. Flammable liquids #2 and #3 must be removed from the cabinet to become OSHA-compliant.
c. The flammable storage cabinet is not OSHA-compliant. Removal of flammable liquid #2 would make the cabinet OSHA-compliant.
d. The flammable storage cabinet is OSHA-compliant, but only a significant amount of flammable liquid #2 could be added to the cabinet.

96. In order to access scaffolding, a 40-foot stairway-type ladder is constructed. Based on the requirements for scaffold access in 29 CFR 1926, how many rest platforms must be constructed?

a. 2
b. 3
c. 4
d. 1

97. A coastal residential community with homes built on elevated platforms to avoid flood damage represents what type of planning?

a. Emergency mitigation
b. Emergency response
c. Crisis management
d. Community development

98. A gas line needs to be installed deep into the ground.

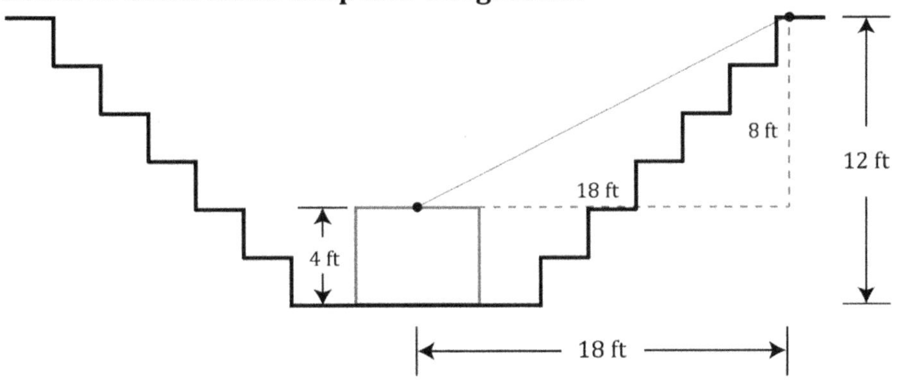

At a depth of 12 feet, the ground is made entirely of hard stone. This requires an excavator to enter the space to break it up and remove the material. The trench is sloped properly for Type C soil, and the very top of the trench is 18 feet wide (1.5 : 1). The excavator is 4 feet high before the crane arm begins and is in the middle of the trench. How long must the crane arm of the excavator be?

a. 23 feet
b. 20 feet
c. 22 feet
d. 21 feet

99. An employee faints during a meeting. Several coworkers are trained in CPR and first aid. The employee is placed in the recovery position and comes to within a minute. The employee states that he feels fine and can continue his day at work. What should the recommended practice be in this situation?

a. If the employee requests to return to work, he should be allowed to return of his own will once he signs a statement.
b. The employee should be taken to the hospital for additional observation.
c. Before returning to work, an official incident report should be filed along with witness statements.
d. No additional treatment or paperwork is needed since the employee was not injured.

100. A safety coordinator identifies a faulty gas monitor on location and determines that the faulty monitor could create an explosive environment, leading to an injury or property damage. The job site is shut down while the gas monitor LEL sensor is replaced by a technician. The new sensor triggers an alarm when it becomes faulty. The job site atmosphere is tested for potential LELs, and when cleared, operations are resumed. This sequence of events represents which safety process?

a. Job safety analysis
b. Process hazard analysis
c. Risk management process
d. Fault tree analysis

101. What is the minimum clearance distance to maintain while operating equipment under overhead powerlines with voltages varying from 50 kV to 1,000 kV?

a. 3 feet
b. 20 feet
c. 10 feet
d. 30 feet

102. You are trying to evaluate a welding arc to determine the UV exposure. If the arc current (I) is 850 amperes and the welder is 25 cm away from the arc (r), what is the effective UV irradiance (E_{eff}) of the welding arc?

a. $0.005\ W/m^2$
b. $170\ W/m^2$
c. $4{,}250\ W/m^2$
d. $54.4\ W/m^2$

103. A hazard assessment reveals an increased risk of exposing workers to an environmental toxin at levels exceeding the permissible exposure limit for the time-weighted average. If the exposure cannot be eliminated, what action should be taken next to protect the workers?

a. The safety team should provide the workers with PPE.
b. A procedure should be put in place to rotate workers in and out of the exposed work area to prevent overexposure.
c. The workers should be given monitors that sound when overexposure takes place so that they can evacuate the workspace quickly.
d. The company should contract the job to a third party.

104. You arrive at your company's construction site and notice that a materials storage shed is located 10 feet from an energized line. The line is energized at 80 kilovolts (kV), and the materials storage shed is in an unrestricted area. According to 29 CFR 1926, is the materials storage shed located sufficiently far enough away from the energized line? Why or why not?

a. No, due to the energization of the line, the materials storage shed needs to be located 11 feet from the energized line.
b. Yes, for lines and equipment energized at 100 kV or less, the distance from any materials storage area is 10 feet, so the materials storage shed is sufficiently far away from the energized line.
c. No, for lines and equipment energized at 100 kV or less, the distance from any materials storage area is 20 feet, so the materials storage shed needs to be located an additional 10 feet away.
d. Yes, for lines and equipment energized at 50 kV or more, the distance from any materials storage area is 10 feet, so the materials storage shed is sufficiently far away from the energized line.

105. How can a company avoid legal ramifications when conducting drug and alcohol testing on employees?

a. Hire third parties to perform all drug and alcohol testing.
b. Provide employees with a detailed written drug and alcohol policy.
c. Have all drug and alcohol testing be conducted through human resources.
d. Only perform random selection testing on a predetermined schedule.

106. The NIOSH equation $RWL = LC \times HM \times VM \times DM \times AM \times FM \times CM$ is used to determine guidance on what safety factor?

a. Airborne contaminant concentration
b. Crane load safety capacity index
c. Time-weighted exposure for a hazardous substance
d. Employee lifting recommendations based on multiple factors

107. A group of construction workers has been assigned the task of transporting pieces of rebar from one location to another. Since you are concerned with the manual material handling hazards of this task, you decide to limit the pieces of rebar that can be moved individually. To do this, you determine a lifting index (LI) value of 1.5 and a recommended weight limit (RWL) of 25. Based on these values, which of the following is the lightest weight that would still require more than one employee to move the rebar safely?

a. 40 lb
b. 10 lb
c. 20 lb
d. 30 lb

108. You need to perform personal sampling for lead on a construction worker who will be welding for 15 minutes. In order to get the minimum air volume (35 liters) in 15 minutes, what flow rate must the air sampling pump have?

a. 1.5345 L/min
b. 525 L/min
c. 2.3333 L/min
d. 2.0132 L/min

109. A workplace management team is challenged with reducing the risk of heat stroke in their facility. Which list accurately describes the order in which changes should be applied?

1. Mandate heat awareness training.
2. Administer lightweight and reflective clothing.
3. Lower the workspace temperature using fans.
4. Implement shorter work periods and require additional breaks.

a. 1, 4, 3, 2
b. 3, 4, 1, 2
c. 2, 4, 3, 1
d. 3, 2, 4, 1

110. Company XYZ needs to train three employees in forklift operations for an upcoming job. The job tasks will include unloading heavy equipment from trailers and stacking it in a construction yard. Which scenario meets the training requirements of OSHA 1926.1427?

a. The employees must pass both a written exam approved by an auditor and a practical evaluation on the forklift doing a series of required skills prior to being certified.
b. The employees complete an online certified forklift driving training program and are paid for the time to complete the course.
c. The employees are monitored on forklift operations by a competent person for at least 40 hours prior to being certified.
d. The employees are trained in a classroom-style setting and then given a forklift driving evaluation by the company safety coordinator.

111. Right before an underground confined space entry was supposed to take place, a sewer pipe burst, releasing a large amount of hydrogen sulfide gas. You have just been informed that the sewer pipe is no longer leaking, and they need to get a crew into that space to repair the damage as soon as possible. You measure the ambient air as having 100 ppm hydrogen sulfide (H_2S) gas (IDLH). With a flow rate of 75 cfm, how long will it take for the H_2S gas in the space (volume $= 1,500\ ft^3$) to lower to 1 ppm?

a. 180 minutes
b. 92 minutes
c. 58 minutes
d. 34 minutes

112. What is the typical current differential that will trip a GFCI?

a. 1–2 mA
b. 4–6 mA
c. 10–15 mA
d. 20–25 mA

113. Noise dosimeter readings for a two-week period establish a work site's steady decibel level at 101 dB. What is the minimum noise reduction rating earplugs would need to reduce a worker's exposure to less than 90 dB per the construction industry permissible exposure limit (PEL)?

$$\text{Decibel Reading} - \frac{(\text{NRR} - 7)}{2}$$

a. 33 NRR
b. 28 NRR
c. 30 NRR
d. 35 NRR

114. What factor is weighted most heavily when determining a company's experience modifier rate (EMR) for workers' compensation premiums?

a. Accident likelihood
b. Accident cost
c. Accident frequency
d. Denied claims

115. You are performing respirator fit testing on workers at your company. After performing four fit testing exercises on the same employee with a full-face respirator, a fit factor for each exercise was determined:

- Exercise #1 fit factor ($\mathbf{ff_1}$): 65
- Exercise #2 fit factor ($\mathbf{ff_2}$): 80
- Exercise #3 fit factor ($\mathbf{ff_3}$): 35
- Exercise #4 fit factor ($\mathbf{ff_4}$): 102

Based on the fit factors for all four exercises, what is the overall fit factor for this employee using a full-face respirator?

a. 60.4
b. 70.5
c. 35.25
d. 282

116. A Geiger counter can be used to detect the presence of what type of ionizing radiation found in concrete and countertops?

a. Microwave radiation
b. Gamma radiation
c. Extremely low-frequency radiation
d. Neutron radiation

117. During the construction of a laboratory, the construction crew needs to lift fume hoods to the second floor of the building. During the risk analysis portion of this hoisting and rigging task, management wants to know the potential destruction that could be caused by the fume hood falling. What is the force (F) of a free-falling fume hood that weighs 300 kg?

a. 300 N
b. 2,940 N
c. 900 N
d. 12,254 N

118. A regulatory agent arrives on a job site where an old school building is being renovated into public office space. The agent asks to see the job scope and hazard mitigation plan for the job site. What safety documentation should the agent be given?

a. Emergency action plan
b. Job safety analysis
c. Site-specific safety plan
d. City and construction work permits

119. You are assigned to perform ventilation testing for a new office building to validate HVAC functionality. Since the office building is located at a high altitude, you need to determine the actual air density (ρ_{actual}). The temperature is 25 °F, or 485 degrees Rankine (°R), and the barometric pressure is 22.28 inHg. What is the actual air density? Assume the following values for standard temperature and pressure (STP) conditions:

- $\rho_{STP} = 0.075 \text{ lb/ft}^3$
- $T_{STP} = 70\ °\text{F or } 530\ °\text{R}$
- $P_{BSTP} = 29.92 \text{ inHg}$

a. 0.075 lb/ft^3
b. 0.092 lb/ft^3
c. 0.061 lb/ft^3
d. 0.071 lb/ft^3

120. While setting a battery tank that weighs approximately 8,000 lb empty, the forklift operator notices the rear tires coming off the ground. The operator lowers the load to the ground to prevent tipping. The forklift is a rental that does not have outriggers, and the boom was extended to 60% during the lift. The forklift lifting capacity is 12,000 lb. What is the most likely reason the forklift is unsteady?

a. The forklift hydraulic lifting cylinders are malfunctioning.
b. The forklift lacking outriggers prevents the forklift from lifting to its full capacity.
c. The forklift is being operated outside of its load rating given the boom extension.
d. The forklift counterweights have not been balanced properly.

121. How much power (P) is in a direct current (DC) circuit that has 240 volts (V) and an electrical current (I) of 0.45 amps?

a. 0.00188 W
b. 533 W
c. 108 W
d. 240 W

122. A construction crew is transporting an overweight load across state lines. What regulatory body would the crew need to file permits with?

a. Occupational Safety and Health Administration
b. Environmental Protection Agency
c. National Highway Traffic Safety Administration
d. Department of Transportation

123. Although there isn't a specific OSHA standard concerning workplace violence, there is a provision that requires all employers to provide a work environment free from recognized hazards that are causing or are likely to cause death or serious physical harm, which would include workplace violence. What is this provision?

a. Title VII of the Civil Rights Act
b. General Duty Clause
c. Employee Assistance Program
d. Safety Act

124. What is the total pressure (TP) of a ventilation system that has a static pressure (SP) of -1.2 inches of water gauge (in. w.g.) and a velocity pressure (VP) of 0.98 in. w.g.?

a. −2.18 in. w.g.
b. −0.22 in. w.g.
c. 2.18 in. w.g.
d. 0.22 in. w.g.

125. You hold a pre-job brief for hot work in a confined space. Two personnel will enter the confined space to perform a welding fix to a 2 ft^2 surface area of the wall. Since all of the surfaces in the confined space are covered with toxic preservatives, what is the cross-sectional area that must be stripped of all toxic coatings in the area of heat application?

a. 6.8 ft^2
b. 8.3 ft^2
c. 4 ft^2
d. 7.2 ft^2

126. An OSHA inspector is conducting an on-site follow-up inspection and finds that the pressure relief valves are not certified. The company was cited for this violation six months prior. What type of violation is the company subject to?

a. De minimis violation
b. Repeat violation
c. Nonserious violation
d. Willful violation

127. You are evaluating a trench for OSHA compliance. The trench is 10 feet deep and 25 feet wide. Based on the sloping angle, what is the least stable soil that this trench could be located in?

a. Type A
b. Type B
c. Type C
d. Type D

128. You are asked by your electric company to determine how far an employee will be from a single-phase arc working with live-line tools and doing live-line barehand work. If the system voltage is 50 kV and the minimum approach distance (MAD) is 25 inches, what would be the reasonable distance between the employee and the electric arc?

a. 15 in
b. 25 in
c. 50 in
d. 75 in

129. A construction company hires a group of new employees. Each employee is assigned a mentor, completes a new hire orientation program, has a checklist of job skills to be trained in and observed on completing successfully, and is given a green hard hat instead of the usual white. What type of program is this an example of?

a. Employee right-to-work program
b. Employee return-to-work program
c. Short service employee program
d. Employee information and training program

130. What equipment design is incapable of producing enough electrical or thermal energy to cause an ignition or explosion when working around potentially hazardous atmospheres?

a. Intrinsically safe equipment
b. Explosion-proof equipment
c. Circuit-protected equipment
d. Battery-only operated equipment

131. According to the base-to-height ratio outlined in 29 CFR 1926.451, a crew working from a supported scaffold that is 16 feet high would need a base width of at least how many feet?

a. 2 feet
b. 4 feet
c. 8 feet
d. 16 feet

132. The corporate safety manager of a construction company is reviewing data and is concerned about an upward trend in incidents related to slips, trips, and falls. Are the accident reports the manager is reviewing examples of leading or lagging indicators, and why?

a. Leading indicators because they provide evidence that indicates more slips, trips, and falls will continue to happen if mitigation does not take place
b. Lagging indicators because the time it takes for the report to reach the corporate office to be reviewed lags far behind the actual event that occurred
c. Lagging indicators because the accidents have already occurred, and the report is a reactive response
d. Leading indicators because future accidents can be prevented by reviewing accident reports to determine the best way to mitigate the hazards

133. You are evaluating a material handling task where a group of construction workers is carrying a box containing 10 bricks (each brick weighs 4.3 lb) from a truck bed to where the bricklayers are working. If the recommended weight limit (RWL) for this task is 29, what is the lifting index (LI)?

a. 1.48
b. 0.148
c. 1,247
d. 1.25

134. A construction crew is transferred to a job site at a much higher elevation with extremely cold conditions. What equipment has the greatest likelihood of causing significant injury because of the new weather conditions?

a. Hand and power tools
b. Scaffolding
c. Batteries
d. Diesel engines

135. Your company asks you to determine airflow requirements for a spray booth using methyl ethyl ketone (MEK), which has a lower explosive limit (LEL) of 1.8%. If MEK evaporates at a rate of 36 ft^3 per gallon (from 29 CFR 1926 Table D-57.8), what is the required dilution volume of air per gallon of MEK?

a. 36,400 ft^3 dilution/gal of MEK
b. 14,140 ft^3 dilution/gal of MEK
c. 8,144 ft^3 dilution/gal of MEK
d. 7,856 ft^3 dilution/gal of MEK

136. According to the NFPA, what concept defines the number of people expected to be present in a building, room, or space at any given time under all possible situations?

a. Occupancy load
b. Tenant availability
c. Assembly factor
d. Maximum occupancy

137. During the process of constructing an aboveground storage tank, you are asked to determine the total rate of venting (ft^3/hr) required based on an exposed wetted surface area of $3,000\ ft^2$. To be OSHA-compliant, what total rate of venting (CFH) is required for this storage tank?

a. 3,000 ft^3/hr
b. 786,000 ft^3/hr
c. 300,000 ft^3/hr
d. 56,900 ft^3/hr

138. An employee walks into the warehouse and sees his coworker lying on the floor, nonresponsive. What should be the first step the employee takes in responding to this situation?

a. Call emergency services.
b. Check for breathing and pulse.
c. Ensure the area is safe before proceeding.
d. Remove the injured person from the area immediately.

139. You work for a company that regularly performs lead-based soldering. In an effort to determine if the local exhaust ventilation systems are effective at capturing the airborne lead produced from soldering, you perform lead wipe sampling according to NIOSH Method 9100. The total amount of lead collected in the six wipe samples was 1.2 mg. What was the average surface concentration of lead for the six wipe samples?

a. 2 $\mu g/cm^2$
b. 56 $\mu g/cm^2$
c. 12 $\mu g/cm^2$
d. 48 $\mu g/cm^2$

140. What would be the most dependable means of communication in an emergency in which cellular and internet power were compromised?

a. Handheld receivers
b. Megaphones
c. Satellite phones
d. HAM radios

141. Your company just purchased a large HVAC system with a sound power of 3.5. What is the sound power level?

a. 35 dB
b. 125 dB
c. 103 dB
d. 87 dB

142. During the construction project of a new facility, the project manager asks you to calculate the lumens needed for the first aid station area. According to 29 CFR 1926.56 Table 3, first aid stations require an illumination intensity of 30 foot-candles (fc). If the first aid station area is 6 feet long, 4 feet wide, and 8 feet tall, how many lumens are necessary to be OSHA-compliant?

a. 960 lumens
b. 720 lumens
c. 5,760 lumens
d. 1,440 lumens

143. You have been asked to determine the force generated when using a fall arrestor. The personal fall arrest system (PFAS) has a free fall distance (FF) of 6 feet and a deceleration distance or fall arrestor distance (FAD) of 3 feet. Using 250 pounds as the weight of the person, what is the force generated (gravitational force) from this PFAS?

a. 500 lb of force
b. 750 lb of force
c. 125 lb of force
d. 1,000 lb of force

144. A personal air sample of a welder reveals an exposure to 125 $\mu g/m^3$ of lead for 180 minutes. In order for the welder's exposure to remain below the OSHA PEL for lead (50 $\mu g/m^3$), what is the maximum concentration of airborne lead that the welder can be exposed to for the remaining 5 hours of his shift?

a. 2.5 $\mu g/m^3$
b. 5 $\mu g/m^3$
c. They are already exposed to lead above the OSHA PEL.
d. 75 $\mu g/m^3$

145. You are asked to determine the hazards associated with a chemical solution based on its molarity (M) concentration. You are told that 95 g of sodium hydroxide, or NaOH (MW = 39.997 g/mol), was dissolved in water to make a solution with a total volume of 1.2 liters. What is the molarity of the NaOH solution?

a. 3.589 M
b. 0.198 M
c. 1.98 M
d. 3,166 M

146. An employee is cited by the safety manager for wearing regular sunglasses instead of approved safety glasses with side shields. What standard would likely be used as the quality requirement for safety glasses?

a. ASSP A10.0
b. ISO 14000
c. ANSI Z87.1
d. ISO 45001

147. Your company needs to lift a 3,000-pound pallet of bricks onto an elevated platform using a crane. If the working load limit (WLL) of the crane is 5,500 pounds and the dynamic amplification factor (DAF) for the lift is 1.89, is this crane sufficient for the lift plan?

a. No, the intended load exceeds the safe working load (SWL) of the crane.
b. Yes, the WLL is greater than the intended load.
c. No, the WLL of the crane is not at least 2x the intended load.
d. Yes, the SWL of the crane exceeds the weight of the intended load.

148. When responding to a medical emergency, which law protects bystanders who provide basic first aid and/or CPR within the realm of their training?

a. Health Insurance and Portability Law
b. Good Samaritan Law
c. False Claims Act
d. Administrative Simplification Compliance Act

149. What is the OSHA recommendation for evaluating established safety programs?

a. These programs should be evaluated at least once annually and whenever opportunities for improvement are identified.
b. These programs should be evaluated biannually and when an incident occurs that is related to a health and safety program.
c. These programs should be evaluated upon the release of industry best practices.
d. These programs should be evaluated every three years and whenever OSHA standards that may affect the program are updated.

150. What is the proper procedure to train employees on when operating a powered industrial truck that strikes a live power line?

a. Immediately exit the truck and call 911.
b. Jump from the vehicle in such a manner that both feet exit the vehicle and hit the ground at the same time.
c. Call 911 and stay put until emergency help arrives.
d. Back the truck as far away from the line as possible and call 911.

151. Before construction workers begin working on an elevated platform, they ask you to perform a fall hazard assessment. As part of the assessment, you need to determine the total fall clearance distance needed for a personal fall arrest system (PFAS) using a shock-absorbing lanyard based on the following factors:

- D-ring shift (C): 1 foot
- Safety factor: (D) 2 feet
- D-ring height (C): 5 feet
- Deceleration distance (B): 3 feet
- Lanyard length (A): 7 feet

You locate the anchor point 3 feet above the D-ring on the employee's harness and measure the employee's height as 6 feet. The D-ring is located at a height of 5 feet. Based on these factors, is 15 feet a sufficient total fall clearance distance, or will the employee make contact with the lower level?

a. The employee would make contact with the lower level. The total fall clearance distance needed is 20 feet.
b. The total fall clearance distance needed is less than 10 feet, so 15 feet is a sufficient clearance distance.
c. The employee would make contact with the lower level. The total fall clearance distance needed is exactly 15 feet.
d. The total fall clearance distance needed is exactly 8 feet, so 15 feet is a sufficient clearance distance.

152. What is the electrical current in a circuit with a voltage of 75 V and a resistance of 950 Ω? What health effects, if any, would the amount of electrical current calculated cause a human being?

a. 71.25 mA, which can cause temporary muscle paralysis
b. 71,250 mA, which can cause temporary muscle paralysis
c. 0.0789 mA, which can cause death
d. 78.9 mA, which can cause death

153. When considering emergency response procedures for hazardous waste, which of the following does NOT classify as hazardous material?

a. Radioactive materials
b. Allergens
c. Flammable materials
d. Sewage

154. Which of the examples below accurately represents direct costs versus indirect costs associated with workplace accidents?

a. Direct cost: workers' compensation payments; indirect cost: property damage
b. Direct cost: training; indirect cost: legal services
c. Direct cost: medical bill; indirect cost: lost productivity
d. Direct cost: accident investigation; indirect cost: OSHA fines

155. You are asked to estimate the radius of a crane (assume a simple geometric model) at a certain configuration. The boom length is 42 feet, and the boom angle is 29°. What is the crane's radius?

a. 28.8 ft
b. 23.3 ft
c. 36.7 ft
d. 20.4 ft

156. Maintenance workers are performing paint stripping using methyl ethyl ketone (MEK) on a large tunnel area defined as a confined space. The tunnel is 8 feet long with a diameter of 20 feet. Based on where they are performing their work, it would take all of them 2 minutes to exit the tunnel area in case of an emergency. If the large container of MEK were to lose its integrity and start spilling MEK, would 2 minutes be enough time for all the workers to exit the area before the ambient concentration reaches the IDLH level for MEK ($\text{IDLH MEK} = 3{,}000 \text{ ppm}$)? The airflow rate in the tunnel is 550 cfm, and the generation rate (G) of MEK is 5 cfm.

a. After 2 minutes, the concentration of MEK is under 1 ppm; 2 minutes is more than enough time to escape.
b. After 2 minutes, the concentration of MEK is around 3,500 ppm; 2 minutes is not enough time to escape.
c. After 2 minutes, the concentration of MEK is just at 3,000 ppm; 2 minutes is just enough time to escape.
d. After 2 minutes, the concentration of MEK is over 5,000 ppm; 1 minute is not even enough time, let alone 2 minutes.

157. Your construction company would like to begin performing radioactive cleanup activities at a site where plutonium-238, which has a half-life of 87.7 years, was deposited 62 years ago. Before the site can be worked on, the current activity of plutonium-238 needs to be determined in order to implement the proper radiological controls. If plutonium-238 has an initial activity of 4,265 GBq/g, what would the current activity be after 62 years?

a. 2,613 GBq/g
b. 2,133 GBq/g
c. 3,208 GBq/g
d. 1,875 GBq/g

158. While working as a contractor on a building site, an employee feels that a new crew of employees is not being treated fairly nor being offered sufficient PPE like respirators. The employee wants to make a complaint but is afraid of retribution. What is the best way for the employee to handle this situation?

a. The employee should make an official anonymous complaint through the OSHA online complaint center or the OSHA complaint call line.
b. The employee should meet with management above their direct supervisor to let them know about the issue so they can decide how to properly handle the complaint.
c. The employee can offer to share PPE with the other crew.
d. The employee should not get involved in another crew's business.

159. An employee fails to notice that a piece of machinery is at its maximum torque pressure. The equipment fails, resulting in a costly repair and significant downtime. Upon investigation, it is documented that the employee not only had to be watching the digital torque pressure but also manually adjusting multiple other parameters and managing his crew. What accident theory would this scenario fall under?

a. Human factors theory
b. Modern causation theory
c. Domino theory
d. Epidemiological theory

160. Employee 1 slips and falls on rubber safety matting placed in front of the operator controls of a generator pack. The employee does not feel injured and brushes the fall off without reporting it. Weeks later, a similar incident occurs, but Employee 2 breaks his wrist and files a report. During the accident investigation, it was determined that condensation was building up on the mat, causing it to become very slick. If Employee 1 had followed protocol and reported the slip, Employee 2's injury may have been avoided. What is the best motivation leadership can provide to encourage employees to report all accidents?

a. Offer support for completing the incident report and make sure feedback is given after the report has been reviewed.
b. Offer rewards for quality incident reports.
c. Ensure employees that there will be no repercussions if a report is filed.
d. Review all incident reports during safety meetings so that employees know they are being read.

161. A fire breaks out while hot work operations are taking place on a construction site. The construction site is located next to a propane storage facility. According to OSHA 1926.35, the emergency action plan must have an alarm system to notify workers of an emergency. The alarm system must follow which of these specifications?

a. The alarm must automatically contact 911 or emergency services when initiated.
b. The alarm should only trigger if there is potential for significant loss of property or life.
c. The alarm should be loud and continuous until the emergency is under control or the injured person(s) has been evacuated.
d. The alarm should have a distinctive signal for each type of potential emergency that can be heard by all workers in the affected area.

162. Preparing for an emergency response exercise, the safety manager questions if the scale of the drill should be referenced as an emergency or a disaster. What criteria are used to define the difference between an emergency and a disaster?

a. Emergencies are usually small-scale, localized incidents that are resolved quickly using local resources. Disasters are large-scale incidents that require a level of response and recovery greater than local communities can provide.
b. Emergencies are only related to an incident in which a life is at stake. Disasters include property and environmental damage.
c. Emergencies and disasters are considered the same, and the terms can be used interchangeably with no confusion.
d. Disasters can only be classified by the federal government. All other incidents are considered emergencies.

163. Occupational safety and health data shows trends in severe injuries occurring in all of the following scenarios EXCEPT:

a. During work performed by new and inexperienced workers
b. During work completed in the first two hours of a worker's shift
c. During routine job tasks completed by the worker
d. During work shifts that occur after 6 pm

164. What is the arc gap, or arc length, of a single conductor in air at 75 kV?

a. 75 mm
b. 190.5 mm
c. 29.5 mm
d. 150 mm

165. Job authorizations, critical task safety analyses, and permits to work are all examples of which of the following?

a. Lagging indicators in hazard identification
b. Reactive approaches to hazard analysis
c. Leading indicators of risk management
d. Management of change procedures

166. During a pre-job safety meeting prior to a critical double crane lift, one crane operator mentions that while inspecting his rigging, he found a frayed wire on the outer strand of a wire inside the lifting sleeve. The wire rope is new and has only been used on one similar lift. The compressor weighs 36,000 lb. The wire rope is rated for 25,000 lb; however, two cranes will be splitting the load between two wire ropes. The single frayed wire does not deem the rigging out of service but does cause reason for question since it is new. Which action is most appropriate prior to conducting the lift?

a. Since the single frayed wire does not deem the wire rope rigging out of service, continue the job but thoroughly inspect the equipment after the lift.
b. Do not risk the lift. Deem the rigging as out of service and replace it with a new wire rope.
c. Note the single frayed wire on the inspection report to make sure the crane operator is covered if an accident takes place.
d. Contact the manufacturer of the wire rope to ensure there are no existing issues with that model of rope and ask about the lifting scenario taking place to ensure proper lifting techniques are being used.

167. Which of the following is NOT a reason for allowing an employee to return to work on light duty after an injury?

a. This allows the employee to stay engaged in the workplace.
b. This lowers workers' compensation costs.
c. This mitigates lost income for the employee.
d. This controls OSHA recordable incident rates.

168. According to the Globally Harmonized System (GHS), a chemical mixture containing greater than or equal to 5% of chemicals categorized as "skin corrosive" will require that the entire chemical mixture be defined equivalently. You are reviewing a chemical mixture that contains only one chemical component defined as "skin corrosive." The chemical concentration (molar mass = 30 g/mol) in the mixture is 1.5 molarity (M). Does this chemical require that the entire chemical mixture be defined as "skin corrosive"? Assume a mass density of 1 kg/L.

a. Yes, the "skin corrosive" chemical makes up 45% (v/v), so the chemical mixture does have to be classified as "skin corrosive."
b. No, the "skin corrosive" chemical makes up 3.0% (v/v), so the chemical mixture does not have to be classified as "skin corrosive."
c. Yes, the "skin corrosive" chemical makes up 30% (v/v), so the chemical mixture does have to be classified as "skin corrosive."
d. No, the "skin corrosive" chemical makes up 4.5% (v/v), so the chemical mixture does not have to be classified as "skin corrosive."

169. Which two standards are designed to improve an organization's occupational health and safety management system and environmental management system?

a. OSHA 1926 and OSHA 1910
b. ISO 45001 and ISO 14001
c. OSHA 18001 and ISO 45001
d. ISO 27001 and ISO 9000

170. Personal noise sampling on a heavy equipment operator resulted in a noise dose of 112%. What is the TWA for this noise dose using OSHA's criteria?

a. 90.8 dBA
b. 112 dBA
c. 106 dBA
d. 84 dBA

171. According to OSHA construction standard 1926.1053, if an employee is accessing a platform from a portable ladder, how many feet shall the ladder extend above the platform?

a. 2 feet
b. 3 feet
c. 1 foot for every 14-foot story
d. 2 feet for every 10 feet of height

172. A company needs to manually access a union that is 25 feet above the subfloor. The crew decides to lift an employee wearing a fall protection harness using a winch. Within a few minutes, the employee becomes dizzy, sweaty, faint, and then unconscious. Based on the scenario, what can be assumed is happening with the employee?

a. No assumptions can be made.
b. The employee is suffering from suspension trauma.
c. The employee has an unknown medical condition
d. The employee's fear of heights has put him in shock.

173. A coastal community's city council implements new construction codes after a hurricane destroys much of the area. The new code requires production facilities that fall under SPCC plans to have enclosed and impermeable secondary containments. This is an example of what stage of emergency planning?

a. Recovery
b. Response
c. Preparedness
d. Mitigation

174. The Incident Command System is designed to allow personnel from different workplaces and agencies to efficiently work together to manage the scene of an emergency. What are the five functions of Incident Command?

a. Treasury, administration, planning, logistics, and reporting
b. Command, reporting, public relations, finance, and director
c. Command, operations, planning, logistics, and finance
d. Liaison, finance, operations, administration, and reporting

175. Before beginning the workday, a group of construction workers asks you to determine the wet bulb globe temperature (WBGT) so that the risk of heat stress can be properly determined for outdoor work. According to a WBGT measuring device, the wet bulb temperature (t_{nwb}) is recorded as 75 °F, the globular temperature (t_g) as 78 °F, and the dry bulb temperature (t_{db}) as 80 °F. What is the WBGT?

a. 80.8 °F
b. 79.5 °F
c. 76.1 °F
d. 77.7 °F

176. Which of the following is NOT a common management function of an Incident Command System (ICS)?

a. Operations
b. Planning
c. Triage
d. Finance/administration

177. From a safety professional's perspective, what is the most beneficial impact of an employee health and wellness program?

a. Fewer workers' compensation claims
b. Reduced health insurance premiums
c. Improved worker health, alertness, and absenteeism
d. Better talent acquisition

178. When an emergency occurs that is beyond the mitigation and aid capabilities of local and state agencies, what organization can be contacted to request additional support?

a. Workers' compensation
b. Temporary Assistance for Needy Families
c. Federal Emergency Management Agency
d. Department of Human Services

179. Which of the following scenarios is an example of mitigating risk through the use of automation?

a. Using backup cameras for forklift operations
b. Using power tools instead of hand tools
c. Using human-machine interface (HMI) screens to operate and set equipment parameters
d. Using computer-based training to deliver safety training

180. At the end of constructing a water tower, water samples need to be taken to determine if Legionella is present. You use 1-liter bottles with Dacron or polypropylene-tipped swabs, along with other necessary equipment, to sample the water tower. The results indicate that 32 colony-forming units (CFU) of Legionella were identified in the 3.5 liters of tested water. What is the concentration of Legionella in the water tower?

a. 9.14 CFU/mL
b. 32 CFU/mL
c. 3.2 CFU/mL
d. 0.009 CFU/mL

181. A company is noticing an increase in incident reports, and although none have yet been recordable injuries, the trend shows a growing likelihood of a severe incident. Due to recent layoffs, increased workloads, and management changes, employee morale is very low. What would be the best option for improving long-term employee morale and reducing incidents?

a. Offer employees rewards and bonuses when there are fewer incident reports.
b. Set clear expectations of employees and managers and build relationships among them.
c. Require employee engagement training and team-building exercises.
d. Enforce penalties such as unpaid time off, demotion, or termination for incident involvement.

182. An employee with a preexisting back condition is hired for manual labor on a construction site. After three weeks on the job, the employee feels a sharp pain in his back when moving a pallet. The employee requests to see a physician. The physician diagnoses the injury as a pulled muscle and prescribes a muscle relaxer. What are the company's reporting requirements in this scenario?

a. The employer is exempt from including this injury on the OSHA 300 log because the injury was preexisting.
b. The employer does not have to include the injury on the OSHA 300 log, but the employer is financially responsible for the injury because the injury was aggravated by the work environment.
c. The employer must record this incident on the OSHA 300 log as a recordable injury.
d. The employer will submit this case to OSHA along with the employee's medical history. OSHA will help determine whether the injury was preexisting or if it was related to the job tasks and must be recorded.

183. A group of construction workers decides to move a large 1,200-kg metal crate by rigging the crate to a dump truck and dragging it on the ground. The rigging used for this makes a straight line from the crate to the dump truck. The coefficient of friction for this activity is 0.8. The rigging used can only withstand a pulling force of 3,000 N. Is this rigging material sufficient for this task?

a. The pull force required to initially move the crate is over 9,000 N. Different rigging material must be used for this task.
b. The pull force required to initially move the crate is 960 N. This rigging material will be more than adequate for this task.
c. The pull force required to initially move the crate is just over 3,000 N. Different rigging material must be used for this task.
d. The pull force required to initially move the crate is just under 3,000 N. This rigging material is likely to fail and should be switched for stronger material.

184. A town handing out sandbags in preparation for significant rainfall from a hurricane is an example of what?

a. Emergency mitigation
b. Emergency recovery
c. Incident response
d. Incident management

185. The employer must provide protective clothing for employees consistently working around open flames and/or arc flash exposure. Which of the following is a true statement?

a. Arc flash-protective clothing is flame-resistant, but not all flame-resistant clothing is arc flash-rated.
b. The terms flame-resistant and flame-retardant can be used interchangeably.
c. Flame-retardant clothing is made of natural fibers that resist flames.
d. Flame-resistant clothing is chemically treated to provide protection against open flames.

186. In order to properly control occupational noise exposure, the frequency of the noise levels must be determined. You measure the wavelength (λ) to be 0.404 meters. The speed of sound (c) is 343 m/s. What is the frequency (f) of the sound wave?

a. 849 Hz
b. 139 Hz
c. 0.0012 Hz
d. 1,486 Hz

187. If you are a supervisor of a crew, why would it be important to keep confidential information such as salaries and performance appraisals private?

a. The employees may get mad about pay ranges and find other jobs.
b. Employees need to be able to trust their supervisors.
c. Sharing salary information is considered taboo.
d. Sharing salary and work performance information is illegal.

188. What is the protocol when subcontractors are working on a site and have different company requirements than those set forth in the site-specific contractor safety plan?

a. The contractors' requirements must be followed.
b. The subcontractors' requirements are fine if they follow OSHA standards.
c. The subcontractors' requirements can be used only if their requirements are stricter.
d. The contractor and subcontractor must find common ground.

189. A coworker complains of tightness in his chest before suddenly collapsing to the ground. You perform a quick assessment and recognize that the man's breathing is shallow, and his pulse is thready. Would using an automated external defibrillator (AED) be proper in this situation? Why or why not?

a. Yes, an AED can help restore natural breathing, which will steady the pulse.
b. Yes, an AED will shock the patient back into a normal sinus rhythm.
c. No, an AED should only be used on someone who has no sign of a pulse and is unresponsive.
d. No, an AED should only be used if the pulse is strong; otherwise, the shock could overpower the natural sinus rhythm and cause it to stop.

190. The weather report states that the highest temperature will be 40 degrees Celsius (°C). What is this temperature in degrees Fahrenheit (°F)?

a. 80 °F
b. 104 °F
c. 95 °F
d. 68 °F

191. You are the health and safety technician for a company that will be performing work near a large aperture circular antenna. You are asked to examine the risk of microwave radiation exposure. The antenna has a diameter (D) of 82 cm and a power rating of 60,000 mW (P). What is the estimated power density level (W) in the antenna's near field (less than 1 wavelength from the source)?

a. 45.4 mW/cm^2
b. 4,920,000 mW/cm^2
c. 49.2 mW/cm^2
d. 731 mW/cm^2

192. What annual OSHA reporting document is required to be posted in a visible location from February 1st through April 30th each year?

a. 300A
b. 300
c. 301
d. 29 CFR 1926

193. You are evaluating scaffolding construction for OSHA compliance. During your inspection, you notice a ramp that seems like it is steeper than the 20° limit OSHA has on ramps and walkways for scaffolding. You measure and draw the following diagram:

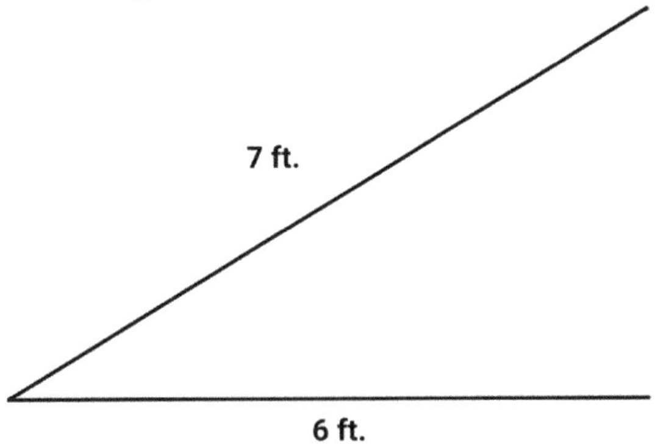

Based on this, does this ramp exceed the 20° limit for ramps and walkways? Why or why not?

a. No, the angle of this ramp is just under 20° and does not exceed the angle limit for ramps and walkways.
b. Yes, the angle of this ramp is 31° and thus exceeds the angle limit for ramps and walkways.
c. No, the angle of this ramp is around 15° and does not exceed the angle limit for ramps and walkways.
d. Yes, the angle of this ramp is over 45° and thus exceeds the angle limit for ramps and walkways.

194. To ensure proper sanitation, how many toilet seats and urinals does OSHA require to be at a temporary workplace with 300 employees?

a. 6 toilet seats and 6 urinals
b. 3 toilet seats and 3 urinals
c. 5 toilet seats and 5 urinals
d. 10 toilet seats and 10 urinals

195. A foreman is responsible for leading his highway crew through a morning safety meeting. On this day, the daylight crew will be pouring asphalt, which is considered a high-risk activity. The job will likely continue over into the night shift as well. Does the first crew's job safety analysis (JSA) suffice for the second shift as well?

a. Yes, if the job scope has not changed, the same JSA can be used.
b. Yes, the same JSA can be used if there are no new workers on the job.
c. No, the same JSA cannot be used. A JSA should be hazard-specific, and the night crew could be exposed to different hazards than the daylight crew.
d. No, a new JSA must be completed each time a job is performed.

196. What federal act performs the primary function of establishing the basic structure for governing and protecting waterways from pollution?

a. Federal Clean Water Act
b. Safe Drinking Water Act
c. Comprehensive Environmental Response, Compensation, and Liability Act
d. North American Wetlands Conservation Act

197. Which of the following accident investigation techniques is primarily reactive in nature, meaning that it is typically used after an accident has occurred rather than used to identify potential causes of accidents in the future?

a. Root cause analysis
b. Failure mode and effect analysis
c. Loss prevention and reduction
d. Fault tree analysis

198. You are asked to evaluate a 0.8-foot diameter abrasive wheel to determine if the maximum angular exposure of the grinding wheel periphery and sides is not more than 90° (29 CFR 1926.300(b)(8)). Measuring the perimeter (outer edge) of the exposed grinding wheel gave a value of 1.62 feet. Does the angular exposure of the grinding wheel periphery exceed the CFR standard?

a. No, the angular exposure of the grinding wheel periphery is 85.2°, which is lower than the 90° limit.
b. Yes, the perimeter of the exposed grinding wheel should be no more than 1.3 feet.
c. No, the perimeter of the exposed grinding wheel needs to be lower than 2.23 inches.
d. Yes, the angular exposure of the grinding wheel periphery is 110.3°, which exceeds the 90° limit.

199. The hazard statement on an SDS for calcium carbide states:

Hazard statement:
Reactive with water
Releases flammable gases that may ignite spontaneously
Causes skin irritation
Causes serious eye damage
May cause respiratory irritation
Toxic to aquatic life

A symbol in which diamond of the NFPA placard would communicate that a substance is hazardous or reactive with water?

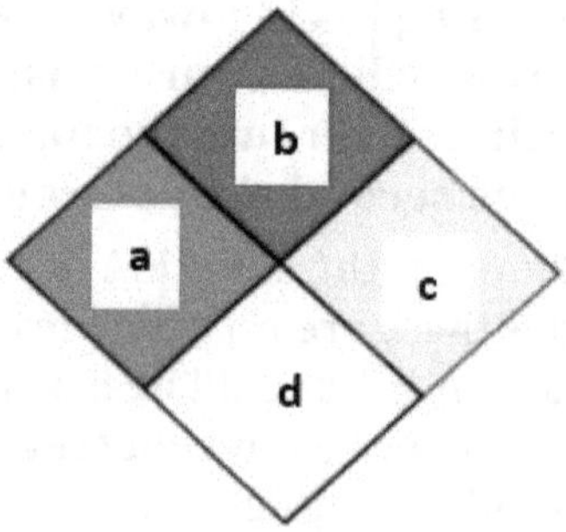

200. A compressed gas cylinder is pressure-filled to 150 atm at 25 °C. You are concerned that excess heat gained during transportation may increase the internal pressure enough to cause the cylinder to explode, which will occur at 155 atm. Which temperature is sufficient to allow the compressed gas cylinder to explode?

a. 32 °C
b. 28 °C
c. 30 °C
d. 34.9 °C

Answer Key and Explanations for Test #1

1. B: When multiple noise sources are emitting simultaneous sound pressure levels, the combined sound pressure level (L_{pt}) is calculated using the following equation:

$$L_{pt} = 10 \log \sum_{i=1}^{N} 10^{\frac{L_{pi}}{10}}$$

Inserting all three sound pressure levels given in the problem will allow us to calculate the L_{pt}:

$$\begin{aligned} L_{pt} &= 10 \log \left(10^{\frac{100}{10}} + 10^{\frac{98}{10}} + 10^{\frac{95}{10}}\right) \\ &= 102.9 \text{ dBA} \end{aligned}$$

2. D: Safety inspections focus on physical safety hazards around people and equipment. Audits focus on safety systems, programs, and processes.

3. D: Emergency wardens or coordinators can be authorized to help fulfill any emergency duties they are capable of and trained for; however, the incident commander position cannot be predetermined, as it is selected based on experience, training, and leadership abilities specific to the emergency.

4. A: Fall protection standards for General Industry and Construction Industry are similar. The main difference between the two standards is the height requirement at which a fall protection system is required. General Industry (1910.28) calls for fall protection at 4 feet. Construction Industry (1926.501) requires fall protection at 6 feet. Both industries require a fall protection system to be in place when working above dangerous equipment or machinery, regardless of height. Fall protection requirements on dock boards for both industries require side walls or for the operator to be able to justify that there is no hazard related to falling from the dock board.

5. C: ABC extinguishers can fight Class A, B, and C fires, therefore eliminating all possible flames in this situation. Class A extinguishers are used on combustibles such as wood, paper, cloth, rubber, trash, and plastics. Class B extinguishers are used for flammable liquids. CO_2 extinguishers can be used for Class B and Class C electrical fires but not Class A.

6. D: Installing an electrical distribution box is the safest way to plug in the air conditioner. This allows the unit to be plugged directly into the power source. OSHA 1926.403 provides that the employer shall ensure that electrical equipment is free from recognized hazards and that certified equipment shall be used in accordance with the instructions included in the listing, labeling, and certification. Electrical cords are not to be longer than 100 feet, and multiple electrical cords cannot be plugged into each other. Power strips are not to be plugged into extension cords.

7. D: Failure Mode and Effects Analysis is a leading risk management strategy used to identify problems or failures in a system, analyze the potential harmful outcomes of the failures, and then apply actions to prevent occurrence.

8. A: Rheumatoid arthritis is an autoimmune disease, not a cumulative trauma disorder. Cumulative trauma injuries occur when repetitive stress on the body leads to damage to the musculoskeletal and nervous systems. Examples of cumulative trauma injuries are Raynaud's syndrome, low back pain, tendonitis, tenosynovitis, nerve entrapment, tennis elbow, and carpal tunnel syndrome.

9. A: A spokesperson during a crisis management situation is responsible for all the options listed but primarily for the delivery of accurate, consistent, and timely information.

10. C: According to 29 CFR 1926 Subpart M App C, the object used to test PFAS should have a girth of 38 inches $\pm$ 4 inches. To calculate girth, the width (6 inches) and the height (8 inches) of the object are added together, and then the sum is multiplied by 2:

$$\text{Girth} = 2 \times (\text{width} + \text{height}) = 2 \times (6 + 8) = 28 \text{ in}$$

Therefore, this object could not be used to test PFAS.

11. D: Silica (SiO_2) does not have a specific PEL assigned by OSHA. Instead, the PEL for SiO_2 is determined per the material using the following equation:

$$\text{PEL}_{silica} = \frac{10 \text{ mg/m}^3}{(\%\text{SiO}_2) + 2}$$

The higher the percentage of SiO_2 the material contains, the lower the PEL_{silica} will be. Plugging 20% SiO_2 into the equation allows us to compute the following:

$$\text{PEL}_{silica} = \frac{10 \text{ mg/m}^3}{20 + 2} = 0.45 \text{ mg/m}^3$$

12. D: Class K fire extinguishers are wet chemical extinguishers that discharge an acidic chemical that reacts with cooking oil and fats to create a substance that sits on top of fire and smothers it. Using a CO_2 extinguisher, ABC extinguisher, or other dry chemical extinguisher may only cool the top layer of oil, and the fire could reignite. Additionally, the discharge pressure could cause the oil to spread, creating a larger fire.

13. B: Universal precautions is a concept introduced by the Center for Disease Control that calls for the assumption that all bodily fluids and blood should be treated as infectious. Not replacing the cap on a used needle reduces the risk of exposure by accidental self-sticking. All contaminated material is recommended to be disposed of in a biohazard bag. Requiring health screens after exposure is a reactive approach and not a precaution. Contaminated scenes should be cleaned up with a bleach and water solution.

14. B: The liaison officer communicates between agencies involved with the emergency and coordinates media messages with the public relations officer. The liaison is considered a neutral point of contact between parties. The liaison officer can help lessen tensions during emergency situations by providing consistent communication amongst all involved in an emergency.

15. D: The job site-specific safety manual would include specific information about the job site, including any additional rules or regulations that would need to be followed per regulatory agencies or city ordinances.

16. B: According to 29 CFR 1926.103, the MUC is defined as the maximum atmospheric concentration of a hazardous substance from which an employee can be expected to be protected when wearing a respirator. The MUC is determined mathematically by multiplying the OSHA PEL of the contaminant and the assigned protection factor (APF) of the respirator being used:

$$\text{MUC} = \text{APF} \times \text{PEL}$$

The MUC for the scenario above is determined by plugging in the APF for a full-facepiece APR (50) and the PEL for the chemical (25 ppm):

$$\text{MUC} = 50 \times 25 = 1{,}250 \text{ ppm}$$

If the concentration of the chemical exceeds 1,250 ppm, then a respirator with a higher APF value must be used.

17. D: In 29 CFR 1926.302(b)(4), OSHA specifies a psi of less than 30 for compressed air used for cleaning purposes. To determine the psi of the compressed air gun for this problem, we must first convert cm^2 to in^2 due to psi being pounds per square inch. Since there are 2.54 cm per 1 in:

$$38 \text{ cm}^2 \times \left(\frac{1 \text{ in}}{2.54 \text{ cm}}\right)^2 = 5.89 \text{ in}^2$$

Now, we can use the values for area and pounds of force given to calculate psi:

$$\text{psi} = \frac{202 \text{ lb (force)}}{5.89 \text{ in}^2} = 34.3 \text{ psi}$$

Because this exceeds OSHA's limit of 30 psi, the pressure must be lowered on this compressed air gun before it is used for cleaning purposes.

18. C: To calculate the $\text{A}_t(8)$ from multiple different operations with differing vibration magnitudes, the partial vibration exposure values are used in the following equation:

$$\text{A}_t(8) = \sqrt{[\text{A}_1(8)]^2 + [\text{A}_2(8)]^2 + \cdots + [\text{A}_n(8)]^2}$$

Inserting the A(8) values for equipment 1 and 2 calculates the $\text{A}_t(8)$ to be:

$$\text{A}_t(8) = \sqrt{[1.2]^2 + [1.4]^2} = 1.84 \text{ m/s}^2$$

19. A: Redesigning the equipment through engineering changes that would eliminate the need for a worker to be near the belt is the best example of the domino theory. All the answer options could reduce the likelihood of injury but do not eliminate the mechanical hazard. Engineering out the hazard is the best option, then administrative controls, such as lockout/tagout and training, can be put into place. PPE to protect the worker is the last line of defense.

20. A: According to the Bureau of Labor Statistics, transportation incidents involving motorized vehicles are currently the leading cause of occupational fatalities.

21. C: The hose diameter needed is determined by first solving for the cross-sectional area associated with the capture velocity (V) and the flow rate (Q) values:

$$Q = VA \rightarrow A = \frac{Q}{V} = \frac{167 \text{ cfm}}{2{,}500 \text{ fpm}} = 0.0668 \text{ ft}^2$$

Once the cross-sectional area (A) is determined, the radius (r) can be calculated:

$$A = \pi r^2 \rightarrow r = \sqrt{A/\pi} = \sqrt{(0.0668 \text{ ft}^2/\pi)} = 0.146 \text{ ft} \times \frac{12 \text{ in}}{1 \text{ ft}} = 1.75 \text{ in}$$

22. A: To solve this problem, you create a right triangle with the hypotenuse being the distance from the anchor point to the convergence point of the two pieces of rope (15 feet) and the base being half of the length of the crate (13 feet):

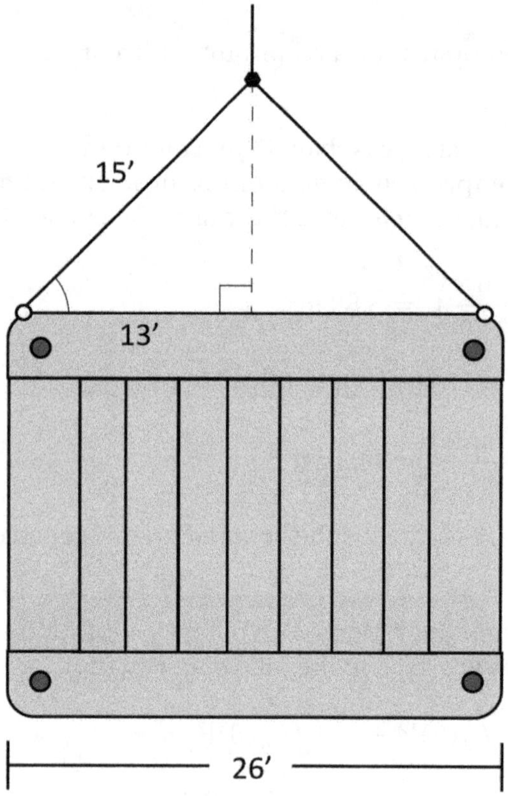

$$\cos\theta = \frac{13}{15}$$

$$\theta = \cos^{-1}\frac{13}{15} = 29.9°$$

Since you have the hypotenuse and the adjacent lengths, you can use the cosine function to determine the angle, which is just below 30° at 29.9°.

23. D: To solve this problem, we will use the inverse square law equation:

$$I_2 = I_1\left(\frac{(d_1)^2}{(d_2)^2}\right)$$

The given values for the intensity (dose rate) at distance 2 (I_2), the intensity (dose rate) at distance 1 (I_1), and the distance at location 1 (d_1) are plugged into the inverse square law equation. We then algebraically rearrange the equation to solve for the distance at location 2 (d_2):

$$133 = 600 \times \left(\frac{(2^2)}{(d_2)^2}\right) \rightarrow (d_2)^2 = \frac{600}{133} \times 2^2 \rightarrow d_2 = \sqrt{\frac{4 \times 600}{133}} = 4.25 \text{ ft}$$

24. D: The flow rate of the eyewash station is computed by dividing the volume of water collected (5 gallons) by the duration of the run time (15 minutes):

$$\text{Flow rate} = \frac{5 \text{ gallons}}{15 \text{ min}} = 0.3333 \text{ gallons/min}$$

Thus, the eyewash station's flow rate does not meet ANSI's standards of 0.4 gallons/min.

25. B: The best option would be to move the injured employee to a first-floor office. This option eliminates the need to exit down the stairs or wait for assistance.

26. B: To determine the daily vibration exposure value [A(8)], the following equation is used:

$$A(8) = a_{hv} \times \sqrt{\frac{T_v}{T_0}}$$

The square root of the quotient of the number of hours worked (T_v) and the reference time of 8 hours (T_0) is multiplied by the acceleration of the object (a_{hv}) to determine the value of A(8). Plugging in all of those values listed yields an A(8) value of:

$$A(8) = 4.2 \times \sqrt{\frac{3}{8}} = 2.6\ \text{m/s}^2$$

27. D: Fall protection should be visually inspected by the employee prior to each use. Additionally, fall protection should be visually inspected after a fall or impact. OSHA also requires a documented fall protection inspection to be completed by a competent person annually.

28. C: The location, amount, and type of flammables or hazardous chemicals stored on site should be communicated with local first responders when an emergency response plan is developed. Having this information prior to an emergency can allow first responders to properly prepare for the types of emergencies they may need to respond to, and it protects the responders by making them aware of extremely hazardous areas.

29. C: Hot work permits are required for any hot work that is conducted within 40 feet of a potential hazard, such as a building or storage container containing flammable material.

30. B: According to 29 CFR 1926.955(b)(1), portable ladders and platforms used on structures or conductors in conjunction with overhead line work shall be capable of supporting without failure at least 2.5 times the maximum intended load. Since the maximum intended load is 250 lb, 2.5 times that would be 625 lb.

31. D: A horizontal lifeline can be installed along the rooftop and can be used as a means for fall restraint or fall arrest. In this scenario, the horizontal lifeline would be installed as a restraint measure, preventing the worker from accessing the edge of the building and eliminating the risk of falling.

32. B: Alcohol and substance abuse training for supervisors informs supervisors what behaviors to look for in an employee with a potential drug or alcohol problem. The training will also cover what actions the supervisor should take, such as notifications, documentation, legalities, and reporting.

33. A: To determine the average flow rate, the average pre-calibration flow rate must first be calculated, and then this value can be used with the post-calibration flow rate to determine the overall average flow rate. This is done because the pre-calibration flow rate is determined before sampling occurs, and its value is independent of the post-calibration flow rate. As such, the pre-calibration flow rate measurements are used to determine the average flow rate prior to sampling:

$$\text{Pre-calibration average flow rate} = \frac{(2.0154 + 1.9843 + 2.0004)}{3} = 2.0000333\ \text{L/min}$$

Next, the average flow rate can be calculated using the pre-calibration average flow rate and the post-calibration flow rate:

$$\text{Average flow rate} = \frac{(2.0000333 + 2.0034)}{2} = 2.0017\ \text{L/min}$$

34. C: XHHW-2 is a standard wire type that indicates:

- X: Cross-linked polyethylene insulation
- HH: Heat-resistant (90°C in dry locations)
- W: Suitable for wet locations
- -2: Rated for 90°C in both wet and dry locations (This designation indicates it is rated beyond the standard 75°C in wet locations)

Some additional standard indicators are:

- T: Thermoplastic insulation
- R: Rubber insulation
- N: Nylon coating
- Romex: The trade name for Thermoplastic-Sheathed Cable (TPS) that is commonly used in residential wiring

35. D: Before responding to any emergency, first aid providers must assess the scene to ensure their own safety prior to engaging in any first aid or CPR efforts.

36. B: When several different chemicals are present at the same time, the chemical exposure from each airborne contaminant must be calculated and then summed together to determine if an overexposure has occurred or not. It is insufficient to say that there is no overexposure just because the PEL of any one contaminant has not been exceeded; rather, all exposures must be considered. To calculate the overall PEL exposure from a chemical mixture, the following equation is used:

$$\text{PEL}_{mixture} = \frac{C_1}{\text{PEL}_1} + \frac{C_2}{\text{PEL}_2} + \cdots + \frac{C_n}{\text{PEL}_n}$$

The exposure fraction is determined by dividing the airborne concentration of the contaminant by the associated PEL. The exposure fractions are then summed together. If the overall value is ≥ 1, an overexposure has occurred.

$$\text{PEL}_{mixture} = \frac{475}{1{,}000} + \frac{62}{100} + \frac{375}{500} = 1.845$$

Since the value of the $\text{PEL}_{mixture}$ is greater than 1, the solvent bath mixture is sufficient to be considered an overexposure.

37. C: OSHA 29 CFR 1926.35 is the OSHA standard for the construction industry that covers emergency action planning. The standard does not mandate how often drills should occur; however, OSHA suggests that employers review emergency response procedures with employees annually as well as when a procedure is put into place, when the scope of a procedure changes, or when job roles and responsibilities change. OSHA regulations in 29 CFR 1910 are related to general industry, therefore eliminating answer options B and D. OSHA 1910.28 is the construction industry fall protection standard.

38. B: The secondary containment should hold 55 gallons, as this represents 10% of the total amount (550 gallons) and the volume of the largest container is 55 gallons.

39. A: H_2S awareness training should be provided whenever a worker may potentially be exposed to hydrogen sulfide. Awareness level training will educate employees on what H_2S is, the hazards that H_2S presents, exposure limits, work practices, and safety equipment. Whenever a worker is required to continue work in an atmosphere where H_2S concentration is greater than 10 ppm, training above the awareness level that includes the use of a positive pressure breathing apparatus and rescue procedures must be provided.

40. B: The Hazardous Communication Standard requires current and accessible Safety Data Sheets for all hazardous chemicals. When a chemical is taken out of use, its SDS should be dated and archived for 30 years.

41. D: The pH of a solution is calculated using the following equation:

$$\text{pH} = -\log(\text{H}^+)$$

The given value of the concentration of hydronium ions $[\text{H}^+]$ is inserted into the equation to determine a pH value of:

$$\text{pH} = -\log(0.0001) = 4$$

Since the pH is below 7, this chemical solution would be considered acidic.

42. A: To determine the number of mold spores, we will use the air sample to extrapolate the data for the entire volume of air in the room. Since 850 mold spores were identified in 75 L of air ($15 \text{ L/min} \times 5 \text{ min} = 75 \text{ L}$), we can establish the number of mold spores per unit of air. First, we must change the units of air to meters in order to match the units of the dimensions of the room:

$$75 \text{ L of air} \times \frac{1{,}000 \text{ mL}}{1 \text{ L}} \times \frac{1 \text{ cm}^3}{1 \text{ mL}} \times \left(\frac{1 \text{ m}}{100 \text{ cm}}\right)^3 = 0.075 \text{ m}^3 \text{ of air}$$

Now, we can establish the number of mold spores per m^3 of air:

$$\frac{850 \text{ mold spores}}{0.075 \text{ m}^3 \text{ of air}} = \frac{11{,}333.33 \text{ mold spores}}{\text{m}^3 \text{ of air}}$$

Next, we need to determine the volume of the room in m^3 and then multiply the volume by the concentration of mold spores in the air:

$$\text{Volume}_{(room)} = 3 \times 4 \times 1.5 = 18 \text{ m}^3$$

$$\frac{11{,}333.33 \text{ mold spores}}{\text{m}^3 \text{ of air}} \times 18 \text{ m}^3 = 204{,}000 \text{ mold spores}$$

43. C: NOAA Weather Radio is a nationwide network of stations that broadcast warning and event information for all types of hazards, including natural disasters, environmental disasters, and public safety alerts.

44. C: The safety manager should first visit the site in person to speak to the employees prior to conducting any audits or inspections. Building relationships is key to forming effective communication and respect among employees and management. Performing audits before communicating expectations and goals could build a lack of trust in a team. Also, reviewing audits with the entire company before reviewing the findings with the site leader or foreman would not build a trusting relationship.

45. A: Joining a recognized national or global professional organization, such as the American Society of Safety Professionals or the National Safety Council, will allow a company to be provided updates to regulations that may affect their industry.

46. C: The emergency action plan does not require procedures for managing hazardous chemicals. The following minimum requirements for emergency action plans are provided in OSHA 1926.35: procedures for reporting a fire or other type of emergency, procedures for emergency evacuation and exit routes, procedures to account for all employees after evacuation, procedures for employees who will perform rescue or medical care, and a list of all employees with their emergency contact information and their job duties.

47. D: Environmental emergency response plans are designed to develop actions to be taken in the event of an environmental emergency that will streamline response, mitigate the spill, and reduce the impact of the spill.

48. C: A safety committee comprising employees from all levels of the organization (entry level to upper management) is a proven way to ensure the buy-in of policies and procedures. Employees are more likely to participate in creating and following procedures they had a say in creating. Also, employees are the most familiar with the hazards that exist and can provide their input on the most efficient ways to avoid these hazards.

49. B: A commercial driver's license (CDL) is needed to drive a vehicle that weighs more than 26,000 lb. This weight includes the weight of the vehicle, the truck, the trailer, and anything being towed.

50. A: Each emergency action plan must be specific to the job and the location. If the job activity and/or location changes, the plan must be updated and readily available on site. Emergency services should always be the first call to avoid delaying response time. Changing protocol to require dispatch or a corporate office to notify emergency services would create an extra step in the notification process. The emergency plan needs to be accessible to everyone on site. If only the supervisor has access to the plan and the supervisor is injured, a delayed response time could result in greater harm.

51. A: The construction company would be responsible for ensuring that BLM permits and additional reporting requirements were approved prior to starting the project because the drilling rig operator would not be accessing the site until after the construction was finished.

52. A: The TL of a material can be calculated from its transmission coefficient by using the following equation:

$$\text{TL} = 10 \log\left(\frac{1}{\tau}\right)$$

Plugging in our value for the transmission coefficient yields a TL of:

$$\text{TL} = 10 \log\left(\frac{1}{0.05}\right) = 13 \text{ dB}$$

53. D: To calculate the concentration of an airborne contaminant from an air sample, the amount of the airborne contaminant is divided by the volume of air collected. The amount of airborne hexavalent chromium (0.125 μg) is divided by 20 liters, which is converted to m^3 to ensure that the units match:

$$20\text{ L} \times \frac{1{,}000\text{ mL}}{1\text{ L}} \times \frac{1^3\text{cm}^3}{1\text{ mL}} \times \frac{1^3\text{m}^3}{100^3\text{cm}^3} = 0.02\text{ m}^3 \rightarrow \frac{0.125\text{ μg}}{0.02\text{ m}^3} = 6.25\text{ μg/m}^3$$

An 8-hour exposure to this ambient air environment would lead to an OSHA overexposure to hexavalent chromium.

54. D: A machine's ground bearing pressure is the amount of pressure exerted on a surface in units of lb/ft^2. This is determined by dividing the gross weight of the machine by the area of the floor it occupies. Convert the dimensions of the scissor lift from inches to feet, then multiply the two measurements together to get the occupied floor area of the machine. This number can then be used to calculate the ground bearing pressure:

$$\frac{2{,}485\text{ lb}}{\left(\left(\frac{65\text{ in}}{12\text{ in/ft}}\right)\left(\frac{30\text{ in}}{12\text{ in/ft}}\right)\right)} = 184\text{ lb/ft}^2$$

55. C: To perform sampling pump flow rate corrections to account for the differences in temperature and/or pressure between the calibration and sampling sites, the following equation is used:

$$Q_{act} = Q_{cal} \times \sqrt{\left(\frac{P_{cal}}{P_{act}}\right) \times \left(\frac{T_{act}}{T_{cal}}\right)}$$

The subscripts *cal* and *act* are used to differentiate between values measured at the calibration site (*cal*) and the sampling site (*act*). Temperature units are in Kelvin, which requires that a value of 273 is added to the temperature in Celsius.

Plugging in the given values for the calibrated average flow rate (2.2 L/min) along with the measured temperatures and pressure values yields an actual flow rate of:

$$Q_{act} = 2.2\text{ L/min} \times \sqrt{\left(\frac{720}{750}\right) \times \left(\frac{(273+0)}{(273+20)}\right)} = 2.08\text{ L/min}$$

Thus, a flow rate of 2.08 L/min must be used in all future calculations and reports.

56. C: Steel-toe boots are rated to withstand 75 foot-pounds, which means they can withstand the impact force of a 75-lb object dropped from a height of 1 foot. This relationship can then be used to determine the weight of an object the boots can withstand at a height greater than or less than 1 foot.

$$\frac{75\text{ ft}\cdot\text{lb}}{12\text{ ft}} = 6.25\text{ lb}$$

Therefore, the maximum weight of an object falling from a height of 12 feet that the boots can withstand is 6.25 lb. The steel-toe boots would not withstand the 8-lb tool.

57. A: OSHA requires employers to offer hepatitis B vaccines at no cost to employees who can reasonably anticipate contact with blood or other potentially infectious materials during their jobs or employees who have potentially contacted infectious material.

58. B: The OSHA PEL for asbestos is 0.1 fibers/cubic centimeter (cc) of air over an 8-hour workday. In order to solve for the concentration of asbestos fibers that the sampled employee was exposed to, we must first determine the volume of air captured during sampling in cubic centimeters:

$$\frac{1\text{ L}}{\text{min}} \times 480\text{ min} = 480\text{ L} \times \frac{1{,}000\text{ mL}}{1\text{ L}} \times \frac{1\text{ cm}^3}{1\text{ mL}} = 480{,}000\text{ cm}^3$$

Next, we will divide the number of identified asbestos fibers collected during the sampling period by the volume of air to calculate the concentration of asbestos:

$$\frac{25{,}000\text{ asbestos fibers}}{480{,}000\text{ cm}^3\text{ air}} = 0.052\ \frac{\text{fibers}}{\text{cm}^3}$$

This is below OSHA's PEL for asbestos.

59. B: When in the recovery position, the injured employee is less likely to aspirate if he throws up. If there is any potential for neck or spinal injury, the employee should not be shifted.

60. B: The emergency response plan is specific to workplaces that fall under the HAZWOPER designation due to operations involving hazardous waste treatments, disposal, or cleanup. OSHA 29 CFR 1926.65 lists the requirements of an emergency response plan, including pre-emergency planning, personnel roles, training, communication, emergency recognition, safe distances, places of refuge, site security and control, evacuation routes and procedures, decontamination, emergency medical treatment and first aid, emergency alerting, PPE, and emergency equipment. Liability insurance and lists of local contractors and PPE vendors are not required components of an emergency response plan.

61. C: If a wheel has a wheel width in between the wheel width values shown in Table D-57.1, extrapolation is necessary in order to calculate the minimum exhaust volume. In this case, the width of 0.46 feet is converted to inches (5.52 in), and then divided by the closest width value that is less than that (5 in). This value is then multiplied by the minimum exhaust volume associated with 5 inches (880 cfm) and rounded up to the nearest increment of 10:

$$0.46\text{ ft} \times \frac{12\text{ in}}{1\text{ ft}} = 5.52\text{ in} \rightarrow \frac{5.52\text{ in}}{5\text{ in}} \times 880\text{ cfm} = 971.52\text{ cfm} \rightarrow 980\text{ cfm}$$

62. A: Requiring asbestos abatement or remediation prior to a remodeling job would protect workers from exposure. Removing the hazard is the highest form of risk reduction and should always be the goal of risk reduction. PPE and training can help reduce risks but are not as effective as eliminating the hazard.

63. A: Employees can conduct pre-job fit-for-duty exams on each other. Performing a pre-job fit-for-duty exam consists of checking someone for signs of fatigue, mental alertness, physical abilities, possible substance abuse, life stress, and any other factors that could prevent someone from performing their job safely. Pre-job fit-for-duty exams are not the same as medical fit-for-duty exams (in which a medical professional in a clinical setting determines workers' health and their ability to perform their jobs).

64. B: Employees that can actively engage in corrective actions are more likely to report incidents. Front-line employees often have the best solutions for operational safety issues.

65. B: During the risk assessment process, once hazards have been assessed, the risk level of the hazards must be determined. Probability is the likelihood of an event occurring based on the current job and conditions. Severity is the degree of undesired consequences that impact the safety of employees or the overall project.

66. C: To calculate ACH, the following equation is used:

$$\text{ACH} = \frac{60\text{Q}}{\text{V}_r}$$

The volume of the room is determined by multiplying the length, width, and height:

$$\text{V}_r = \text{L} \times \text{W} \times \text{H} \rightarrow \text{V}_r = 50 \times 25 \times 100 = 125{,}000 \text{ ft}^3$$

Plugging in V_r and the flow rate (Q) yields an ACH of:

$$\text{ACH} = \frac{60 \times 7{,}000}{125{,}000} = 3.36 \approx 3.4$$

67. D: According to 29 CFR 1926.250(b)(6), a stack of bricks cannot be higher than 7 feet. To solve this problem, we must first determine the height of the bricks in feet:

$$\frac{2.25 \text{ inches height}}{\text{brick}} \times \frac{1 \text{ foot}}{12 \text{ inches}} = \frac{0.1875 \text{ feet height}}{\text{brick}}$$

Now, we will calculate how many bricks it will take to get to a height of 7 feet:

$$\frac{1 \text{ brick}}{0.1875 \text{ feet height}} \times 7 \text{ feet height} = 37.333 \text{ bricks}$$

Since using 38 bricks would exceed 7 feet, the number of bricks is rounded down to 37.

68. C: OSHA 1926.652 requires excavations greater than 5 feet deep to have cave-in protection. Shoring, sheeting, shields, sloping, and benching are all means used to prevent cave-ins.

69. A: Toeboards, handrails, and certified tie-off points are required on scaffolds, aerial lifts, scissor lifts, and other elevated working platforms. If the work being performed is greater than six feet above the ground, fall protection requirements are put into place. The requirements vary depending on the type of elevated platform, but at a minimum, toeboards, handrails, and tie-off points are required.

70. D: Sweat is not a carrier for bloodborne pathogens. According to the CDC, the most common bloodborne pathogens are HIV, hepatitis B, and hepatitis C. Bloodborne pathogens are transmitted through infected blood or bodily fluids that potentially contain blood. Saliva is not considered a means of transportation for bloodborne pathogens, but saliva contaminated with blood could potentially transmit bloodborne pathogens.

71. C: Span of control is used by FEMA to define the number of individuals or resources that one supervisor can manage effectively during an incident.

72. C: Lockout/tagout does not require the use of a multimeter or any other device to measure voltage. It is recommended to try out the equipment after the lockout/tagout procedure is in place to ensure that the equipment is turned off and any residual energy is dissipated.

73. A: Respiratory protection training includes the donning and doffing of equipment and training employees on how to ensure the respirator is a good fit. This training cannot be effectively taught online.

74. D: Designating an official spokesperson to represent an organization is part of a crisis management plan. Crisis management and emergency response plans are both critical during emergency and problematic situations. Crisis management involves the mitigation and response to an emergency to protect the organization, whereas emergency response is the immediate actions implemented during the emergency to protect people and property. Answer choices A, B, and C are all components of an emergency response plan.

75. D: The OSHA PEL is established for an 8-hour work shift. When work shifts are longer than 8 hours, a reduction factor (RF) must be calculated using the longer shift length (i.e., 10 hours):

$$\text{RF} = \frac{8}{h} \times \frac{24-h}{16} \rightarrow \text{RF} = \frac{8}{10} \times \frac{24-10}{16} = 0.7$$

The RF value is then multiplied by the PEL of the contaminant of interest to calculate the shift-adjusted PEL:

$$\text{PEL}_{adj} = \text{RF} \times \text{PEL} = 0.7 \times 50 = 35\ \mu\text{g/m}^3$$

Therefore, employees working a 10-hour shift can only be exposed to 35 $\mu\text{g/m}^3$.

76. A: The static coefficient of friction is the ratio of the frictional force (F_s) and the normal force (F_N). F_N is the product of the object's mass (3.5 kg) and gravitational acceleration (9.81 m/s^2):

$$\mu_s = \frac{\text{F}_s}{\text{F}_N} = \frac{20\text{ N}}{(3.5\text{ kg} \times 9.81\ \text{m/s}^2)} = 0.58$$

77. D: According to 29 CFR 1926.451(a)(4), each suspension rope used on adjustable suspension scaffolds shall be capable of supporting, without failure, at least 6 times the maximum intended load. By multiplying the maximum intended load by 6, you can determine the test weight required:

$$750\text{ lb} \times 6 = 4{,}500\text{ lb}$$

78. C: A lack of proper fall protection, including failing to wear proper fall protection, is the most common cause of fatal falls in the construction industry. Lack of planning contributes to this key underlying cause of falls.

79. C: Kinetic energy (KE) is calculated as half the product of an object's mass (m) and the square of its velocity (v). Inserting the given values for mass (4,000 kg) and velocity (4 m/s) yields:

$$KE = \frac{mv^2}{2} = \frac{4{,}000 \times 4^2}{2} = 32{,}000\text{ N}$$

80. C: Potential fines and citations should not be considered when preparing an emergency contingency plan. The protection of people and the environment should be placed at the forefront of all emergency operations. Muster areas, evacuation procedures, and rescue plans must all be included in an emergency action plan.

81. D: The sound pressure level (L_p) is calculated using the sound pressure reference (P_0) value of 0.00002 pascals and the sound pressure (P) in the following equation:

$$L_p = 20 \log\left(\frac{P}{P_0}\right)$$

Plugging in the values for the sound pressure and the sound pressure reference yields a sound pressure level of:

$$L_p = 20 \log\left(\frac{1.4}{0.00002}\right) = 97 \text{ dB}$$

Therefore, a sound pressure of 1.4 pascals is equivalent to a L_p of 97 dB.

82. A: OSHA 1926.651 states that a stairway, ladder, ramp, or other safe means of egress shall be in trench excavations that are 4 feet deep or more, and no more than 25 feet of lateral travel can be between egress locations. A 52-foot trench would have to have 2 means of egress to meet this standard.

83. B: Behavior-based safety is a proactive approach to increasing workplace safety by observing behaviors, analyzing the consequences, and reinforcing desired safety behaviors.

84. C: Human factors engineering evaluates whether a proposed design, product, or system will function safely and effectively in real-world conditions, taking into account the possibility of human error. The goal of human factors engineering is to design solutions that minimize or eliminate the potential for human error.

85. B: To maintain compliance with fire and building codes, two means of egress are required. The emergency action plan evacuation procedures would need to be updated immediately to reflect the temporary rearrangement of employee office space.

86. B: OSHA conducts inspections based on the six priorities. First, imminently dangerous situations are a top priority and are inspected immediately. A faulty crane would be considered an imminent danger. Second are severe injuries and illnesses. An amputation is an example of a severe injury. Third are worker complaints, such as allegations of wrongdoings or mistreatment. Fourth are referrals from violations reported by other regulatory agencies. Fifth are targeted inspections of high-hazard industries with high levels of injury reports, such as high-rise building projects. Finally, follow-up inspections to check for abatement are the lowest priority of OSHA inspections.

87. B: Multiple dose rates (D) and time (T) events can be calculated using the following equation:

$$\text{Dose} = (D_1 \times T_1) + (D_2 \times T_2) + \cdots + (D_n \times T_n)$$

Plugging in the dose rate (D) and the time (T) of exposure for all three of the different periods yields a total radiation dose of:

$$\begin{aligned}\text{Dose} &= (320 \text{ mrem/hr} \times 3 \text{ hr}) + (25 \text{ mrem/hr} \times 2.5 \text{ hr}) + (104 \text{ mrem/hr} \times 2.5 \text{ hr}) \\ &= 1{,}282.5 \text{ mrem}\end{aligned}$$

88. C: The volumetric flow rate (Q) needed in an LEV system is calculated by multiplying the area (A) by the velocity (V). The cross-sectional area of a circle is πr^2. Since the diameter is 14 inches, the radius is 7 inches. Converting the radius from inches to feet yields 0.58333 ft (7 ÷ 12). This is then plugged into the previously identified equation:

$$A = \pi(0.58333^2) = 1.069 \text{ ft}^2$$

The cross-sectional area (A) is then inserted along with the given value for V (1,200 fpm) to calculate Q:

$$Q = VA = 1.069 \times 1{,}200 = 1{,}283 \text{ cfm} \rightarrow 1{,}290 \text{ cfm}$$

89. D: No, this product is reactive to water and should not be stored in the flammable cabinet. OSHA 1926.152 states that materials that react with water and create a fire hazard shall not be stored in the same area as other flammable liquids.

90. D: To convert from mg/m^3 to ppm, the following equation is used:

$$\text{ppm} = \frac{\text{mg/m}^3 \times 24.45}{\text{MW}}$$

For sample 1, 2.89 is plugged in along with the molecular weight (MW) for sulfur dioxide (64 g/mol) to convert to ppm:

$$\text{ppm} = \frac{2.89 \times 24.45}{64} = 1.1 \text{ ppm}$$

For sample 2, 8.65 is plugged in along with the molecular weight (MW) for sulfur dioxide (64 g/mol) to convert to ppm:

$$\text{ppm} = \frac{8.65 \times 24.45}{64} = 3.3 \text{ ppm}$$

91. C: It is the duty of the authorized forklift operator, not the spotter, to inspect the forklift daily or at the beginning of each shift. Inspecting the forklift identifies maintenance issues before they become a hazard.

92. D: Optical density (OD) is a measure of the attenuation of laser radiation at a specific wavelength through a material. OD determines the specifications of laser safety eyewear that must be used to protect employees from laser radiation that is not controlled. OD is calculated by taking the log of the ratio value between E and MPE:

$$\text{OD} = \log\left(\frac{\text{E}}{\text{MPE}}\right) = \log\left(\frac{650}{2.55}\right) = 2.4$$

Therefore, this would require laser safety eyewear marked for an OD of 3 covering the wavelength of 500 nm.

93. B: On average, a fully charged 5-lb extinguisher can last 10–15 seconds. This is important to cover in emergency response training so that employees understand the capabilities of the equipment available to them in an emergency.

94. D: According to 29 CFR 1926.1436(e)(2)(ii), the employer must ensure that new hoists are load tested to a minimum of 110% of rated capacity. The minimum weight that the manufacturer should use to load test the hoist can be calculated by multiplying the decimal form of the percentage by the rated capacity:

$$1{,}250\text{ lb} \times \left(\frac{110}{100}\right) = 1{,}375\text{ lb}$$

The manufacturer must use a minimum weight of 1,375 lb to validate a rated capacity of 1,250 lb.

95. D: Flammable liquids are categorized into four different categories based on the Globally Harmonized System (GHS):

- Category 1: liquids with a flash point below 73.4 °F (23 °C) and a boiling point at or below 95 °F (35 °C)
- Category 2: liquids with a flash point below 73.4 °F (23 °C) and a boiling point above 95 °F (35 °C)
- Category 3: liquids with a flash point at or above 73.4 °F (23 °C) and a boiling point at or below 140 °F (60 °C)
- Category 4: liquids with a flash point above 140 °F (60 °C) and at or below 199.4 °F (93 °C)

According to 29 CFR 1926.152(b)(3), not more than 60 gallons of Category 1, 2, and/or 3 flammable liquids or 120 gallons of Category 4 flammable liquids shall be stored in any one storage cabinet. Using the GHS, the three flammable liquids are categorized as the following:

- Flammable liquid #1: Category 2
- Flammable liquid #2: Category 4
- Flammable liquid #3: Category 1

To determine whether either or both of the OSHA requirements for storage have been met, we can calculate the ratio of the amount of flammable liquids present divided by the storage limit. Flammable liquids #1 (25 gallons) and #3 (25 gallons) are added together and compared to the storage limit of 60 gallons, and flammable liquid #2 (25 gallons) is compared to 120 gallons:

$$\frac{(25\text{ gal } + 25\text{ gal})}{60\text{ gal}} = 0.833\text{ storage capacity used}$$

$$\frac{25\text{ gal}}{120\text{ gal}} = 0.208\text{ storage capacity used}$$

Since neither ratio exceeds 1, this flammable cabinet is OSHA-compliant.

96. B: According to 29 CFR 1926.451(e)(3)(ii), rest platforms must be provided for every 12 feet of vertical height of a stairway-type ladder. With 40 feet total, we can calculate the number of rest platforms needed:

$$\frac{1\text{ rest platform}}{12\text{ feet vertical height}} \times 40\text{ feet vertical height} = 3.33\text{ rest platforms}$$

Thus, only 3 platforms are required.

97. A: Structures built on stilts are an effort to prevent structural flood damage during hurricanes. Mitigation focuses on reducing the loss of life and property by mitigating (reducing) the overall damage caused by an incident.

98. C: To solve this problem, a right triangle is created. Factoring in the excavator height of 4 feet, the height is equal to 8 feet with 18 feet of width. The formula $a^2 + b^2 = c^2$ is then used to determine the length of the excavator arm to reach the top of the trench (see image):

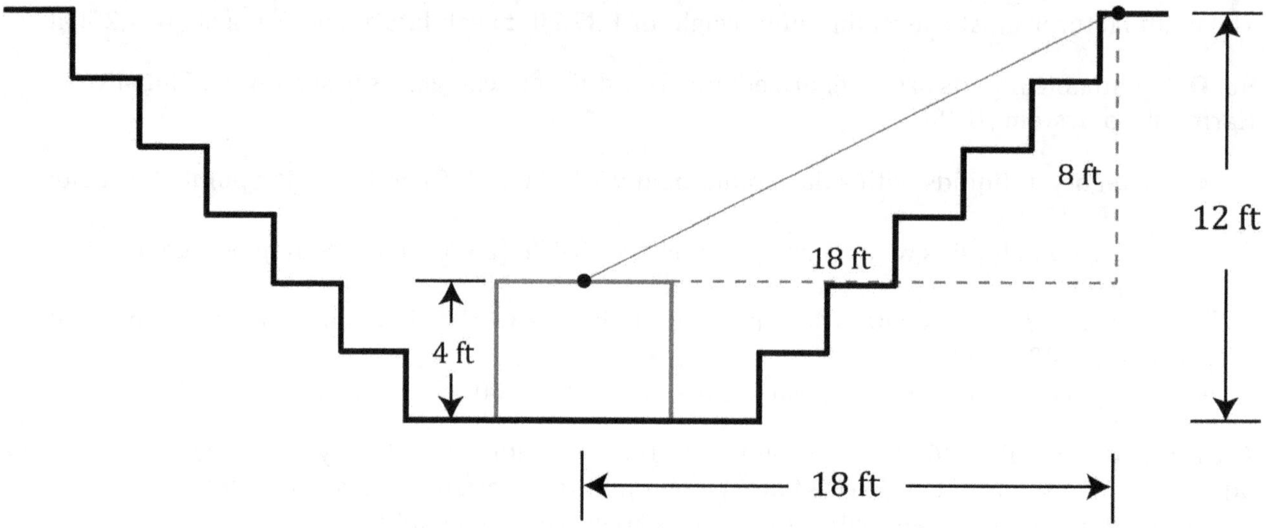

$$a^2 + b^2 = c^2$$

$$\sqrt{18^2 + 8^2} = c = 20 \text{ ft}$$

In addition to reaching the top of the trench, excavated material (spoils) must be placed at least 2 feet from the edge of the trench; therefore, the length of the excavator arm needed is 22 feet.

99. B: Fainting could potentially not be medically related, but since fainting is a common symptom of heart problems, the employee should be taken to the hospital for additional observation.

100. C: The risk management process has five components: identify the risk, analyze the risk, reduce or eliminate the risk, monitor the risk, and review the risk. The faulty gas monitor was identified, the risk was analyzed, the sensor was replaced, the atmosphere was checked, and a process was put in place for future faults.

101. B: OSHA 1926.1411 identifies safe distances for working under powerlines. Higher voltages require greater distances. A clearance of 20 feet would safely cover all voltages for 50 to 1,000 kV.

102. D: For welding arcs, the effective UV irradiance (E_{eff}) at a given distance (r) can be estimated if the arc current (I) is known using the following equation:

$$E_{eff}\left(\frac{W}{m^2}\right) = \frac{\left(\left(0.004 \times \frac{W}{\text{amperes} \cdot m^2}\right) \times I\,(\text{amperes})\right)}{r^2\,(m)}$$

Plugging in the given values for the distance from the source (25 cm = 0.25 m) and the arc current (850 amperes) yields an E_{eff} value of:

$$E_{eff} = \frac{(0.004 \times 850)}{(0.25)^2} = 54.4 \text{ W/m}^2$$

103. B: A procedure should be put in place to rotate workers in and out of the exposed work area to prevent overexposure. This kind of procedure would be a type of administrative control. Engineering out the hazard is the first line of defense. After that, administrative controls can be put in place. PPE is always the last line of defense.

104. A: According to 29 CFR 1926.958, materials stored in unrestricted areas near lines and equipment energized above 50 kV must be located at least 10 feet away plus 4 inches for every 10 kV over 50 kV. To determine the distance beyond 10 feet due to the 30 kV above 50 kV, first determine the number of feet of distance required per kilovolt above 50 kV:

$$\left(4 \text{ in} \times \frac{1 \text{ ft}}{12 \text{ in}}\right)/10 \text{ kV} = 0.0333 \text{ ft/kV}$$

The calculated value can then be used to determine the distance required for 30 kV above 50 kV:

$$0.0333 \text{ ft/kV} \times 30 \text{ kV} = 1 \text{ ft}$$

Therefore, the materials storage shed must be located 11 feet from the energized line.

105. B: Providing employees with a detailed written drug and alcohol policy upon employment and having employees sign the document stating that they agree to adhere to the testing requirements is the best way for an organization to avoid legal conflicts related to drug and alcohol testing.

106. D: RWL = LC × HM × VM × DM × AM × FM × CM is the NIOSH equation used to define the acceptable weight that healthy employees can lift over an eight-hour shift without increasing their risk for a musculoskeletal disorder of the lower back. The factors associated with the formula are how often the lift takes place, whether the lift requires bending or twisting, how heavy the lift is, where the lift begins to exert force upon the body, how close the load is to the body, and how long the load is held.

107. A: To determine the load weight (L) associated with an LI of 1.5 and an RWL of 25, the NIOSH Lifting Equation is used and rearranged to solve for L:

$$\text{LI} = \frac{\text{L}}{\text{RWL}} \rightarrow \text{L} = \text{LI} \times \text{RWL} = 1.5 \times 25 = 37.5 \text{ lb}$$

Thus, a piece of rebar weighing 40 pounds would require more than one employee to move it, as the maximum load weight of 37.5 pounds is exceeded.

108. C: To determine the flow rate needed for the air sampling pump, the air volume needed (35 liters) is divided by the sampling time (15 minutes):

$$\text{Flow rate} = \frac{35 \text{ (L)}}{15 \text{ (min)}} = 2.3333 \text{ L/min}$$

109. B: The first step should be to remove the hazard through engineering controls (i.e., lowering the temperature of the workspace by using fans). The next step is applying administrative controls, such as implementing shorter work periods and requiring additional breaks and training. Then, management should provide PPE that reduces heat retention, such as lightweight and reflective clothing.

110. A: OSHA 1926.1427 requires formal training to become forklift certified. The training must be a combination of written and practical exams. The formal forklift training program must be approved by an auditor. The driving evaluation can be performed by a competent person and must be documented.

111. B: To solve for the amount of time needed for an enclosed space to reduce (dilute) the airborne contaminant(s) present, the following equation is used:

$$t_2 - t_1 = -\frac{V_r}{Q}\ln\left(\frac{C_{g2}}{C_{g1}}\right)$$

The volume of the room (V_r, ft^3), the flow rate in the room (Q, $\frac{ft^3}{min}$), and the concentration of the contaminant at the beginning (C_{g1}) and end (C_{g2}) of the dilution period are used to determine t_2 ($t_1 = 0$). Since we are going from 100 ppm to 1 ppm, $C_{g1} = 100$ ppm and $C_{g2} = 1$ ppm. Plugging in all the given values yields a t_2 of:

$$t_2 = -\frac{1{,}500}{75}\ln\left(\frac{1}{100}\right) = 92 \text{ min}$$

Therefore, it will take 92 minutes for the concentration of H_2S gas to be diluted from 100 ppm to 1 ppm. Entry into this space should not be encouraged until the time necessary to dilute the contaminant gas has passed.

112. B: Ground fault circuit interrupters (GFCIs) are fast-acting devices that instantly cut off the electrical current when a ground fault occurs, typically for a current differential of 4–6 mA. The National Electrical Code recommends that GFCI protection be provided for all 20-amp receptacles in high-water-risk areas of the home.

113. C: Earplugs with a 30 NRR rating would reduce the dB level to 89.5, which is below the construction industry standard PEL of 90 dB.

114. C: Accident frequency is factored most heavily when determining EMR because it is hard to predict the cost of an accident in advance; however, it is reasonable to assume that employers with frequent incidents will cost more to an insurer in the long run. Denied claims and accident likelihood are not factors used in determining EMR.

115. A: The overall fit factor using individual exercise fit factors is calculated using the equation:

$$\text{Overall fit factor} = \frac{\text{Number of exercises}}{\left(\frac{1}{ff_1} + \frac{1}{ff_2} + \frac{1}{ff_3} + \frac{1}{ff_4} + \cdots + \frac{1}{ff_n}\right)}$$

$$= \frac{4}{\left(\frac{1}{65} + \frac{1}{80} + \frac{1}{35} + \frac{1}{102}\right)} = 60.4$$

116. B: A Geiger counter measures ionizing radiation such as alpha, beta, and gamma radiation. Radioactive material can be found in building materials such as sandstone, concrete, brick, natural stone, gypsum, and granite. Exposure to radioactive material can result in a number of negative health effects such as mutations and increased risk of cancer.

117. B: To determine the amount of force of the falling fume hood, the equation F = ma is used. When an object is free-falling, only gravity is acting upon it; therefore, its acceleration is equal to 9.8 m/s^2 (the same acceleration for any falling object). Inserting the acceleration value as well as the mass value of 300 kg equates to a force of:

$$F = 300 \times 9.8 = 2{,}940 \text{ N}$$

118. C: The site-specific safety plan should include the scope of the construction project, contact information for management and emergency services, the project location, risks associated with the job scope, and hazard mitigation procedures. A job safety analysis (JSA) breaks down job tasks to help employees understand each hazard associated with a job. Emergency action plans are designed to prepare organizations for how to respond during an emergency. City and construction work permits are relative to the compliance regulations in the area where a job is completed. These permits can include regulatory compliance documentation, city zoning permits, and building requirements.

119. C: At conditions different than STP, the actual air density can be calculated by first determining the air density correction factor (d_c):

$$d_c = \frac{T_{STP}\ (°R)}{T_{actual}\ (°R)} \times \frac{P_B\ (\text{inHg})}{P_{BSTP}\ (\text{inHg})}$$

$$\rho_{actual} = d_c \times \rho_{STP}$$

Plugging in the STP and actual values for temperature and pressure yields a d_c of:

$$d_c = \frac{530\ °R}{492\ °R} \times \frac{22.28\ \text{inHg}}{29.92\ \text{inHg}} = 0.8137$$

This can then be multiplied by the air density at STP (ρ_{STP}) to find ρ_{actual}:

$$\rho_{actual} = 0.8137 \times 0.075\ \text{lb/ft}^3 = 0.061\ \text{lb/ft}^3$$

120. C: In this scenario, the forklift operator is probably not trained in proper forklift lifting and rigging operations. Considering the boom is extended to 60%, the weight of the load would increase well beyond the forklift weight rating of 12,000 lb. The forklift operator has not properly read the load rating chart that is required to be on the forklift. The weight on the extended boom would cause the forklift to tilt.

121. C: The power in DC circuits is the product of the voltage and the current. Inserting the given values for volts and current calculates the power in the DC circuit to be equal to:

$$P = VI = 240 \times 0.45 = 108 \text{ W}$$

122. D: Department of Transportation regulations are set by the federal government but can vary from state to state; therefore, companies must review the regulations for each state they enter. The National Highway Traffic Safety Administration is more involved with safety standards and recalls. Choices A and B would not be relevant in this instance.

123. B: The General Duty Clause from the OSHA Act of 1970 requires all employers to provide a work environment free from recognized hazards that are causing or are likely to cause death or serious physical harm. The General Duty Clause covers all hazards not specifically addressed in a standard.

124. B: The TP of a ventilation system is the sum of SP and VP. Plugging in the given values from SP and VP yields:

$$\text{TP} = \text{SP} + \text{VP} = (-1.2) + 0.98 = -0.22 \text{ in. w. g.}$$

125. C: According to 29 CFR 1926.354(c)(1), "In enclosed spaces, all surfaces covered with toxic preservatives shall be stripped of all toxic coatings for a distance of at least 4 inches from the area of heat application."

Since the area of heat application is 2 ft^2, the overall surface area will be this area plus an additional 4 inches all around. To determine the cross-sectional area that must be stripped, the radius of the area of heat application can be solved for, and then the additional 4 inches (0.333 feet) can be added:

$$2 \text{ ft}^2 = \pi r^2 \rightarrow r = \sqrt{\left(\frac{2}{\pi}\right)} = 0.798 \text{ ft}$$

Adding the additional 0.333 feet (4 inches) gives us an overall radius and cross-sectional area of:

$$0.798 \text{ ft} + 0.333 \text{ ft} = 1.13 \text{ ft} \rightarrow \text{A} = \pi(1.13^2) = 4 \text{ ft}^2$$

126. D: Willful violations occur when the employer knowingly takes no reasonable steps to correct a previously revealed offense. The company was aware of the lack of certification for the pressure relief valves but still failed to have them inspected.

127. B: To calculate the sloping angle of the trench, you form a right triangle with one of the sides of the trench and half of the overall width of the trench. Because it's a right triangle, you can now use the two known sides of the right triangle (opposite and adjacent) and the tan function to calculate the sloping angle:

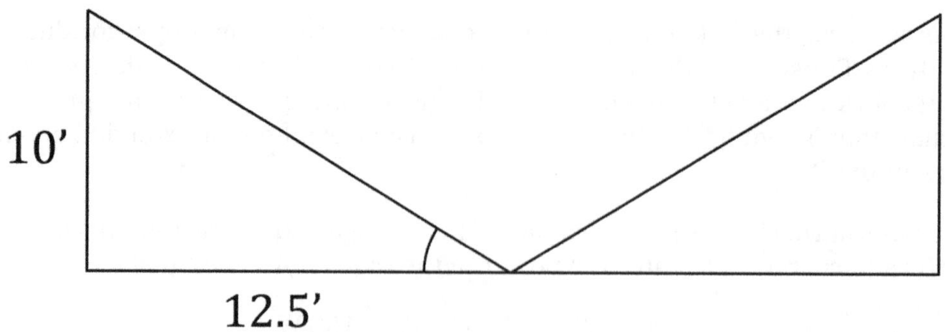

$$\tan(\theta) = \frac{10}{12.5} \rightarrow \theta = \tan^{-1}\left(\frac{10}{12.5}\right) = 38.66°$$

Since trenches in Type C soil require a sloping angle of 34° or less, this trench can only be used in Type B soil.

128. A: According to Table 4 (*Selecting a Reasonable Distance from the Employee to the Electric Arc*) in Appendix E to Subpart V of 29 CFR 1926 Appendix E to Subpart V Table 4—*Selecting a Reasonable Distance from the Employee to the Electric Arc*, the employer can determine how far the employee will need to be from an electric arc for single conductors in air working with live-line tools and doing live-line barehand work using the following equation:

$$D\ (\text{in}) = \text{MAD} - \left(2 \times \left(\frac{\text{kV}}{10}\right)\right)$$

Plugging in the voltage of the system (50 kV) along with the MAD value of 25 inches allows us to calculate a reasonable distance (D) that an employee will be from the electric arc of this system:

$$D\ (\text{in}) = 25 - \left(2 \times \left(\frac{50}{10}\right)\right) = 15\ \text{in}$$

129. C: A short service employee program is designed to guide new hires through an onboarding process that protects new workers from on-the-job hazards through mentorship from a seasoned employee and training specific to their jobs.

130. A: Intrinsically safe equipment is designed to prevent explosions by ensuring the equipment cannot release sufficient energy to ignite in a hazardous environment. Explosion-proof equipment is designed to contain an explosion within its housing, preventing the ignition of a hazardous environment. Battery-operated equipment and power tools operated on a circuit are not rated to work in hazardous environments.

131. B: To comply with OSHA 1926.451, the scaffold base would need a minimum width of 4 feet to meet its 4:1 ratio requirement.

132. C: Accident reports are examples of lagging indicators because the accidents contained in them have already occurred. Other examples of lagging indicators are worker compensation claims, environmental spills, and equipment damage. Leading indicators are measures taken that prevent accidents and incidents. Examples of leading indicators are safety training, best practice reviews, behavior-based safety programs, and hazard assessments.

133. A: The lifting index (LI) is calculated by dividing the object weight (L) by the RWL:

$$\text{LI} = \frac{\text{L}}{\text{RWL}}$$

Since each box contains 10 bricks and each brick weighs 4.3 lb, the overall object weight (L) is equal to 43 lb. Using the given RWL value of 29 and 43 pounds for L yields an LI of:

$$\text{LI} = \frac{43}{29} = 1.48$$

134. B: Scaffolding involves working from an elevated platform. During extremely cold conditions, the metal can weaken, become brittle, and fracture, leading to a fall. Hand and power tools, batteries, and diesel engines may all work less efficiently due to the cold temperatures, but they do not pose as severe of a threat.

135. D: According to 29 CFR 1926.57(h)(6), the total air volume exhausted through a spray booth shall be such as to dilute solvent vapor (i.e., MEK) to at least 25% of the LEL of the solvent being sprayed. The ft^3 dilution per gallon of solvent required to do this is calculated using the following equation:

$$\frac{\text{dilution air } (\text{ft}^3)}{\text{gal of solvent}} = \frac{4 \times (100 - \text{LEL}) \times \left(\frac{\text{ft}^3 \text{of vapor}}{\text{gal}}\right)}{\text{LEL}}$$

Plugging in the given values for LEL and the rate of evaporation per gallon for MEK determines:

$$\frac{4 \times (100 - 1.8) \times 36}{1.8} = 7{,}856 \text{ ft}^3 \text{ dilution/gal of MEK}$$

136. A: Occupancy load refers to the number of people expected in a building at a time based on the building's floor space and function. The maximum occupancy refers to the maximum number of people legally permitted in a room based on means of egress.

137. B: According to 29 CFR 1926.152(i)(2)(v)(D), for tanks and storage vessels, the CFH shall be calculated using the exposed wetted surface area (A) in the following formula:

$$\text{CFH}\left(\frac{\text{ft}^3}{\text{hr}}\right) = 1{,}107 \times \text{A}^{0.82}$$

Using 3,000 ft^2 as our exposed wetted surface area yields a CFH of:

$$\text{CFH}\left(\frac{\text{ft}^3}{\text{hr}}\right) = 1{,}107 \times 3{,}000^{0.82} = 786{,}000 \text{ ft}^3/\text{hr}$$

138. C: The first step in an emergency is to ensure your own safety before proceeding. Once safety is confirmed, the next step should be to contact emergency services. Once help is on the way, the injured person can be assessed, and if a properly trained individual is present, basic first aid and/or CPR can be performed.

139. A: To solve for the average surface concentration of lead, the total amount of lead collected is divided by the total surface area sampled. NIOSH Method 9100 requires that each wipe samples an area of 10 cm × 10 cm (surface area = 100 cm^2). Based on this information, we can compute the average surface concentration as follows:

$$\frac{\left(1.2 \text{ mg} \times \frac{1{,}000 \text{ µg}}{1 \text{ mg}}\right)}{\left(\frac{100 \text{ cm}^2}{\text{wipe}} \times 6 \text{ wipes}\right)} = 2 \text{ µg/cm}^2$$

140. D: HAM radio is a form of radio that does not require the need for cell phones or the internet. A permit is required through the Federal Communications Commission (FCC) to operate a HAM radio. HAM radios can be used over short distances and across the world.

141. B: The sound power level (L_w) is calculated using the sound power (W) in the following equation:

$$L_w = 10 \log\left(\frac{W}{W_0}\right)$$

Plugging in the sound power reference value (W_0) of 10^{-12} watts and the sound power of the HVAC system yields a sound power level of:

$$L_w = 10 \log\left(\frac{3.5}{10^{-12}}\right) = 125 \text{ dB}$$

Therefore, a piece of equipment with a sound power of 3.5 W will produce a sound power level of 125 dB.

142. B: The number of lumens is calculated based on the room's square footage and the required number of foot-candles for the room. Since the first aid station area is in a room measuring 24 ft^2 (6 ft long by 4 ft wide) and has a requirement of 30 fc (1 lumen per ft^2), the number of lumens can be calculated as follows:

$$24 \text{ ft}^2 \times 30 \text{ fc} = 720 \text{ lumens}$$

143. B: To calculate the force (gravitational force) generated when using a fall arrestor, the following equation is used:

$$\text{Force} = \left[\text{Weight (lb)} \times \frac{\text{FF (ft)}}{\text{FAD (ft)}}\right] + \text{Weight (lb)}$$

The ratio between the FF and FAD is multiplied by the weight of the user and then added to the weight of the user. Plugging in the given values, we can calculate the force generated as:

$$\text{Force} = \left[250 \text{ lb} \times \frac{6 \text{ ft}}{3 \text{ ft}}\right] + 250 \text{ lb} = 750 \text{ lb of force}$$

144. B: To calculate the employee's TWA exposure when more than one sample is consecutively taken during his work shift, the following equation is used:

$$\text{8-hr TWA} = \frac{C_1 \times T_1 + C_2 \times T_2 + \cdots + C_n \times T_n}{480}$$

C_n are sample results, and T_n are times in minutes for each respective sample. Plugging in 50 $\mu g/m^3$ (the OSHA PEL) for the 8-hr TWA as well as 125 $\mu g/m^3$ (C_1), 180 minutes (T_1), and 300 minutes (T_2) allows us to solve for the maximum ambient air concentration (C_2):

$$50 = \frac{125 \times 180 + C_2 \times 300}{480} \rightarrow C_2 = \frac{((50 \times 480) - (125 \times 180))}{300} = 5 \ \mu g/m^3$$

Therefore, if the welder is exposed to more than 5 $\mu g/m^3$ over the last 5 hours (300 minutes) of his shift, his exposure exceeds the OSHA PEL. (Note that this is the theoretical max and more conservative PELs are often used in these types of calculations for safety.)

145. C: Molarity is the moles of solute divided by the liters of solution:

$$M = \frac{\text{moles of solute}}{\text{L of solution}}$$

For this problem, grams of NaOH must be converted to moles using the molar mass equation:

$$95 \text{ g NaOH} \times \frac{1 \text{ mole NaOH}}{39.997 \text{ g}} = 2.375 \text{ moles NaOH}$$

Plugging in the given value for liters of solution (1.2) along with the number of moles of NaOH yields a molarity value of:

$$M = \frac{2.375 \text{ moles NaOH}}{1.2 \text{ L}} = 1.98 \text{ M}$$

146. C: American National Standards Institute (ANSI) Z87.1 is the standard for occupational and educational personal eye and face protection devices. ANSI sets forth testing requirements for safety eyewear related to impact, heat, chemical splashes, and dust.

147. A: To determine if a piece of lifting equipment is sufficient for the intended load, the safe working load (SWL) must be determined, which is the maximum load that a piece of lifting equipment may raise, lower, or suspend under operating conditions. The SWL can be found by dividing the WLL by the DAF:

$$\text{SWL} = \frac{\text{WLL}}{\text{DAF}}$$

The WLL is the maximum load a piece of lifting equipment is designed to raise, lower, or suspend, and the DAF is a variable factor that represents the dynamic effects that the working load is exposed to during lifting off or lowering/grounding the load. Plugging in the given values for WLL and DAF yields an SWL value of:

$$\text{SWL} = \frac{5{,}500 \text{ lb}}{1.89} = 2{,}910 \text{ lb}$$

Therefore, under the current operating conditions, this crane is not sufficient for the intended load.

148. B: The Good Samaritan Law protects individuals who provide care to those who are injured or ill. The law was created to encourage bystanders to provide care for those in need without fear of being sued. The care provided must not be compensated.

149. A: The OSHA recommendation for reviewing health and safety programs is to review them at least once annually to determine what is working and what is not. A program can be evaluated by tracking lagging and leading indicators, verifying that the program is being fully implemented, and correcting any shortcomings.

150. C: Calling 911 and staying put until emergency help arrives and shuts the power down is the safest option. The only time an employee should try to exit the vehicle is if they are located in an area where help will not arrive for a period long enough to be life-threatening and there is no way to contact 911.

151. C: Since the total fall clearance distance needed is exactly 15 feet, the employee would make contact with the lower level. Total fall clearance distance is the sum of all of the following factors:

- Free fall distance
- Deceleration distance (can't be greater than 3.5 feet)
- D-Ring shift (assumed to be 1 foot)
- Back D-ring height (usually assumed to be 5 ft)
- Safety factor (value of 2 feet)

When the D-ring is above the anchor point (like in this scenario), the free fall distance is calculated by subtracting the lanyard length from the distance from the D-ring to the anchor point. Since the lanyard is 7 feet long and the distance from the D-ring to the anchor point is 3 feet, our free fall distance for this scenario is 4 feet. Adding this value to our deceleration distance of 3 feet, along with the other given/assumed values, brings the total fall clearance distance to exactly 15 feet.

152. D: The electrical current in a circuit (I) can be calculated by dividing the voltage (V) by the resistance of the circuit (R). This is known as Ohm's Law. Plugging in a voltage of 75 V and a resistance of 950 Ω yields:

$$I = \frac{V}{R} = \frac{75}{950} = 0.0789 \text{ A}$$

Since there are 1,000 mA in 1 A, the current is equal to 78.9 mA. As little as 75 mA can cause death.

153. D: The Environmental Protection Agency does not consider sewage a hazardous material. Hazardous materials are classified as toxic, corrosive, flammable, explosive, radioactive, gas, medicine, human tissue (e.g., blood), and allergen.

154. C: Direct costs associated with an accident are considered insured costs. These include medical expenses, workers' compensation payments, legal services, and medical bills. Indirect costs are uninsured and are viewed as additional costs. Examples of indirect costs are lost productivity, training new employees, accident investigation, OSHA fines, damage to company reputation, and equipment repair.

155. C: The radius of a crane the product of the boom length and the cosine of the boom angle. Plugging in the values for the boom length and the boom angle equates to a crane radius of:

$$\text{Radius}_{crane} = (\text{boom length}) \times \cos(\theta_{boom}) = (42) \times \cos(29°) = 36.7 \text{ ft}$$

156. B: To determine the concentration of MEK after a period of time (2 minutes), we will use the following equation:

$$C_{g2} = \frac{G}{Q}\left(1 - e^{-\left(\frac{Q \times \Delta t}{V_r}\right)}\right)$$

The first step is to calculate the volume of the room (V_r), which, in this case, is the tunnel the workers are in. To find the volume of a cylinder, the cross-sectional area is multiplied by the length of the cylinder. In this case, the diameter of 20 is halved to get the radius (10 feet), and the length of 8 feet is plugged into the following equation:

$$V_{cylinder} = \pi r^2 \times L \rightarrow V_{cylinder} = \pi(10^2) \times 8 = 2{,}513 \text{ ft}^3$$

With the volume of the workspace, we can then plug in the given values for the generation rate (5 cfm), time (2 minutes), and the flow rate in the tunnel (550 cfm) to solve for C_{g2}:

$$C_{g2} = \frac{5}{550}\left(1 - e^{-\left(\frac{(550\times 2)}{2{,}513}\right)}\right) = 0.0035$$

It is important to remember that the value of C_{g2} is not in units of ppm. To convert to ppm, the calculated value must be multiplied by 10^6. Doing that produces an MEK ambient concentration value of 3,503 ppm, which is well above the IDLH value of 3,000 ppm.

157. A: To calculate radioactive decay, the following equation is used:

$$N_t = N_0 e^{-\left(\frac{0.693t}{t_{1/2}}\right)}$$

The current activity (N_t) is calculated by plugging in the given values for the initial activity (N_0), time of decay (t), and the half-life ($t_{1/2}$):

$$N_t = 4{,}265 \times e^{-\left(\frac{0.693(62)}{87.7}\right)} = 2{,}613 \text{ GBq/g}$$

Therefore, after 62 years, the current activity (also known as the number of atoms remaining) is 2,613 GBq/g.

158. A: If an employee fears retribution for making a complaint, the OSHA complaint hotline or online complaint center are both anonymous ways to report incidents or issues. OSHA will not release the name of the individual who files the complaint. Respirators are fitted for each employee and should never be shared.

159. A: Human factors theory describes the phenomenon of accident causation as a result of work tasks that are beyond the capability of a typical human worker. This can be due to environmental, internal, or situational factors. In this case, the worker was being asked to manage too many tasks. His capacity was overloaded, resulting in an incident. Modern causation theory revolves around unsafe acts and positions. Domino theory posits that accidents are the result of a chain of events, like the sequential pattern of falling dominos, because of careless persons or poorly designed and improperly maintained equipment. Epidemiological theory focuses on industrial hygiene practices.

160. A: Often, the paperwork to complete an incident report is overwhelming, and employees do not want to take the time to complete it. If employees are never given feedback on incidents that have been filed or do not see corrective actions, they will stop reporting.

161. D: OSHA 1926.35 requires an emergency notification system to be in place to alert employees of a potential emergency. Emergency notification alarms will be specific for each emergency action plan, and employees must be trained to recognize all alarm sounds and their usage. The alarm system does not have to directly contact emergency services. Not all situations that require the alarm to be sounded will require emergency services. For example, an injury could be a reason to sound the emergency alarm system, but the injured party may not need care beyond first aid given on site. Alarms can be initiated for many different hazards, but there doesn't have to be potential for significant loss of property or life.

162. A: According to FEMA, emergencies are usually small-scale, localized incidents that are resolved quickly using local resources. Disasters are typically large-scale and cross geographic and political boundaries. Disasters require a level of response and recovery greater than local communities can provide. An emergency can escalate into a disaster if mitigation is not handled properly. Most emergency response exercises within a company will be classified as emergencies rather than disasters.

163. C: High-severity injuries are not associated with job tasks that are completed routinely by the worker.

164. B: For a single-phase arc in air, the electric arc will almost always occur when an energized conductor approaches too close to the ground. According to Table 5 (*Selecting a Reasonable Arc Gap*) in Appendix E to Subpart V of 29 CFR 1926, an employer can determine the arc gap, or arc length, by multiplying the dielectric strength of air by the voltage on the line. Using the voltage of 75 kV for the single conductor yields:

$$\text{Arc gap (mm)} = \text{Voltage (kV)} \times 2.54 = 75\text{ kV} \times 2.54 = 190.5\text{ mm}$$

165. C: Job authorizations, critical task safety analyses, permits to work, and job safety analyses are all examples of leading indicators in the risk management process. These procedures require formal documentation to be completed prior to a high-risk task taking place. Policies and procedures, job tasks, and potential hazards are reviewed with employees prior to the job taking place.

166. D: Contact the manufacturer of the wire rope to ensure there are no existing issues. A new sling should not have frayed wires, and although a single frayed wire does not mean the rigging needs to be taken out of service, it is a cause of concern. The lifting angles could be causing a disproportionate weight distribution, resulting in one sling taking on more weight than it can handle. Calling a manufacturer or subject matter expert is always advised when questions about equipment viability arise.

167. D: Allowing an employee to return to work on light duty can help the employee stay engaged and active in the workplace. Light duty can also help the employee maintain income and prevent the company from rising workers' compensation costs. Allowing a worker to return to work on light duty does not affect OSHA incident reporting requirements, as light duty is not considered days away from work.

168. D: To convert chemical concentration in M to a percentage (v/v) in a chemical solution, multiply M by the molar mass and then divide by 10 times the mass density:

$$\frac{1.5\text{ mol}}{\text{L of solution}} \times \frac{30\text{ g}}{1\text{ mol}} \times \frac{1\text{ L}}{10 \times 1{,}000\text{ g}} = 4.5\%\text{ (v/v)}$$

169. B: ISO 45001 and ISO 14001 are international standards that provide frameworks for organizations to implement to improve occupational safety and health and environmental compliance.

170. A: To convert noise dose of 112% to TWA using OSHA's noise criteria, the following equation is used:

$$\text{TWA} = 16.61\log\left(\frac{\%\text{D}}{100}\right) + 90\text{ dBA} = 16.61\log\left(\frac{112}{100}\right) + 90\text{ dBA} = 90.8\text{ dBA}$$

Answer Key and Explanations for Test #1

171. B: The OSHA standard states, "When portable ladders are used for access to an upper landing surface, the ladder side rails shall extend at least 3 feet above the upper landing surface to which the ladder is used to gain access."

172. B: Suspension trauma, also known as harness hang, occurs when blood pools in the legs due to the restriction of blood flow caused by the person's body weight and the position of the fall protection harness around the legs. Suspension trauma can be deadly if not resolved quickly. Certified lifting devices are designed for scenarios such as this. A regular fall protection harness should never be used for lifting.

173. D: Mitigation focuses on reducing the impacts of future emergencies. Enclosed containments would reduce environmental contamination in the event of a hurricane. Recovery is the process of rebuilding. Response refers to the initial measures put in place to protect life and property during or after an emergency. Preparedness is the planning process for how to deal with different types of potential emergencies.

174. C: The Incident Command System establishes five functional areas under which all other operations can be assigned. The command function has overall responsibility for the scene. The operations function manages staff and resources. The planning function gathers information about the incident and determines how to meet the objectives of the command. The logistics function obtains and delivers resources to the command and the emergency. The finance function manages the financial details of the incident and maintains documentation related to the incident. Planning, reporting, public relations, liaison, and treasury are all duties that can fall under the umbrella of the five main functions of incident command.

175. C: The WBGT for the outdoors is calculated using the following equation:

$$\text{WBGT} = 0.7 \times t_{nwb} + 0.2 \times t_g + 0.1 \times t_{db}$$

Inserting the given values for t_{nwb}, t_{db}, and t_g yields a WBGT value equal to:

$$\text{WBGT} = 0.7 \times 75 + 0.2 \times 78 + 0.1 \times 80 = 76.1\ °\text{F}$$

176. C: Incident Command Systems are typically categorized into five major management functions:

- Incident command
- Operations section
- Planning section
- Logistics section
- Finance/administration section

177. C: From the safety perspective, the greatest benefit of an employee health and wellness program is the overall improvement of employee health. A healthy, alert, and sufficiently staffed workforce reduces the likelihood of accidents.

178. C: The Federal Emergency Management Agency (FEMA) is an organization that provides federal assistance in disaster management. FEMA states that its mission is to support citizens and first responders to build, sustain, and improve our capability to prepare for, protect against, respond to, recover from, and mitigate all hazards. Temporary Assistance for Needy Families is a government welfare program for low-income families. Workers' compensation is insurance that employers purchase that provides medical benefits and wage replacement to injured employees.

The Department of Human Services provides assistance in programs such as social security, health benefits, and child protective services.

179. C: The use of HMI screens can eliminate the hazard of a human entering a hazardous work area by using touchscreens and keyboards to engage and interact with machines.

180. D: To calculate the concentration of Legionella in the water tower, the number of CFU is divided by the volume of tested water in mL:

$$\frac{32\ \text{CFU}}{\left(3.5\ \text{L} \times \frac{1{,}000\ \text{mL}}{1\ \text{L}}\right)} = 0.009\ \text{CFU/mL}$$

Thus, the concentration of Legionella in the water tower is 0.009 CFU/mL. As a general guideline, the acceptable level of Legionella in water is < 1 CFU/mL.

181. B: The best option for improving morale and reducing incidents in the long term is to establish relationships among managers and employees. Employees need clear expectations from management, and management must be present in the workplace to communicate and demonstrate the safety culture that is expected from employees.

182. C: The OSHA 1926.1091 recordkeeping requirement states that employers must consider an injury or illness to be work-related if an event or exposure in the work environment either caused or significantly aggravated a preexisting injury or illness. In this scenario, although there is a preexisting injury, the employer would still have to count the injury on the OSHA 300 log as a recordable injury. In some cases, when an injury results solely from a medical condition such as seizures or diabetes, the injury would not be recorded.

183. A: To solve this problem, the amount of force necessary to overcome the forces of gravity acting on the crate, as well as the coefficient of friction between the surface and the crate, is calculated as follows:

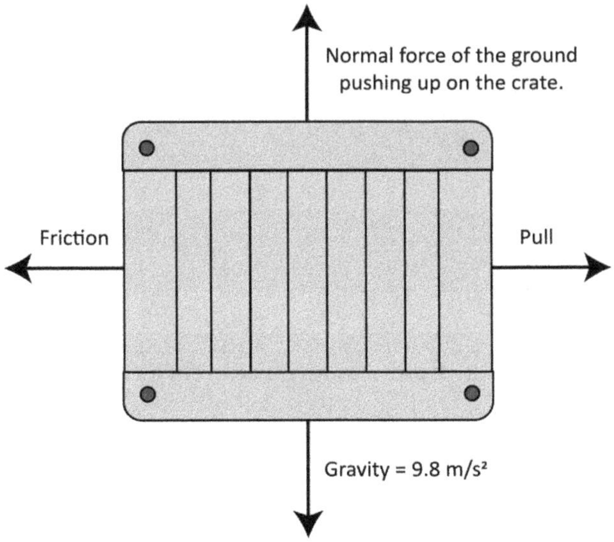

$$\begin{aligned} F &= \mu m g \\ &= (0.8)(1{,}200\ \text{kg})(9.8\ \text{m/s}^2) \\ &= 9{,}408\ \text{N} \end{aligned}$$

Since a force of 9,408 N is needed to move the crate, different rigging material is necessary to perform this task.

184. A: Emergency mitigation describes the actions taken to reduce the loss of life and property by lessening the impact of disasters and emergencies. Mitigation can occur before and after a disaster. Preemptive sandbag placement is meant to minimize or altogether prevent flooding during an emergency.

185. A: The only true statement is that arc flash-protective clothing is flame-resistant, but not all flame-resistant clothing is arc flash-rated. The term flame-resistant refers to natural fibers that resist flames. The term fire-retardant means the fabric has been chemically treated to resist flames. Arc flash-rated clothing provides more insulation and goes through more stringent testing than flame-resistant and flame retardant clothing.

186. A: The frequency of a wave can be found by dividing its speed by its wavelength. Inserting the given values for the speed of sound (343 m/s) and the wavelength (0.404 meters) yields:

$$f = \frac{c}{\lambda} = \frac{343}{0.404} = 849 \text{ Hz}$$

Therefore, when choosing material or equipment to perform noise control, be sure that it is effective at a frequency of around 850 Hz.

187. B: Employees need to be able to trust their supervisors. Sharing salary and work performance information could cause tension within the crew, leading to discontent, poor attitudes, and a lack of trust in the supervisors, which could all affect work performance and safety.

188. C: Subcontractors are required to meet the requirements of the site-specific safety plan unless the subcontractor has more stringent policies than the site-specific safety plan calls for. In this case, the subcontractor's policies should be reviewed and approved by the contractor.

189. C: An AED should only be used on someone who has no sign of a pulse and is unresponsive. An AED is used to shock a person's heart back to a normal rhythm after cardiac arrest. A patient can have gasping breaths as a result of natural reflexes yet have no pulse. These gasps should not be considered breathing if there is no pulse, and the AED should be applied.

190. B: To convert a temperature in °C to °F, the following equation is used:

$$°\text{F} = 32 + \left(\frac{9}{5}\right) \times °\text{C}$$

Inserting the given value of 40 °C yields a temperature of:

$$°\text{F} = 32 + \left(\frac{9}{5}\right) \times 40 = 104\ °\text{F}$$

191. A: To calculate the power density level in the near field of a large aperture circular antenna, use the following equation:

$$\text{W} = \frac{16\text{P}}{\pi \text{D}^2}$$

Plugging in the given values for the diameter (D = 82 cm) and the antenna power (P = 60,000 mW) equates to a power density (W) of:

$$W = \frac{16 \times 60{,}000}{\pi(82^2)} = 45.4 \text{ mW/cm}^2$$

This power density for microwave radiation exceeds OSHA's limit of 10 mW/cm^2 (29 CFR 1926.54(l)).

192. A: OSHA Form 300A is a summary of work-related injuries and illnesses required to be completed annually and visibly posted in the workplace from February 1st through April 30th.

193. B: The angle of this ramp is computed by creating a right triangle and using the cosine function to solve for the angle:

$$\cos(\theta) = \frac{6}{7} \rightarrow \theta = \cos^{-1}\left(\frac{6}{7}\right) = 31°$$

Therefore, as it is presently constructed, this ramp is not in compliance with OSHA's incline limits outlined in 29 CFR 1926.451(e)(5)(ii).

194. A: According to Table D-65.2 in 29 CFR 1926.65, for a temporary workplace consisting of more than 200 employees, there must be 1 toilet seat and 1 urinal per 50 employees. The number of toilet seats and urinals needed for 300 employees can be determined by calculating the ratio of toilets per employee and then multiplying that value by 300:

$$\frac{\text{1 toilet seat and 1 urinal}}{\text{50 employees}} = \frac{0.02}{\text{employee}} \times \text{300 employees} = \text{6 toilet seats and 6 urinals}$$

195. C: JSAs should be updated each time the hazards could potentially change. The JSA is a tool that helps employees identify hazards and risks associated with a job task and develop a work procedure to mitigate those hazards. JSAs are specific to each job task and may need to be modified as the job scope or the environment changes.

196. A: The Federal Clean Water Act regulates the discharge of pollutants into the waters of the United States through wastewater standards. The act makes it illegal to discharge any pollutant into a source of navigable water unless a permit is approved.

197. A: Root cause analysis is a formal process that is conducted through an investigation in which the root causes for an accident are revealed. Root cause analysis is designed to eliminate the problem, not just its symptoms.

198. B: To solve this problem, we need to compare the perimeter measurement of the exposed grinding wheel to the circumference (entire perimeter) of the grinding wheel. The circumference (C) of a circle is calculated by first finding the radius of the circle by dividing the diameter in half and then using the following equation:

$$C = 2\pi r = 2\pi\left(\frac{0.8}{2}\right) = 5.23 \text{ ft}$$

The exposed perimeter of the grinding wheel should be no more than 25% of the circumference. Dividing the perimeter of the exposed grinding wheel by the circumference of the wheel gives:

$$\frac{1.62}{5.23} = 0.31 \text{ or } 31\%$$

Thus, this abrasive wheel is not in compliance with 29 CFR 1926.300(b)(8).

199. D: The white diamond on the NFPA placard represents special hazards from three areas: reactivity with water, oxidizer, or asphyxiant gas.

200. D: To determine mathematically how a change in temperature affects the pressure of an enclosed vessel (i.e., gas cylinder), or vice-versa, Gay-Lussac's law is used:

$$\frac{P_1}{T_1} = \frac{P_2}{T_2}$$

Since we are trying to determine what temperature (T_2) will cause the pressure (P_2) in the cylinder to increase to 155 atm and thereby explode, the given values for P_1 (150 atm) and T_1 (25 °C) are inserted into the equation to solve for T_2. Additionally, it is important to remember that the temperature measurements must be converted to Kelvin prior to being inserted into the equation. To convert °C to Kelvin, simply add 273 to the °C value:

$$\frac{150}{(25 + 273)} = \frac{155}{T_2} \rightarrow \frac{155}{\left(\frac{150}{(25 + 273)}\right)} = T_2 = 307.9 \text{ Kelvin}$$

To convert T_2 from Kelvin to °C, 273 is subtracted from 307.9 to yield 34.9 °C.

CHST Practice Test #2

1. Which of the following is a method of risk transfer when painting the exterior of a multistory building?

a. Using properly installed tie-off points for body harnesses
b. Renting a boom lift for the maintenance staff
c. Hiring a painting subcontractor
d. Putting off the paint job for another year

2. Which specific OSHA regulation addresses Safety Training and Education?

a. 29 CFR 1926.21
b. 29 CFR 1921.35, Appendix A
c. 29 CFR 1916.48
d. 29 CFR 1910.727, Subparts B & C

3. With regard to soil or other granular material, the angle of repose can be defined as:

a. The angle at which underground load-bearing columns are placed relative to the trench angle
b. The steepest angle that soil will form when it is piled up
c. The difference between the angle of an installed footing and its collocated boring
d. The difference between the angle of an installed pile and its collocated footing

4. Per OSHA 29 CFR 1904.33 (Recordkeeping Standard), how long must the OSHA 300 Log be retained?

a. Six months
b. Two years
c. Five years
d. Seven years

5. Which of the following is NOT an effective method of reducing the impact of vibration on construction workers?

a. Keeping equipment properly maintained
b. Ensuring guards are in place on the equipment
c. Reducing the length of time a worker uses vibrating equipment
d. Purchasing vibration-dampening gloves

6. Which of the following is NOT a safety measure associated with scaffold use?

a. Selecting a scaffold that is rated to hold four times the maximum load to be carried
b. Regularly checking structural members for damage
c. Tying, bracing, or using guy-wires on scaffolds with a height-to-base ratio exceeding 4:1 to prevent tipping
d. Requiring fall protection when working on scaffolds at a height of 15 feet or more above ground level

7. Which of the following is regarded as a structured brainstorming technique to identify hazards of a systemic nature that can potentially lead to an accident?

a. A failure modes and effects analysis
b. A hazard and operability analysis
c. An event-tree analysis
d. A workplace hazards assessment

8. Which of these severity ratings would MOST aptly categorize the following conditions after a facility accident event: (1) considerable, but not totally irreparable, damage to equipment or systems; (2) two killed workers and 14 injured workers in a workforce of 200; (3) operational status or capabilities likely delayed for months; (4) roughly $5 million in property damage; and (5) quantities of chemical releases to the environment which exceed EPA NESHAP limits by 40 to 50 percent?

a. Critical
b. Catastrophic
c. Acute severe
d. Beyond design basis

9. Identify which of the following records is NOT maintained as part of a safety program.

a. Training logs
b. Injury and illness logs
c. List of subcontractors
d. Confined space entry permits

10. At a construction site, which of the following activities has the risk of exposing a worker to ionizing radiation?

a. Conducting a land survey with equipment that uses laser sights
b. Inspecting concrete using radiographic techniques
c. Installing fluorescent emergency exit signs
d. Performing ultraviolet curing of moisture-resistant coatings

11. What is the minimum number of AEDs required to be at a job site having 30 total employees?

a. There is no minimum number
b. One
c. Two
d. Three

12. What is the required PPE for employees who are removing asbestos-containing material during a demolition?

a. Hard hat and coveralls
b. Hard hat, particulate protective suit, and air-purifying respirator
c. Hard hat, chemical protective suit, and self-contained breathing apparatus
d. Full-encapsulating chemical protective suit

13. Which of the following is NOT a type of ionizing radiation?

a. Gamma rays
b. Alpha particles
c. X-rays
d. Microwaves

14. The National Fire Protection Association (NFPA) standard system for identifying hazards of materials includes a diamond hazard, as shown below. What hazard does the yellow portion of the diamond represent?

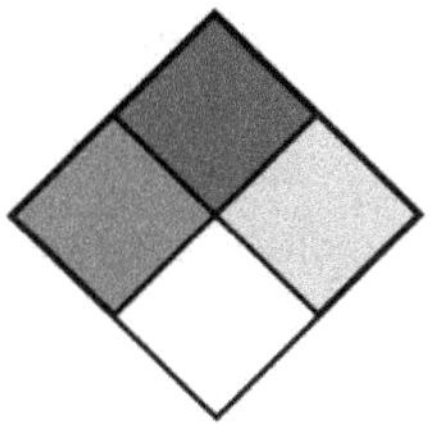

a. Health
b. Flammability
c. Special hazard
d. Instability

15. Lifting frequency is the average number of lifts per minute performed over a _____-minute period.

a. 60
b. 45
c. 30
d. 15

16. Routinely scheduled __________ are a vital component of an effective occupational safety and health program. As such, the frequency of such __________ should be aligned with the degree of risk posed by a subject operation in conjunction with associated regulatory requirements.

a. NQA-1 audits
b. Inspections
c. Industrial hygiene benchmark reviews
d. ISO-9001 evaluations

17. As a project advances, what generally happens to the cost of the builder's risk insurance?

a. It remains constant throughout the project.
b. It decreases over the life of the project.
c. It increases over the life of the project.
d. It increases and decreases depending on the number of employees and equipment on site.

18. Which of the following is NOT a recognized benefit of computer-based safety training (CBT)?

a. Employees have more control over the learning process.
b. CBT is easier to understand than in-person lectures.
c. Training can occur on the employee's schedule.
d. Employees can review the material without fear of embarrassment.

19. Which of the following is NOT an outcome from the misuse or poor maintenance of a hand tool?

a. Eye injuries caused by using a screwdriver as a chisel
b. Pieces of wooden handles on hammers or axes breaking off and striking an individual
c. Stripping a pair of pliers' set-screw while in use
d. Springing a wrench's jaws

20. What is the first step in conducting a job safety analysis?

a. Scoping out potential hazards associated with a particular function or task
b. Verifying that worker competence is commensurate with the associated responsibilities of the function or task
c. Identifying the job to be analyzed
d. Reviewing prior-incident data historically associated with a particular function or task

21. A safety inspection looks to identify ___________, while an audit evaluates ___________.

a. hazardous conditions, adherence to safety programs
b. hazardous conditions, risky behavior
c. risky behavior, hazardous conditions
d. control methods, training programs

22. Which of the following is NOT true regarding personal protective equipment (PPE)?

a. Fit-testing for respirator PPE is a necessity.
b. PPE is often considered a vital element within the construct of a safety plan.
c. Nitrile gloves are considered a type of PPE.
d. PPE should be utilized as a primary means for controlling hazards.

23. When assessing alternatives to mitigate risk related to a fall hazard, which option best describes options in order of priority (best to worst)?

a. Fall arrest system, safety nets, fall restraint system
b. Elimination of exposure, fall arrest system, guardrail system, fall restraint
c. Elimination of exposure, fall restraint, fall arrest system, guardrail system
d. Elimination of exposure, guardrail system, fall restraint, fall arrest system

24. What is the required frequency for inspecting portable fire extinguishers?

a. At the beginning of the project
b. Daily
c. Weekly
d. Monthly

25. Effective safety training courses based on adult learning theories should incorporate which of the following?

a. Presentations, tests, and feedback
b. Presentations, exercises, and scenarios
c. Videos, notes, and stories
d. Roundtables, exercises, and tests

26. OSHA's Bloodborne Pathogens standard requires employers to enforce "universal precautions." Which choice best describes universal precautions?

a. All body fluids should be treated as infectious.
b. Only blood should be treated as potential infectious material.
c. Precautions outlined in the United Nations Universal Precautions standard shall be observed.
d. ANSI's standards on universal precautions shall be implemented.

27. You are the new safety director for ABC Protection, which is a small fall-protection solutions company. The company employs 10 employees; two are office/managers, and the remaining are laborers. The previous director maintained every injury and illness record for the last 30 years. You decide to reorganize the department's recordkeeping and retention program because the accumulated injury and illness records are taking up too much space. Which answer below describes the best means for handling these training records?

a. Injury and illness records are maintained for 1 year following training and then are irrelevant.
b. Injury and illness records are maintained for 30 years after the employee leaves the company.
c. Injury and illness records do not need to be retained for this number of employees.
d. Injury and illness records are discarded when the safety program's retention period elapses.

28. For a remodel project that will involve replacing the facade on a three-story building, which of the following safety programs will likely NOT need to be in the site safety plan?

a. Fall protection
b. Scaffold erection and inspection
c. Electrical safety
d. Materials handling

29. Identify the principle below that can reduce the risk of injury before relocating a material laydown area.

a. Job hazard analysis
b. Management of change
c. Plan-do-check-act
d. Hierarchy of controls

30. Which of the following hazards accounts for the highest fatality rate in the construction industry?

a. Exposure to harmful substances
b. Falls
c. Fires and explosions
d. Contact with objects and equipment

31. Identify which one of the following site conditions would NOT negatively impact worker safety.

a. Elevation changes
b. Absence of the superintendent
c. Delivery vehicle traffic
d. Excessive noise

32. The company has been awarded a job to expand a laboratory space for a college. Although the crew is experienced in remodels, this is the first laboratory for the company. Identify the best approach for conducting the site-safety training.

a. Prepare an in-person training at the company office, including a representative from the college, to discuss hazards, systems, and alarms, as well as answer questions from the crew.
b. Brief the crew by way of a "tailgate" meeting just before starting the demolition.
c. Have the employees review the Hazard Communication standard online.
d. No additional training is required, as the crew has done several remodel projects.

33. Which of the following benefits of good housekeeping contributes to fire prevention?

a. Reduction of fuel available to start or sustain a fire
b. Better appearance of work area for visitors
c. Easy location of tools in the event of a fire
d. Lower insurance premiums due to clean workspaces

34. The theory of "safety by design" incorporates the __________________ in the design phase of a project to increase construction worker safety.

a. cost of protective equipment
b. job site organizational structure
c. final product
d. hierarchy of controls

35. Identify the definition for standard threshold shift.

a. A change in a foundation slab that results in a trip hazard
b. A medical test for blood lead levels
c. A change in hearing due to noise exposure
d. A change in an OSHA regulation that requires rewriting a safety program

36. Which of the following extinguishing agents would NOT be appropriate for a fire involving flammable liquids?

a. Halon
b. Water
c. Carbon dioxide
d. Dry powder

37. The only time a body belt may be used as part of a personal fall arrest system is when a construction worker is using a(n) _________.

a. safety sling
b. arrestor
c. positioning device
d. braking mechanism

38. An ANSI-compliant first aid kit for a construction site would NOT be required to contain which of the following?

a. Disposable gloves
b. Mouth barrier
c. Two 50 mL syringes
d. Scissors

39. Which of the following would NOT be considered a type of surcharge load?

a. Heavy equipment working close to the trench
b. Spoil piles deposited adjacent to the trench
c. Traffic passing near the trench
d. A trench box that exceeds the crane's lifting capacity

40. The safety professional notes that a subcontractor has brought on a day laborer to meet a deadline. What is the responsibility of the CHST?

a. No action is required, as the individual is the responsibility of the subcontractor.
b. Check the contractor agreement to determine if day laborers are allowed on site.
c. Check with the subcontractor's on-site supervisor to verify that the individual has undergone all required safety training.
d. Get a list of all day laborers at the next project meeting.

41. If a five-foot-deep trench is dug in clay material, what angle must the trench wall not exceed?

a. 22.5 degrees
b. 53 degrees
c. 63 degrees
d. 72 degrees

42. Which of the following is NOT considered a repetitive motion disorder (RMD)?

a. Bursitis
b. Carpal tunnel syndrome
c. Stenosis
d. Tendonitis

43. What is the maximum height for a lumber stack from which workers will be pulling boards by hand?

a. 4 feet
b. 8 feet
c. 10 feet
d. 16 feet

44. Fill in the blanks. Employers can implement the following prevention strategies related to bomb threats:

promoting open ______________ among managers, employees, and union representatives;
maintaining mechanisms for complaints and concerns and allowing them to be expressed in a non-judgmental forum that includes timely feedback to the initiator;
promoting importance of work-life balance;
maintaining ______________ and consistent discipline for employees who exhibit improper conduct and poor performance.

a. discipline; subjective
b. communication; impartial
c. communication; subjective
d. punishment; impartial

45. What is the incident rate of a company that had 15 recordable injuries and the total hours worked by all employees was 745,000 hours?

a. 3.63
b. 4.96
c. 2.42
d. 4.03

46. A Hearing Conservation Program is required under OSHA regulation 29 CFR 1926.52 (Occupational Noise Exposure) when workers are exposed to noise levels in excess of an eight-hour time-weighted average of______.

a. 60 decibels (dB)
b. 75 decibels (dB)
c. 90 decibels (dB)
d. 95 decibels (dB)

47. Classroom-based safety training benefits the attendees because:

a. Most adults learn best by hearing information.
b. Having everyone in the room at once is best for business operations.
c. Those who don't like asking questions do not have to participate.
d. Content and delivery can be adjusted to fit the audience.

48. Which of the following could best be described as a lagging indicator?

a. Reviewing injury records from the past quarter
b. A pre-employment physical that indicates higher risk to a worker due to a pre-existing health condition
c. The measure of the number of employees who have successfully completed proper lifting technique training
d. An event that occurs soon after a risk assessment is conducted

49. What does PII stand for as it relates to employees?

a. Project information iteration
b. Planned information interchange
c. Personally identifiable information
d. Protected idea information

50. A multimeter can be used to measure which of the following electrical parameters?

a. Voltage, current, and capacitance
b. Inductance, amplitude, and resistance
c. Voltage, impedance, and resistance
d. Voltage, current, and resistance

51. The process of fit-testing is used for what type of personal protective equipment?

a. Full-body suits
b. Respirators
c. Fall arresting harnesses
d. Cardiopulmonary pulse monitors

52. Which of the following is NOT a condition related to cold stress?

a. Trench foot
b. Chilblains
c. Hyperthermia
d. Frostbite

53. If an employee is hospitalized because he or she was overcome with fumes while standing in a newly dug pit when a sewer line was punctured by the excavator, which standard would likely NOT be cited in the related violation?

a. 29 CFR 1926.28 (Personal Protective Equipment)
b. 29 CFR 1926.59 (Hazard Communication)
c. 29 CFR 1926.1204 (Permit-Required Confined Space Program)
d. 29 CFR 1926.651 (Specific Excavation Requirements)

54. Identify the most effective method for changing the behavior of an adult who is not wearing earplugs while working with a jackhammer.

a. Pause the work, provide them with earplugs, and inquire why they weren't wearing them.
b. Send them home immediately without pay for the day for violating a rule.
c. Ask the other employees what the worker is doing wrong and suggest a correction.
d. Notify the employee's supervisor of the violation.

55. As a safety professional, you review the project plan for repainting the lane markings of a loading dock area for a warehouse in an industrial park. Which of the following topics will NOT need to be part of the site-specific safety plan?

a. Vehicle-worker interaction
b. Hearing protection
c. Hazard communication
d. Personal protective equipment

56. Identify the standard that outlines the requirements for emergency alarms.

a. 29 CFR 1910.165
b. 29 CFR 1926.35
c. 29 CFR 1910.1200
d. 29 CFR 1926.59

57. In a construction area, rigging is best defined as which of the following?

a. Ropes, chains, or slings that attach a load to a lifting device
b. Hand-operated derricks or winches
c. Cables affixed between the jib and trolley of a crane
d. Aerial baskets

58. What does the acronym CBT stand for?

a. Construction build teams
b. Current budget trajectory
c. Computer-based training
d. Concern brainstorm team

59. Which leadership philosophy prioritizes the needs of the employees and encourages employee development?

a. Matrix organization
b. Servant leadership
c. Top-down
d. Flat organization

60. Identify the incident investigation technique that is least subject to bias.

a. Photographs
b. Witness interviews
c. Written notes
d. Sketches

61. A crisis management plan should address:

a. Any potential emergency, including both natural and human-caused events
b. When employees can contact their families during an emergency
c. How to contact emergency services
d. The location of fire extinguishers at the work site

62. To get HVAC ductwork onto the third level, _____________ has the highest risk of injury, while ___________ has the lowest risk.

a. hiring day labor, carrying up a ladder
b. lifting using a pulley system, tossing up from worker to worker
c. manually carrying, lifting with a crane
d. pulling the sections up using a rope, manually carrying

63. All of the following should be a part of a site safety plan EXCEPT:

a. A description of the project timeline
b. Job-specific hazards
c. Training requirements
d. Chain of command

64. How can a crew relocating ventilation ductwork on the job site by crane reduce the risk of striking other workers with the load?

a. Wear hard hats
b. Use a radio system to stay in contact with the crane operator
c. Move the load slowly
d. Use tag lines to control the load

65. A drum of adhesive that is punctured by a forklift and spilled on the ground can be cleaned up by:

a. The forklift operator, to avoid disciplinary action
b. Properly trained and outfitted employees
c. The supervisor who is responsible for keeping the project on schedule
d. Any employee in the area, to prevent the spill from spreading

66. The preferred protective ensemble for responding to a cleanup of 300 pounds of dry cement would be _________.

a. level A
b. level B
c. level C
d. level D

67. Which alternative best describes risk transfer?

a. An employer does not allow employees to enter permit-required confined spaces and hires a contractor to perform such work.
b. An employer requires hot work to only be performed by trained individuals.
c. An employer transfers the risk of flammable liquid exposure from the finishing department to the quality assurance department.
d. An employer moves flammable liquids from the main production area to a warehouse.

68. Identify the most effective method for controlling atmospheric hazards in a confined space.

a. Mechanical ventilation
b. Multi-gas meter
c. Self-contained breathing apparatus
d. Mechanical retrieval system

69. What is the importance of keeping an inspection log?

a. The activity can be charged to the client as an extra income source.
b. It allows employees to check the condition of the equipment.
c. It validates the need for a safety professional to be on site.
d. It demonstrates to an inspector that all necessary items are checked prior to use to reduce the risk of failure.

70. A site-specific safety plan should include which of the following elements?

a. Location of the restroom facilities
b. Name of the prime contractor
c. Legally mandated meal break schedule
d. Employee training records

71. An OSHA inspector is on the site of a concrete repair project that is using caulking to seal the seams adjacent to the building. The inspector requests training records for Hazard Communication. The company has not performed the training, and the superintendent is asking the safety professional to create records to avoid a fine. How should a CHST respond?

a. Create the records to avoid the fine as directed by the superintendent.
b. Do not respond to the demand of the inspector.
c. Stop the project and conduct a focused training on the hazards of the caulking.
d. Report the superintendent to OSHA for trying to falsify records.

72. A Type-A fire extinguisher is recommended for which type of fire?

a. Flammable liquids
b. Trash, wood, or paper
c. Electrical equipment
d. Burning metal

73. Which of the following is NOT a control for preventing a materials-handling accident?

a. Creating linear, narrow passageways or lanes for movement of materials outside of work areas
b. Training workers to properly use hand signals
c. Training workers on proper lifting techniques
d. Implementing traffic controls in materials-handling areas

74. Which of the following is required whenever welding will be conducted away from a designated area?

a. Authorization from the superintendent
b. Approval from the project supervisor
c. Fire permit issued by a Fire Marshal
d. Hot work permit

75. Identify which of the following is NOT a component of a behavior-based safety program.

a. Observations
b. Checklists
c. Consequences
d. Feedback

76. How does a site-specific safety plan (SSSP) differ from an emergency action plan (EAP)?

a. They are the same plan, just different names.
b. The SSSP addresses chemical hazards, while the EAP relates to injuries.
c. The SSSP addresses hazards from the topography, location, and local environment, while the EAP outlines general strategies.
d. The SSSP is the management-level plan, while the EAP is for employees.

77. What protective ensemble is appropriate during lead paint removal using a grinder?

a. Hard hat, jeans, long-sleeve shirt
b. Hard hat, protective coveralls, safety glasses, half-face respirator with HEPA filter cartridges
c. Hard hat, chemical splash protection suit, face shield, air-line respirator
d. Full-encapsulating chemical protective suit

78. Which of the following individuals would be most qualified to measure hazards in the workplace and provide recommendations to mitigate such hazards?

a. A certified health physicist (CHP)
b. A certified industrial hygienist (CIH)
c. A qualified person
d. An NSPE-registered industrial engineer

79. All of the following must be checked on a monthly inspection of a portable fire extinguisher EXCEPT that the:

a. Safety pin is in place
b. Exterior paint is not chipped
c. Hose/nozzle is in good condition
d. Unit contains extinguishing media via the "lift test"

80. An employer recently went through an OSHA inspection for an internal asbestos complaint. During the inspection, OSHA sampled several locations that they reasonably assumed to contain asbestos and sent them to a lab for analysis. An employer representative has requested OSHA to release each tangible sample taken so identical samples can be sent to their third-party lab. What are the required subsequent steps?

a. Under the Freedom of Information Act (FOIA), tangible samples are exempt requests.
b. Under FOIA, requests shall be disclosed promptly.
c. Under FOIA, material requests are confidential.
d. Under FOIA, samples are withheld until initial lab analysis.

81. At a job site having several subcontractors, who is primarily responsible for developing emergency plans?

a. The safety representative for each subcontractor shall develop a plan for their employees
b. The prime contractor
c. The employer with the largest number of employees on site
d. The subcontractor performing the most hazardous work

82. Identify which of the following is NOT a benefit of on-the-job training?

a. The training is specific to the job the employee will be doing.
b. Company-specific culture and methods are passed on to the new employee.
c. The employee is guaranteed to be trained correctly.
d. It is cost-effective.

83. What is the primary purpose of a lanyard in a fall-protection system?

a. Distributing the fall arrest force to multiple parts of the body
b. Connecting safety harnesses to anchorage points
c. Being a point of attachment for fall arrest systems
d. Controlling the fall of an employee to a lower level

84. How can a safety professional determine if a safety training focusing on proper lifting techniques was effective?

a. Review the annual OSHA 300 Log for musculoskeletal injuries.
b. Ask trade supervisors if employees are using proper lifting techniques.
c. Walk around the site to see if proper lifting is being used.
d. Ask the subcontractor supervisors if there were any complaints about the training.

85. Which of the following is NOT a hazard associated with the use of manually powered material handling equipment such as dollies, carts, wheelbarrows, and hand trucks?

a. Striking walls, equipment, or other people
b. Muscular injuries
c. Falling from heights
d. Shifting, tipping, or falling loads

86. The project manager has included scaffold erection in the bid. The company does not employ a scaffold-competent person, so the PM asks the certified CHST to sign off on the scaffold tag so that the work can begin. Which provision of the BCSP Code of Ethics does this violate?

a. Acting with integrity by being fair and honest in all dealings
b. Being objective and truthful, and providing all relevant facts
c. Acting in the interest of safety and the environment
d. Only performing work for which the professional is qualified by education or training

87. What should be the first step implemented when developing a facility emergency action plan?

a. Submitting a plan license mandate request to a regional FEMA office
b. Assembling a team from a variety of organizational functions and disciplines
c. Creating a first-tier disaster-response outline or template
d. Constructing a charter for the plan to be approved by the facility's general manager

88. A trainer is preparing for a confined-space training class, and instruction on how to use a gas meter will be covered. The training will also incorporate trainee-to-trainee discussions and role-play activities. The trainer has the choice of four different classrooms to present the training, but each classroom offers a different seating arrangement. Select the seating arrangement most suitable for the scheduled activities and learning needs.

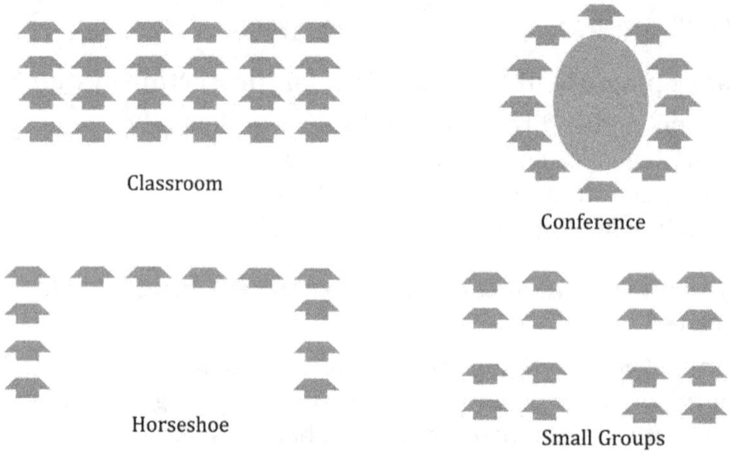

a. Classroom
b. Conference
c. Horseshoe
d. Small groups

89. For rotating parts, what is the most effective control to prevent hand injuries caused by the hand being caught between the rotating part and the housing?

a. Guards
b. Cut-resistant gloves
c. Safety glasses
d. Welding gauntlets

90. Which of the following is a typical mode of respiratory protection used in the workplace?

a. Vacuum-fitted respirators
b. Roughing-filter respirators
c. Supplied-air respirators
d. Free-02 respirators

91. Air sampling is conducted during a concrete cutting operation to evaluate silica dust exposure. The cyclone pump runs at 1.7 L/min for 4 hours. The lab report indicates that the sample disc contained 0.022 mg of silica dust. Assuming the exposure is continuous for 8 hours, what is the occupational exposure to silica dust?

a. 22 ug/m^3
b. 34 ug/m^3
c. 44 ug/m^3
d. 54 ug/m^3

92. Who is responsible for ensuring that all hand and power tools used on a job site are well-maintained and operable?

a. The employee
b. The employer
c. OSHA
d. Contracted labor unions

93. Which of the following is NOT true regarding OSHA inspections?

a. The employer may or may not be advised by the compliance officer of the reason for the inspection.
b. The employer must accompany the compliance officer on the inspection.
c. The compliance officer must show official identification upon an inspection.
d. The compliance officer must assure the employer that any trade secrets observed during the inspection will remain confidential.

94. What are four elements of a successful PPE program?

a. Selection, management, use, maintenance
b. Identification, procurement, review, hygiene
c. Need, use, mitigation, stewardship
d. Designation, protocol, ownership, compliance

95. If an employer's emergency contact system for medical assistance does not automatically provide the location of the job site, the employer must:

a. Post the latitude and longitude in a common area
b. Erect a sign at the entrance with the address of the job site
c. Send an employee to the job site entrance to flag down emergency vehicles
d. Transport the injured worker to the hospital

96. Which of the following is NOT a hazard associated with welding, cutting, and brazing operations?

a. Air contaminants
b. Potential eye damage from UV light
c. Asphyxiation
d. Explosions

97. According to MIL-STD-882E risk assessment, "frequent" is to "likely to occur often" as "occasional" is to "_________".

a. will occur several times
b. likely to occur sometimes
c. unlikely to occur
d. can be assumed will not occur

98. What type of OSHA violation can result in a fine of up to $145,027?

a. Willful violation
b. Contributory negligence
c. Minor violation
d. Errors and omissions

99. A warning line system is described as which of the following?

a. An interlock system that prevents the decoupling of a lanyard from a snap hook
b. A series of concentric floor demarcations that depict incrementally increasing radiation levels within a radiation area
c. An erected barrier on a rooftop to warn workers they are approaching an unprotected side or edge
d. A pressure-differential gauge on a respirator system that indicates it is time for a filter-cartridge replacement

100. Which of the following is typically NOT a tool used during an accident investigation at a work site?

a. Soil sampling equipment
b. Rope or security tape
c. Camera
d. Employee interviews

101. Which of the following is NOT a significant factor in determining the impact of an electrocution event on a worker?

a. The amount of electrical current passing through the body
b. The amount of electrically resistant material present in work boots and work gloves
c. The length of time an exposure to electricity occurs
d. The path through the body the current takes

102. What is the most common injury that results from electric shocks?

a. Electrocution
b. White finger syndrome
c. Head injuries due to loss of consciousness
d. Burns

103. Foot protection for construction employees shall meet the requirements of:

a. American National Standard for Men's Safety-Toe Footwear, Z41.1-1967
b. OSHA Standard 29 CFR 1910.136
c. ASTM F 2413-2005
d. ISO Personal Protective Equipment Footwear Standard — ISO 20345:2011

104. Which of the following is NOT typically regarded as a control method to reduce the risk of machine and tool accidents?

a. Implementing a regular inspection and maintenance schedule
b. Using proper lockout and tagout protocol while machines or tools are being used
c. Requiring that operators have no loose clothing, jewelry, or hair near machines or tools that are being used
d. Ensuring proper functioning of guards and guard devices

105. Which of the following would be activated in response to a potential local emergency resulting from a tornado or flood?

a. A facility contingency charter
b. A facility emergency action plan
c. A US NOAA mitigation response beacon
d. A US FEMA emergency condition notice

106. Which of the following is NOT a method for the delivery of safety training?

a. Needs-analysis training
b. Instructor-led training
c. Self-paced training
d. Structured on-the-job training

107. Which of the following is NOT an appropriate method for notifying employees of new hazards at a job site?

a. Posting daily safety sheets on each trade's bulletin board
b. Attaching job hazard analysis sheets to invoices
c. Telling the subcontractor representatives at daily project meetings
d. Sending out electronic notifications via employee mobile phones

108. A new employee has gone through a brief training on tool safety. During the first hour on the job, they used an air-powered nail gun and discharged the tool accidentally twice, resulting in near misses. Which of the following would be the most appropriate method to address this situation?

a. Allow the employee to keep working since there are no other workers available.
b. Immediately fire the employee and remove him or her from the site.
c. Send the employee home to review the nail gun manufacturer tool safety video online.
d. Immediately stop the employee so that he or she can receive additional on-the-job training.

109. What is the primary function of the Code of Ethics adopted by BCSP?

a. Establishes standards for member behavior, best practices, and decision making
b. Is a requirement in order for BCSP to issue certifications
c. Provides a framework for disciplining members
d. Establishes a legal framework in the event a member is sued

110. Which of the following is NOT a shoring system that can be used to prevent cave-ins in excavations?

a. Aluminum hydraulic shoring
b. Timber shoring
c. Pneumatic shoring
d. Dead shoring

111. What is the definition of an Incident Commander?

a. A person who is responsible for all aspects of an emergency response
b. The lead federal representative responsible for disaster preparedness
c. The individual responsible for evaluation, mitigation, and remediation after an event
d. An OSHA-certified individual who is responsible for properly allocating all appropriate resources in response to an event

112. Where are OSHA standards published when they are considered to be in effect?

a. The OSHA website
b. The Federal Register
c. The Library of Congress
d. Local newspapers

113. If a hospital is not within a "reasonable" distance from the work site, how can an employer ensure an injured worker has access to proper medical care?

a. Post emergency contact numbers in the superintendent's trailer
b. Make sure that every supervisor has a mobile phone to call 911 when necessary
c. Ensure an individual certified in first aid is on the job site
d. Have the safety professional train employees on first aid techniques

114. A ______________ is a regimented and methodical evaluation of a planned or existing undertaking or process that attempts to identify and assess potential issues that may ultimately cause a risk to workers, equipment, and/or operational efficiencies.

a. hazard and operability study (HAZOP)
b. failure modes and effects analysis (FMEA)
c. job safety analysis (JSA)
d. fishbone analysis

115. The proper management of hazardous waste is regulated by which agency?

a. OSHA
b. CDC
c. EPA
d. DOT

116. When comparing scores from a risk assessment, which jobs should be given priority?

a. Jobs that have a high frequency and high severity of hazards
b. Jobs that have a high frequency and low severity of hazards
c. Jobs that have a low frequency and high severity of hazards
d. Jobs that have a low frequency and low severity of hazards

117. Which is the most common control used to prevent workers from being struck by vehicles on a job site?

a. Having all vehicles check in with the superintendent prior to driving on site
b. Requiring all workers to wear high-visibility clothing
c. Designating laydown areas for materials
d. Prohibiting the use of earbuds or headphones

118. Which of the following is the LEAST effective method for identifying injury trends at a job site?

a. Reviewing medical billing records
b. Reviewing workers' compensation claims
c. Reviewing injury reports
d. Speaking with employees

119. During an OSHA inspection, your facility is fined for the violations shown in the picture below. What are those violations?

a. Improper storage of combustibles materials
b. Blocked exit discharge area and fire extinguisher
c. Improper fire extinguisher type for materials
d. Exit sign placement too high for discharge area

120. According to OSHA's standard on emergency action plans (EAPs), an employer must designate and train employees to assist in safe and orderly evacuation of other employees. Which method best meets this criterion?

a. Identifying an emergency coordinator and providing him or her with classroom training on evacuation routes
b. Posting evacuation routes in the hallways
c. Identifying an emergency coordinator and conducting evacuation drills
d. Identifying an emergency coordinator, requiring employees to review the EAP, and conducting evacuation drills

121. Based on the hierarchy of controls, which should be the FIRST measure implemented to reduce a painter's exposure to volatile organic compounds (VOCs) found in oil-based paint?

a. Reduce the amount of time the worker paints on any given day.
b. Provide the painter with an air-purifying respirator with organic vapor cartridges.
c. Restrict oil-based paints to exterior application only.
d. Replace the paint with a water-based formulation.

122. What would be the preferred plan of action for a company that accepts its first exterior-painting job for a three-story building?

a. Buy scaffold planks that can be supported by ladders the company already owns.
b. Purchase taller ladders for the painters.
c. Paint the upper level from the roof, using ropes as safety lines.
d. Include the cost of a scaffold company in the bid.

123. The demolition subcontractor hired for an office building project has worked together with the prime contractor for a few years and specializes in this type of work. As the prime contractor's safety professional, you review their safety program, which is comprehensive, has current training logs, and shows less than one first aid injury for every two jobs. Which of the following methods would be most appropriate to deliver the site-specific safety training?

a. A three-hour computer-based training
b. A "tailgate" review of site-specific hazards
c. An eight-hour in-person comprehensive safety program review
d. No additional training is required

124. Drenching showers and eye washes must be available at the job site whenever _______________ materials are used.

a. hazardous
b. corrosive
c. flammable
d. toxic

125. An emergency plan should always include procedures for sheltering in place, which need(s) to specifically address which of the following?

a. How to extinguish a small (i.e., manageable) fire during a shelter-in-place event
b. How to evacuate a building during a shelter-in-place event
c. How to shut down all ventilation systems and elevators during a shelter-in-place event
d. How to don appropriate personal protective equipment (PPE) during a shelter-in-place event

126. Identify the topic below that is NOT required to be included in a work site's emergency action plan.

a. Evacuation procedures
b. Systems used to report emergencies
c. Method to account for evacuated workers
d. Workers' compensation carrier for employee injuries

127. Calculate the DART rate from the data presented below:

Number of first aid cases = 12
Number of OSHA recordable cases = 5
Number of sick days taken by all employees = 45
Number of injuries resulting in days off = 1
Number of injuries resulting in days with job transfer = 2
Number of injuries resulting in days with job accommodation due to injury = 1
Number of employees = 72
Number total employee hours worked = 127,500

a. 7.8
b. 6.3
c. 26.7
d. 18

128. What are the three variables that are evaluated as potential root causes using the behavior-based safety model?

a. Superintendent, supervisor, employee
b. Person, work environment, behavior
c. Behavior, actions, consequences
d. Safety, security, planning

129. How long must an employer keep information for a chemical that an employee was exposed to and received medical treatment for?

a. Until the chemical is no longer used by the company
b. Three (3) years
c. Ten (10) years
d. Thirty (30) years

130. A bloodborne pathogen is:

a. Blood that contaminates a tool, rendering it inoperable
b. The medical term for scraped skin that exposes the epidermis
c. Microorganisms present in blood and other body fluids that can cause diseases in humans
d. A first aid treatment for cuts and scrapes

131. An electrician is installing a new lighting system and will be wearing a personal fall arrest system to protect them from a fall hazard. The system places the D-ring height at 4 feet, and the free fall distance is 6 feet. What is the required fall clearance? Assume that the deceleration distance is 3.5 feet, the D-ring shift is 1 foot, and the safety factor is 2 feet.

a. 14.5 ft
b. 16.5 ft
c. 18.5 ft
d. 20.5 ft

132. An employee serves as a volunteer building emergency coordinator. The building in question contains multiple hazardous processes that could make an emergency worse if quickly abandoned. The employee writes the building's emergency action plan (EAP). You are tasked with reviewing the plan. The EAP contains information on emergency reporting procedures, evacuation procedures, roll-call procedures, rescue procedures, and emergency coordinator contact information. Which OSHA-required element is missing?

a. Procedures for directing emergency response upon arrival
b. Procedures for accounting for all employees upon evacuation
c. Procedures for those who must fulfill critical duties before evacuating
d. Procedures for calling OSHA upon safe evacuation

133. An employee is working on an elevated work surface that is 13 feet from the lower level. The employee is using a fall arrest system with an anchor point at head-level. The connecting device is a 6-foot self-retracting lanyard with a 3-foot free-fall distance. The worker is 6 feet tall. Which best describes this fall arrest arrangement?

a. This fall arrest arrangement is not recommended. The employee will likely hit the ground.
b. The fall arrest system will be effective. The employee will not hit the ground.
c. More information is required to assess the effectiveness of the fall arrest system.
d. The fall arrest system will be effective, but the employee will hit the ground.

134. A painting company is hired to repaint the interior of a theater that has ceilings 30 feet above ground level. Which of the following standards would NOT apply to this job?

a. 29 CFR 1926.502 (Fall Protection Systems Criteria and Practices)
b. 29 CFR 1926.57 (Ventilation)
c. 29 CFR 1926.59 (Hazard Communication)
d. 29 CFR 1926.1204 (Permit-Required Confined Space Program)

135. Wearing disposable gloves while rendering first aid to a worker who has been cut is an example of implementing:

a. Best business practices
b. Liability reduction
c. Best medical practices
d. Universal precautions

136. Which of the following is NOT a requirement during the re-energization of lines and equipment?

a. Ensuring that no one initiates action to re-energize lines or equipment at a point of disconnection until all subject components are verified as grounded
b. Ensuring that all crews working on lines or equipment release their clearances
c. Ensuring that all employees are physically clear of lines and equipment
d. Ensuring that all protective tags are removed from the point of disconnection

137. Per OSHA regulation 29 CFR 1926.1053, the interval between landing platforms on a fixed ladder that uses a cage or well with multiple ladder sections must be ________.

a. no greater than 24 feet
b. no greater than 30 feet
c. no greater than 36 feet
d. no greater than 50 feet

138. Who is responsible for carrying liability insurance during a project?

a. All parties
b. The prime contractor
c. All subcontractors
d. The property owner

139. Which of the following is NOT a control employed for exposures to hazardous chemicals?

a. Personal protective equipment
b. Length of exposure
c. Safety Data Sheets
d. Personal hygiene

140. Recommended weight limit (RWL) is most accurately defined as which of the following?

a. The maximum weight that healthy workers could safely lift for up to eight hours
b. A lifting weight of 50 pounds for healthy male workers and 30 pounds for healthy female workers
c. The maximum weight associated with a body mass index of 25 (threshold of "overweight")
d. The weight at which a back-support apparatus should be worn

141. In the GHS system of chemical labeling, what is a pictogram?

a. A statement that describes the nature and severity of the hazard
b. One of nine (9) standard images that convey hazard classes
c. A measure to minimize the effects of the chemical
d. A description of the severity of a chemical hazard on a scale from 1 to 4

142. What additional respirable crystalline silica dust exposure controls are required for a worker using a walk-behind concrete saw outfitted with a water delivery system for six hours outdoors?

a. No additional controls are required
b. A dust collection system
c. An air-purifying respirator with HEPA filters
d. Both a dust collection system and an air-purifying respirator with HEPA filters

143. The three fundamental categories of risk management (process) controls are ____________, ____________, and ____________.

a. physical, educational, avoidance
b. visual, procedural, mitigative
c. sensory, tactile, reflexive
d. precursory, process, serial

144. Improperly stored and stacked materials can fall, resulting in ____________.

a. electrical hazards from damaged cords and lines
b. noise hazard caused by containers hitting the ground
c. struck-by hazards causing cuts, bruises, or fractures
d. vibration hazard if several containers fall at once

145. Training construction employees on the OSHA Hazard Communication Standard is required:

a. Semiannually, with initial training being performed within one month of the employment start date
b. When a new hazardous material is introduced into the workplace
c. For a work site to potentially qualify for VPP Star status
d. After an incident occurs with a hazardous material

146. Site safety plans need to be updated with which frequency?

a. At the beginning of the project
b. Quarterly
c. At the start and midpoint of the project timeline
d. Whenever there is a change in site conditions

147. Which choice represents the key components of a training or learning objective?

a. Thesis, body, hook, conclusion
b. Antecedent, behavior, motivation, measurement
c. Achievable, consequence, measurement, result
d. Audience, behavior, condition, degree

148. The ISO 45001:2018 occupational health and safety management system standard includes:

a. Proactively identifying and responding to hazards before an accident occurs
b. Advising workers on safe work practices
c. Summarizing all construction site standards as outlined in 29 CFR 1926
d. Providing strict managerial and supervisory response guidelines in the event of an accident

149. All of the following are hazards presented by powder-actuated tools EXCEPT:

a. Hazardous noise that can affect an employee's hearing
b. Long-term nerve damage from hand vibration
c. Flying debris that can strike a worker's eye
d. A fire that could start when the device is fired in the presence of flammable vapors

150. Per OSHA's 29 CFR 1904.33 (Recordkeeping Standard), how long must a document such as an OSHA Form 301 be retained?

a. Five years following the end of the calendar year that the record covers
b. Three years following the end of the calendar year that the record covers
c. One year following the end of the calendar year that the record covers
d. Seven years following the end of the calendar year that the record covers

151. The multilinear events sequencing (MES) method uses a(n) __________ to illustrate the sequence of events or occurrences that contributed to or was the direct cause of an event.

a. flow chart
b. iterative algorithm
c. time line
d. hierarchy

152. During a crisis or emergency event, the media responses should be provided by ______________.

a. any available supervisor
b. the employee nearest the emergency when it happened
c. a representative of the subcontractor involved in the incident
d. a pre-designated person

153. Which three components make up an emergency exit route?

a. Exit access, exit, exit discharge
b. Exit egress, exit, assembly point
c. Exit access, stairway, exit door
d. Exit ingress, exit stairway, discharge point

154. Which of the following should be consulted when preparing a safety inspection checklist for a portable power generator?

a. Manufacturer's instructions
b. Rental contract
c. Supervisor
d. ASME guidelines

155. An asphyxiant is defined as:

a. A material that causes irritation to the eyes or upper respiratory tract
b. A material that displaces oxygen in an environment or interferes with oxygen transport in the body
c. A substance that affects the genetic material within cells
d. A form of compressed nitrogen that can cause the lungs to bleed if inhaled

156. Hazards for a maintenance task were identified and evaluated on the risk assessment matrix below. It was determined that eye injuries from flying debris had a probability of "possible" and a severity of "critical'. What is the risk ranking of this task?

		Severity			
		Negligible	Marginal	Critical	Catastrophic
Probability	Improbable	Low	Medium	Medium	High
	Possible	Low	Medium	High	Extreme
	Probable	Medium	High	High	Extreme

a. Low
b. Medium
c. High
d. Extreme

157. Risk can be defined as which of the following?

a. Cost basis × fatalities
b. Occurrence × cost basis
c. Remedial cost × frequency
d. Frequency × severity

158. As the year progresses, what changes, if any, should be incorporated into a pre-job safety training program?

a. No change to the program is required if there haven't been any serious injuries.
b. The training program should be shortened as the year goes on because staff have heard the presentation multiple times.
c. Different topics should be covered at each meeting to keep the attendees interested.
d. Environmental risks, such as heat, rain, and cold should be inserted as the seasons change.

159. Which of the following is typically NOT a means of reducing construction ergonomic risk factors?

a. Adjustable scaffolding
b. Reduced-vibration power tools
c. Dolly aligners
d. Motorized concrete screeds

160. During a rigging inspection, if the throat opening of a hook has increased in excess of _____ percent of the original throat opening, it should be replaced.

a. 5
b. 10
c. 22.5
d. 30

161. Which of the following is NOT required by the BCSP Code of Ethics?

a. Protect property and the environment
b. Act with integrity
c. Avoid misrepresenting professional qualifications
d. Charge fair rates that increase over time

162. When developing a maintenance checklist, which of the following should be consulted when determining what items need to be checked and at what frequency?

a. ANSI/ASTM standards
b. Manufacturer specifications and product literature
c. Applicable OSHA standards
d. ASME recommended maintenance schedules

163. During a safety inspection, a worker is observed preparing to use a ladder on the second level of a scaffold. How should this situation be addressed?

a. Make a note to mention the observation during the next scheduled safety meeting.
b. Immediately ask the employee to cease their activities.
c. Prepare a bulletin to be sent out to all trade supervisors at the beginning of the next shift.
d. Schedule scaffold safety training for the next quarter.

164. Temporary circuits of _____ volts and _______ amps at construction sites must be protected by ground-fault circuit interrupters.

a. 110; 10 and 15
b. 120; 20 and 30
c. 120; 15 and 20
d. 110; 20 or more

165. Which of the following regulations applies to the demolition of a building having asbestos-containing ceiling tiles?

a. Asbestos Hazard Emergency Response Act (AHERA)
b. Comprehensive Environmental Response, Compensation, and Liability Act (CERCLA)
c. Clean Air Act (CAA)
d. Resource Conservation and Recovery Act (RCRA)

166. Identify the three parts of a communication that can result in a misunderstanding during a safety training.

a. Presentation, speaker, AV equipment
b. Sender, receiver, slide presentation
c. Superintendent, supervisor, employee
d. Sender, information, receiver

167. Which of the following tactics is NOT typically employed as an enforcement measure for ensuring compliance with safety program protocols?

a. Disciplinary action for blatantly disregarding safety rules
b. Incentives for proactively following safety rules
c. Training regarding safety requirements
d. Mandatory participation in safety program activities and initiatives

168. OSHA's Voluntary Protection Program (VPP) is a program that encourages a company to:

a. Implement an integrated safety management system in its corporate bylaws
b. Have zero recordable incidents over a period of two years
c. Systematically improve its health and safety management system in a partnership between the employer, OSHA, and the workforce
d. Proactively utilize behavior-based initiatives and corrective action metrics

169. Electrical bonding is best defined as:

a. The attraction of polarized charges within an open circuit
b. The transfer of electrons from one atom to another through an ionic bond
c. The inability to detach from an electrified object during an electrocution event
d. Connecting two or more objects with a conductor

170. Which of the following is NOT a typically employed safety measure associated with portable ladder use?

a. Purchasing ladders with slip-resistant rungs
b. Placing ladders far enough from a wall so that toes can be comfortably placed on the rungs
c. Surveying the area overhead for nearby electrical conductors
d. Anchoring or tying ladders to a support structure

171. Which types of records does HIPAA regulate access to?

a. Employment
b. Payroll
c. Medical
d. None of the above

172. How are permit-required confined spaces identified to prevent unauthorized entry?

a. Color-coded markers
b. Signage
c. Blinking visual indicators
d. Permit-required spaces are not required to be specifically identified

173. Excavation cave-in protection is required under which of the following conditions?

a. The excavation mainly comprises gravel or sand and is three feet in depth.
b. The excavation mainly comprises clay or silt and is five feet in depth.
c. The excavation is four feet in depth, regardless of material.
d. The excavation requires heavy equipment to remove excavated material.

174. Of the following, which is NOT a requirement for an emergency spill response plan?

a. Pre-planning
b. Notification systems
c. Fire risk reduction strategies
d. Evacuation procedures

175. During an audit, findings can be organized into which of the following categories?

a. Violations, infractions, notes
b. Deficits, actions, recommendations
c. Programmatic fault, marginal compliance, enhancement actions
d. Nonconformances, areas of concern, opportunities for improvement

176. Whenever there is a safety concern, crane operators must have the authority to ____________ until a qualified person has determined that the concern has been addressed.

a. evacuate a work area
b. preliminarily examine such a concern
c. file a corrective-action request
d. stop and refuse to handle loads

177. Which of the following is NOT a hazard related to poor housekeeping?

a. Noise hazards from material being dropped on the floor
b. Fire from the presence of combustible material near ignition sources
c. Injuries caused by tripping over scrap materials on the floor
d. Bites from pests living in debris piles

178. Which of the following design features reduces the risk of injury related to using hand tools?

a. Shorter handles on axes and hatchets can enhance user safety.
b. Bent handles can help prevent injuries caused by awkward positions.
c. Metallic handles with low coefficients of friction can help reduce the chance of the tool slipping.
d. Non-powered hand tools must include a lockout or tagout ability similar to that for power tools.

179. Which of the following is NOT generally considered an effective control to prevent stored materials from falling?

a. Tapering
b. Blocking
c. Stepping-back
d. Storing pipes and bars in racks that face main aisles

180. "White finger" in the workplace is typically caused by which of the following?

a. Long-term use of vibrating hand-held tools
b. Prolonged reaching over the shoulder level
c. Keeping fingers in the same position for an extended period of time
d. Recurring blisters in the same location over an extended period of time

181. A leading indicator is best defined as which of the following?

a. Metrics for accidents that have already happened
b. The most likely action or event responsible for an accident
c. An objective measure that is used to assess proactive actions taken to improve organizational performance
d. An identified root cause for an accident

182. Each job should have emergency equipment on-site for:

a. Every potential emergency, even if the equipment may not be used
b. Events identified during a risk assessment based on the project and work site location
c. Fires and explosions only
d. Heavy equipment failures

183. An emergency evacuation plan should include the location of ________, alarm pull stations, and assembly areas.

a. fire extinguishers
b. landline telephones
c. fire sprinklers
d. evacuation routes

184. Which of the following is a TRUE statement regarding occupational safety training in the workplace?

a. Experienced workers should be given priority over new hires to receive training.
b. Accidents are more likely to occur when workers have recently started a new job.
c. OSHA standards typically do not outline required training.
d. Accidents always increase soon after accident-prevention training.

185. When researching regulatory requirements, where is the most reliable source of current OSHA standards?

a. Crowd-sourced references (e.g., Wikipedia)
b. Technical journals that contain safety-related articles
c. The Department of Labor's website
d. The binders in the superintendent's office

186. Rubber-tired self-propelled scrapers, rubber-tired front-end loaders, rubber-tired dozers, wheel-type agricultural and industrial tractors, crawler tractors, crawler-type loaders, and motor graders that are used in construction work shall all be equipped with:

a. Transmission interlocks
b. Compartmentalized ballast
c. An independent emergency cutoff switch
d. Rollover-protective structures

187. As it relates to conducting a job hazard analysis, what is a hazard?

a. A situation that has no obvious solution
b. An object or condition that can cause harm
c. An injury that requires more than first aid
d. A health condition that does not manifest itself for several years

188. Identify which of the following can be used to identify an impending at-risk condition at a job replacing underground water pipes.

a. Soil sample tests that could indicate if the trench will fail
b. The weight of the replacement pipes, to evaluate the risk of being dropped by the crane
c. A weather report indicating that a rainstorm will be arriving
d. Traffic pattern data from the city, to avoid workers getting hit by cars

189. Which of the following has the lowest risk of injury for a project that requires installation of new roof supports?

a. Purchasing pre-built trusses and lifting them into place
b. Building trusses on the ground and lifting them into place
c. Building trusses in place using nail guns
d. Building trusses in place using manual tools

190. The ISO 14001 environmental management system is built upon which of the following models?

a. Strategize, conduct, verify, conclude
b. Scope, order, efficiency, accountability
c. Plan, do, check, act
d. Scope, evaluate, reassess, learn

191. Identify the safety program below that would NOT be necessary for a job to replace an underground water main located eight feet under a concrete slab.

a. HAZWOPER
b. Emergency response plan
c. Excavation safety
d. Airborne silica

192. The Globally Harmonized System refers to which of the following?

a. A standard traffic flow pattern for all contractors entering a work site
b. A method of safety training to increase employee retention
c. A universal labeling system for chemicals
d. A communication system that allows for quick project updates

193. Which of the following is NOT a type of respiratory protection used in the workplace?

a. Free-oxygen respirators
b. Air-purifying respirators
c. Supplied-air respirators
d. Self-contained breathing apparatuses

194. Which of the following root-cause analysis methods is implemented AFTER an accident occurs?

a. The Five Whys
b. Job hazard analysis
c. Failure mode and effects analysis
d. Hazard and operability analysis

195. In a group safety training session, the facilitator must be careful to avoid all the following EXCEPT:

a. One person dominating the discussion
b. Introverts not participating
c. Groupthink
d. Slide format

196. For the concrete-cutting portion of a water line repair project, which of the following lists the controls for reducing exposure to airborne crystalline silica in order from MOST effective to LEAST effective?

a. Wearing respirators, reducing the number of employees in the area, using a wet saw
b. Using a wet saw, rotating workers operating the saw, wearing respirators
c. Wearing safety glasses, wearing respirators, wearing gloves
d. Using a jackhammer, wearing respirators, rotating workers operating the jackhammer

197. Calculate the RIR from the data presented below:

Number of first aid cases = 8
Number of OSHA recordable cases = 7
Number of sick days taken by all employees = 35
Number of injuries resulting in days off = 2
Number of injuries resulting in days with job transfer = 2
Number of injuries resulting in days with job accommodation due to injury = 1
Number of employees = 90
Number of total employee hours worked = 180,000

a. 7.8
b. 5.5
c. 16.7
d. 12.9

198. During a project that involves replacing the roof, which of the following activities does NOT have an increased risk during a rainstorm?

a. Using a rotary saw with an extension cord
b. Carrying materials by hand while walking on a concrete foundation
c. Working while wearing a T-shirt
d. Using a gasoline-powered air compressor

199. Any safety program that reduces the _____________ accidents will likewise reduce workers' compensation claims.

a. liability for
b. number of fatal
c. frequency and severity of
d. high-deductible cost of

200. According to the adult learning model, which of the following is NOT a mode of instruction that should be incorporated in any effective safety training program?

a. Visual
b. Audio
c. Comedic
d. Kinesthetic

Answer Key and Explanations for Test #2

1. C: Risk transfer is the act of shifting the responsibility for managing a risk from one party to another. The risk is not eliminated or controlled but simply moved. Subcontracting out a painting operation at heights moves the risk from the building owner to the subcontractor. That subcontractor may then elect to manage the risk using other methods, such as acceptance or control.

2. A: OSHA 29 CFR 1926.21 establishes the requirements for an employer whose employees conduct construction, demolition, remodeling, or other activities that fall under the scope of 29 CFR 1926 to educate employees on the hazards they may encounter on the work site.

3. B: The angle of repose is defined as the steepest angle that soil (or other granular material, such as gravel) forms when it is piled up. If the angle of the pile is steeper than the angle of repose, the pile will slide or slump. It is a function of the density and shape of the particles, surface area, coefficient of friction, and moisture content.

4. C: Per OSHA 29 CFR 1904.33, the OSHA 300 Log must be kept for five years following the end of the calendar year that the record covers.

5. B: Equipment that is not properly maintained may result in increased vibrations that can impact a worker's capillaries and nerves, resulting in permanent damage. Vibration injuries are also a function of time; thus, the lower the exposure time, the lower the risk of injury. For tools that vibrate as a normal part of their operation, vibration-dampening gloves may be another way to reduce the risk of vibration injuries.

6. D: 29 CFR 1926.451 requires that fall protection be used when working on scaffolds that are more than 10 feet above the next lower level. Fall protection systems include guardrails and personal fall arrest systems.

7. B: A hazard and operability analysis (HAZOP) is a structured technique to identify hazards of a more systemic or operational nature that can potentially lead to unexpected or undesired products or outcomes, such as injuries or accidents. It principally relies on a regimented brainstorming technique that guides analysts into considering all modes of deviation that can potentially lead to a failure.

8. A: In risk-assessment space, a severity-rating of critical is typically associated with the following conditional scenarios: (1) major damage to equipment or systems; (2) permanent partial impairment or temporary total impairment to workforces (likely includes injuries and/or fatalities); (3) appreciably disabled operational capabilities; (4) a major degree of overall property damage; and/or (5) high degrees of environmental releases or impacts.

9. C: In addition to a visual inspection, an OSHA inspector will rely on logs and records as a means to demonstrate compliance. To that end, several standards have recordkeeping requirements. Injury and illness logs must be kept for five (5) years after being posted, while confined space entry permits must be retained for one (1) year after the permit was canceled. Training logs should be kept to demonstrate that the training occurred, as well as who attended. It is recommended that training logs be kept for the duration of an employee's time with the company.

10. B: Ionizing radiation, such as x-rays and gamma rays, can cause damage to cells and genetic material. Concrete inspections using radiographic techniques can use an x-ray device to examine the internal structure of finished concrete for voids or reinforcement members.

11. A: Under OSHA's medical standard for construction (29 CFR 1926.50), there is no prescribed minimum number of automated external defibrillators (AEDs) required to be located at a job site. Safe practices would suggest that at least one unit be available in the event of a cardiac emergency.

12. B: The primary hazard associated with asbestos is inhalation of the airborne fibers, which can cause lung scarring and cancer. Protecting the worker's clothing will also reduce the risk of being exposed to fibers after the job is completed. Thus, a Level C ensemble consisting of a hard hat, particulate protective suit, and air-purifying respirator is appropriate PPE for asbestos remediation. Although a Level B ensemble could be used for asbestos removal, the splash protection of the chemical protective suit and the higher level of respiratory protection are offset by the risk of heat stress and the limited work duration associated with the SCBA.

13. D: Microwaves are not a form of ionizing radiation. Forms of ionizing radiation include alpha particles, gamma rays, and x-rays. Ionizing radiation is the most dangerous type of radiation because it can damage cells and genetic material, which can result in mutations and cancer.

14. D: The yellow portion represents the instability of a material. The rating ranges from 0 to 4, increasing in severity from "stable" to "may detonate."

15. D: One of the factors in the NIOSH lifting equation in determining the recommended weight limit (RWL) is the frequency multiplier. That factor takes into consideration how often a worker performs a lift, termed "lifting frequency." This value is tallied as the average number of lifts per minute a worker performs over a 15-minute period.

16. B: Routinely scheduled inspections are a vital part of an effective occupational safety and health program. As such, the frequency of inspections must be aligned with the degree of risk imposed by a subject operation in conjunction with regulatory requirements. Inspections that are not carried out at regular intervals are likely indicative of an insufficient occupational safety and health program.

17. C: Builder's risk insurance is a policy that covers the project during construction. The policy protects against fire, flood, and certain types of damage. The cost of the policy generally increases as the project progresses because the overall value of the project typically increases (e.g., a framed building has less value and material than a structure in the finishing stage).

18. B: Depending on the type of CBT deployed by an organization, the training may not necessarily be easier than in-person training, but employees can control the pace of the training and the learning process, reviewing information as often as needed. This also avoids any potential embarrassment for asking questions in a group setting. Additionally, employees can take the training as their schedules allow, reducing the need to schedule a time for an entire group to be pulled out of the workplace for training.

19. C: Injuries can result from the misuse or improper maintenance of hand tools. Such misuse can include using these tools in manners they were not fashioned or designed for. For example, using a screwdriver as a chisel can break the screwdriver and result in flying debris. Improper maintenance can cause pieces of wooden handles on hammers or axes to break off and strike an individual. Aging tools, such as sprung wrench jaws, can create a poor grip on an object, causing the worker to slip, resulting in injury.

20. C: The first step in a job safety analysis (JSA) is to identify the job that is to be evaluated for hazards. The job may be the one that has resulted in the most injuries, has had the highest medical costs, has caused the most serious injuries, or has the potential for the most serious injuries. OSHA publication 3071 outlines the process for conducting a JSA.

21. A: Although the terms are often used interchangeably, inspections and audits serve different purposes. An inspection is a method to proactively identify hazardous conditions or risky behaviors in an effort to prevent injuries or illnesses. An audit is a formal process of evaluating an organization's compliance with its own safety programs. Inspections observe people, places, and equipment, whereas audits examine systems, programs, and processes.

22. D: Personal protective equipment (PPE) serves and only serves as a barrier between a worker and a hazard, but it does not remove the hazard. PPE should never be considered as a means for removing a hazard or for altogether perceiving a hazardous environment as a safe environment.

23. D: Based on the hierarchy of controls, elimination of a hazard is always the priority. A guardrail system is an engineering control that prevents a fall from occurring. A fall restraint system is an administrative control but still prevents an employee from reaching the fall hazard. Fall arrest relies on people wearing personal protective equipment and following the rules. Fall arrest does not prevent a fall but provides protection in the event of a fall.

24. D: NFPA 10, Standard for Portable Fire Extinguishers, requires that they be inspected monthly. Fire extinguishers must have a durable tag attached that records the inspections. Monthly inspections are required to ensure that the equipment is available and ready if needed.

25. B: Safety training courses that are based on adult learning theories should incorporate presentations, exercises, and scenarios to best leverage the common ways in which adults learn.

26. A: Universal precautions means that employees should treat any bodily fluids as potentially infectious. Employees may never assume any individual is healthy and will not have infectious blood or other bodily materials. This results in taking standard (universal) safety measures (precautions) when handling any type of material.

27. C: According to OSHA 1904.1, employers with 10 or fewer employees during the previous calendar year are exempt from retaining injury and illness records unless OSHA or the Bureau of Labor Statistics (BLS) informs the employer in writing that he or she must keep records.

28. C: This project will require workers to be at locations at least six feet above the next level, requiring a fall protection program. The use of scaffolding will increase the safety of the workers and allow for staging of materials at different levels. The scaffolding will require proper erection and required inspections. Materials and tools will need to be either mechanically lifted or hand-carried to the various levels during the project, requiring safe lifting, proper ergonomics, and safe storage of those materials.

29. B: Management of change is a systematic process of managing transition in an organization to reduce the risk to workers. Whenever there is a change on the job site that impacts workers, it must be conveyed deliberately so that workers can understand what the new process and expectations are to reduce errors that can cause injuries. For example, if a material laydown area is moved, workers need to know that traffic patterns and volume will increase in a different area than they are accustomed to. This will increase worker awareness in the new laydown area and reduce the risk of vehicle-employee incidents.

30. B: Falls at construction work sites represent the most frequent cause of fatalities annually (about one-third). Other major sources include transportation-related fatalities (about 25 percent), contact with objects and equipment (about 20 percent), and exposure to harmful substances (about 15 percent).

31. B: In addition to the hazardous activities that are involved with the work, the site itself can present various hazards. Elevation changes can present a risk of heavy equipment tipping over or stacked materials falling. The constant movement of vehicles on and off the site can present a hazard to workers walking or working in the area. With all the tools, vehicles, and activities, the noise level can interfere with a worker hearing a vehicle, instructions, or warning systems.

32. A: The employer is required to provide training whenever new hazards are introduced into the workplace. In this case, the systems present in a college laboratory, such as specialized gas lines, fume ventilation systems, and purified water lines, may be unfamiliar to the remodel crew. Thus, the required training will focus more on the systems present in the building, as well as the specific alarm systems used for those hazards. A facility or maintenance representative from the college would be beneficial so that employees can ask specific questions prior to beginning the project.

33. A: Housekeeping entails cleaning and organizing a work area. By sweeping and cleaning up scrap, workers can reduce the amount of fuel available to start or sustain a fire. Placing flammable or combustible materials in proper storage containers reduces an ignition source's access to fuel to start a fire.

34. D: The hierarchy of controls is a method of mitigating hazards by (in order of preference) eliminating the hazard, engineering the hazard away, creating administrative controls to reduce exposure to the hazard, or using personal protective equipment to protect the workers. If tradespeople are engaged in the design phase of the project, they can provide input on how to create safer work environments, such as allotting space for laydowns or installing appropriate anchor points for fall protection. This reduces delays caused by redesign, procuring unexpected equipment, and injuries during the construction.

35. C: A standard threshold shift is when an employee's ability to hear sounds during an audiogram shifts by more than 10 dB in one or both ears. The change is measured relative to an employee's initial physical and hearing test, referred to as a baseline audiogram. This means that an employee is unable to hear test signals at low levels compared to the baseline. If this change is 25 dB or more and is attributed to workplace noise exposure, then the injury becomes an OSHA-recordable incident.

36. B: A fire caused by flammable liquids may be effectively extinguished via the use of halon, carbon dioxide, or dry powder extinguishing agents. Any of those three will interrupt the fire tetrahedron by starving the fuel of the oxygen necessary to support combustion. Water should never be used to extinguish a fire involving flammable liquids, as the water will actually cause the fire to spread.

37. C: Under 29 CFR 1926.502, the only time a body belt may be used as part of a personal fall arrest system is when a construction worker is using a positioning device. A positioning device is a system that allows the worker to use both hands for an activity.

38. C: Appendix A to 29 CFR 1926.50 references the ANSI standard for first aid kits (ANSI Z308.1-1978). The ANSI standard, updated in 2014, establishes minimum performance standards for first aid kits by outlining recommended equipment based on the anticipated hazards. A construction site would require a Class B first aid kit for high-risk activities. Such kits are not required to include syringes.

39. D: A surcharge load is defined as any additional weight on the ground that adds stress to excavation walls. Examples of surcharge loads are heavy equipment working close to trench walls, spoil piles deposited too close to the trench walls, and nearby traffic.

40. C: It is the responsibility of any contractor or subcontractor to ensure that the individuals they bring on site under contract have undergone all required safety trainings. Even short-term contract workers must be made aware of the hazards of the site, hazard communication systems, alarm systems, muster points, and any requirements of the prime contractor.

41. B: The first step in determining the proper slope angle is to determine the soil type by way of 29 CFR 1926 Subpart P, Appendix A. Once the soil type is determined, the maximum slope angle is cross-referenced in 29 CFR 1926 Subpart P, Appendix B. Clay is a "cohesive" Type-A material and requires that a 53-degree slope not be exceeded, with an associated height-to-depth ratio of ¾ to 1. Thus, a five-foot trench requires, at most, a 3.75-foot angle, or 53 degrees. Any angle greater than this can result in a cave-in.

42. C: A repetitive motion disorder results from long-term, cumulative trauma or stress to ligaments, tendons, muscles, nerves, and joints caused by repeating the same motion without necessary rest. Typical conditions associated with repetitive strain injuries include tendonitis, carpal tunnel syndrome, trigger finger, and bursitis. Hands, arms, shoulders, and necks are usually the most susceptible to such injuries.

43. D: Under 29 CFR 1926.250, lumber that will be handled manually cannot exceed a stack height of 16 feet.

44. B: Promoting communication throughout the organization can make for a more open work environment, which may allow issues to be resolved before they reach a point of violence or threats of any kind. Employers must receive issues in an objective and impartial manner. They must use facts and evidence to make decisions and treat all employees the same way using the same progressive disciplinary actions.

45. D: To calculate the incident rate, the following formula is used:

$$\text{Incident rate} = \text{number of injuries and illness} \times 200{,}000 \,/\, \text{employees total hours worked}$$

$$\text{Incident rate} = (15 \times 200{,}000) \,/\, 745{,}000$$

$$\text{Incident rate} = 4.03$$

46. C: An eight-hour time-weighted average of 90 dB is considered the threshold (per 29 CFR 1926.52) at which a Hearing Conservation Program, including the use of engineering and administrative controls, must be implemented.

47. D: Classroom-based training programs can be modified based on the experience, ability, or makeup of the group. If the group is experienced, more time can be spent on advanced concepts, whereas a group of inexperienced workers can spend more time on basic information. Specific questions can also be answered in real time by the facilitator, which may increase retention, as the information is associated to real-life situations, one of the pillars of the adult learning model.

48. A: Lagging indicators are data that evaluates incidents that have already occurred. Reviewing data from a certain time period will give the safety professional areas to focus preventative actions to prevent similar injuries from recurring.

49. C: As a matter of business operations, each employer will have access to an employee's personally identifiable information. PII includes various data, such as full name, email address, phone number, home address, Social Security number, driver's license number, biometric information, and medical records. Although PII is not regulated by any single law, the employer must be aware of the requirements for managing various types of PII (such as HIPAA for medical information). The employee entrusts the employer with protecting the information and preventing unauthorized access to the data that could result in a loss to the employee, either financial loss (e.g., if a checking account number or Social Security number is stolen) or loss of privacy.

50. D: A multimeter is a device that can be used to measure electrical voltage, current, or resistance. It can be used to troubleshoot electrical problems in a wide array of industrial systems, such as electronic equipment, motor controls, power supplies, and wiring systems, as well as to determine whether a circuit is live.

51. B: The term "fit-testing" is used when referring to a respirator. A fit-test tests the seal between the respirator's face-piece and the user's face. It is used to ensure proper sizing and fit of the device. An employee can only use a respirator if they have passed a fit-test for that specific make and model.

52. C: There are several cold-stress-related conditions that may arise in the workplace, including trench foot, chilblains, frostbite, and hypothermia. Hyperthermia, on the other hand, is the elevation of core temperature caused by high environmental temperatures.

53. B: An employee standing in a pit is technically in a confined space. The presence of an atmospheric hazard converts the space into a permit-required confined space. An employee must be provided the appropriate personal protective equipment for the hazards anticipated to be encountered on the job. Finally, any underground utilities must be identified prior to any installation, as required by the Excavation Requirements standard.

54. A: Coaching is an effective method for adult behavior modification. Punishment, public ridicule, or a reprimand will not reinforce good behavior. It is common that employees do not follow a safety rule for a reason other than choosing not to — for example, because the box of earplugs was empty, they didn't know they were required, or the protective gear is uncomfortable and they need another solution. Once the root cause of the behavior is identified, the employee will no longer have a reason not to comply and can work in a safe manner.

55. B: The primary elements for a project involving the painting of loading dock surface markings will include struck-by hazards presented by moving vehicles (including limited visibility of the vehicle operators), the hazards presented by the paint removal and application processes, and proper protective equipment (such as high-visibility vests, proper foot protection, and chemical protective equipment). In an exterior loading dock area, the need for hearing protection is likely minimal.

56. A: The construction standard for emergency action plans (29 CFR 1926.35) references a since-removed standard for alarm systems used to alert employees of the need to evacuate. Thus, OSHA has directed employers to reference 29 CFR 1910.165 for emergency alarm requirements.

57. A: Rigging is the means for attaching a load to a lifting device, such as a crane. Rigging can include ropes, chains, cables, and synthetic slings.

58. C: CBT stands for computer-based training, which is a method of delivering information using an online platform. CBT has the advantages of flexible scheduling, remote delivery, consistent information delivery, and typically lower cost than in-person training. The disadvantages are that individuals may do other things while the training is playing, the training does not allow for questions, and it does not alter the delivery based on the experience of the trainee.

59. B: Under the servant leadership philosophy, the goal of the leader is to serve the employees. The leader empowers subordinates and prioritizes their development by putting the needs of employees first, allowing them to perform at the highest level possible.

60. A: During an incident investigation, photographs are critical. The images will record the state of the site at the time of the incident, including lighting, status of control measures (engaged or absent), and overall location of various items involved. Photographs do not suffer from problems of interpretation, bias, or subjectivity. Although interviews, notes, and sketches are critical to any investigation, they must be considered under the understanding that they are subject to bias, omitted information, and lack of details relevant to the incident.

61. A: A crisis management plan establishes roles, chain of command, and actions that the organization should take to minimize loss during natural and human-caused emergencies. Different than an emergency action plan, the Crisis Management Plan addresses economic events, threats of violence, or injuries to the public. The Plan should be written so that it can be referenced in the event of crisis.

62. C: Whenever materials must be moved from one location to another, the removal of human power from the equation will reduce the risk of injury. Mechanical lifting reduces the risk for muscle strains, fall injuries, and dropped materials. Carrying ductwork not only impairs an individual's ability to see where they are walking but also changes the shape of the body, requiring additional space to walk through doorways and stairwells. Finally, a crane or lift can move more pieces in a single lift, thereby reducing the time required to move the materials.

63. A: A site-specific safety plan should be used for every job as a method to demonstrate efforts to mitigate known or expected hazards to protect workers. Such plans should include hazards specific to the jobs that will be performed during the project, the training provided or required to inform workers of those hazards, and who is to be contacted in the event an unsafe condition is identified.

64. D: Tag lines can be attached to overhead loads to control the movement during transport. The lines are held by workers on the ground to prevent the load from swinging, swaying, and spinning, which — if not controlled — can strike workers passing by or result in a dropped load that can land on pedestrians.

65. B: Personnel who respond to hazardous-material spills or releases must be properly trained and provided with appropriate PPE. The levels of training are outlined in 29 CFR 1926.65. Employees must be aware of the hazards of the material, how to properly use specialized PPE, and how to prevent spreading the contamination.

66. C: In this scenario, the primary hazards are a particulate respiratory hazard and minor skin irritation. Thus, a Level C ensemble consisting of an air-purifying respirator outfitted with particulate (e.g., N95) cartridges to protect against airborne silica, a barrier garment (that is not required to be liquid-proof) to prevent skin irritation, and disposable chemical gloves would be the most appropriate. A higher level of respiratory protection would not be warranted by dry cement dust.

67. A: Risk transfer is when the employer transfers risk to another organization. Examples include outsourcing, contracting, and insurance policies.

68. A: Mechanical ventilation is used both to bring fresh air into a confined space and to remove atmospheric hazards. Toxic gases and flammable vapors are removed from the space, reducing both to acceptable levels. Additionally, since clean air is being brought into the confined space, the oxygen content is maintained at safe levels, further reducing the risks to workers in the space.

69. D: Inspection logs act as a record that necessary inspections have been completed. The log should include the items that are required (by both the manufacturer and any applicable OSHA standards) to be inspected. The log should include the date, time, and identity of the inspector as proof that the inspections were completed in accordance with regulations. Inspection logs will likely be requested by any OSHA inspector to verify compliance.

70. D: A good site-specific safety plan should include training records for all employees on site related to the work they will be doing. This will demonstrate to any inspector that the individuals working have been properly and adequately trained in the hazards they may encounter during the job being performed.

71. C: The BCSP Code of Ethics requires that certified safety professionals act with integrity. When confronted with a deficiency, the CHST must admit the mistake and work to correct it as quickly as possible. Forgery is a violation of the BCSP Code of Ethics and could result in loss of certification.

72. B: A Type-A fire extinguisher is recommended for a fire involving trash, wood, or paper, and is symbolically denoted by a green triangle. Extinguishers rated as Type-A are likely water-based, making them inappropriate for flammable liquids or electrical equipment.

73. A: Typically, controls employed for preventing materials-handling accidents include creating wide (not narrow) passageways and lanes for movement of materials, training workers to properly use hand signals, training workers on proper lifting techniques, and implementing traffic controls to keep materials-handling areas clear of people.

74. D: Under NFPA 51B, Standard for Fire Prevention During Welding, Cutting, and Other Hot Work, a hot work permit must be issued whenever a "spark-producing activity" (such as welding, cutting, brazing, or grinding) is conducted outside of a designated area. The permit is a checklist used to prevent fires that could be initiated by the sparks generated by the work.

75. C: A behavior-based safety program is designed to proactively address unsafe behaviors to prevent injuries or accidents. Employees are observed while performing a task to identify potential hazards and any behaviors that may increase risk. Checklists are helpful to observers to maintain consistency and to avoid missing important elements. Feedback should be given in real time to reinforce good behaviors and correct risky behaviors.

76. C: A site-specific safety plan (SSSP) is a plan that takes into consideration the hazards presented by the specific work location, such as topography, proximity to roads, weather, and the businesses in the area that can pose a hazard (such as a refinery). While the EAP outlines general response actions for several types of situations, the SSSP is specific to the conditions and environment of that exact work site.

77. B: Using a grinder to remove lead paint will generate lead-containing dust. Thus, the primary route of entry for lead is inhalation. To protect the worker from breathing in lead dust, a particulate filter respirator, such as a half-face unit with a HEPA filter, will capture the dust and reduce the risk of exposure. Additionally, protective coveralls should be worn to keep the dust off the worker's clothes. This will prevent the worker from being exposed when they take off their work clothes, as well as preventing potential exposure to family members.

78. B: A certified industrial hygienist (CIH) would be the most qualified individual to measure hazards in the workplace and provide recommendations on how to effectively mitigate such hazards. A certified health physicist (CHP) is primarily a subject-matter expert regarding radiological hazards and doses; a qualified person is one who has the knowledge and experience to resolve problems related to a specific subject; and an industrial engineer's main area of focus is usually innovating processes and systems that improve quality and productivity.

79. B: Under NFPA Standard 10, the monthly inspection should include a check of the gauge (indicating "full"), presence of the safety pin, presence of the breakaway loop, and condition of the hose or nozzle, as well as lifting the unit to determine if extinguishing media is present.

80. A: The Freedom of Information Act (FOIA) provides a way for the public to request access to records from federal agencies such as OSHA. Under FOIA, agencies must release information that is requested, but there are items listed as exempt, such as tangible materials, spoken requests, and questions, which are defined as not FOIA requests.

81. B: When there is an injury at a multi-employer job site, OSHA will evaluate the organizational structure of the project to determine who is ultimately responsible for the health and safety of the employees at the site. Thus, the prime contractor bears the ultimate responsibility for ensuring that all safety plans are in place, all controls are installed, and safe practices are being implemented. While the prime contractor may direct the subcontractors to develop trade-, task-, or job-specific safety plans for their own activities, the prime contractor is responsible for making sure those plans are created, implemented, and enforced. However, if an unsafe condition exists, the subcontractor responsible for the task can also be held responsible. For example, in the event of an injury related to failed scaffolding, OSHA can cite both the prime contractor (work site owner/operator, or "Controlling Employer") and the scaffold contractor since the unsafe condition was due to its negligence ("Creating Employer") for violations.

82. C: On-the-job training (OJT) is a cost-effective method of training in that it is typically completed by an employee who is already getting paid for their work. The culture and processes specific to the company are transferred to new employees, and the training is specific to what the employee will be doing. One negative aspect of OJT is that bad habits or shortcuts can also be passed on to the next individual performing the task, laying the foundation for a potential injury or loss of quality.

83. B: A lanyard's primary function within a fall-protection system is to connect a safety harness to an anchorage point. Lanyards absorb energy, so they reduce the forces on a person when a fall is arrested. The forces are distributed across the body by way of a harness.

84. C: Effective training will result in behavior modification. In order to determine if a behavior has changed, observations can be made by walking the site to see if employees are integrating new skills into their everyday activities. The safety professional should then provide feedback to the person who conducted the training.

85. C: Manually handling materials can involve the use of dollies, carts, wheelbarrows, and hand trucks. However, these devices carry risks that must be addressed. Equipment that is overloaded or improperly loaded can present a struck-by risk to walls, equipment, or other people. Even though the device provides a mechanical advantage, users must take caution not to injure themselves while loading the device, and to avoid overloading the device, which can result in muscle injuries. Loads that are not properly placed can shift during transit, resulting in tipping over or falling. Injuries can also result from workers trying to stop a load from tipping or falling.

86. D: Per 29 CFR 1926.45, scaffolds should only be erected, moved, dismantled, or altered under the supervision and direction of a competent person. Thus, if the CHST is not qualified as a scaffold-competent person, they are performing work outside the scope of their qualifications and knowledge.

87. B: The first step that should be implemented when developing a facility emergency action plan is selecting team members from a variety of organizational functions and disciplines. This scenario inevitably supplies a diverse set of viewpoints and areas of expertise that will ultimately result in the development of a stronger overall plan.

88. C: Instructional communication theory suggests that in-classroom learning is impacted by seating arrangements. Engagement, motivation, and focus can all be affected by how desks and chairs are designed and arranged. For training that involves learning to use equipment and trainer instruction, the horseshoe seating arrangement is best.

89. A: Rotating and spinning components of power tools (such as boring drills) and machinery (such as axles, powertrains, or conveyor wheels) can catch loose clothing, hair, or jewelry, pulling an employee into the mechanism and causing an injury. The injury is caused by the body part getting caught between the rotating mechanism and the housing of the tool or machinery. The most effective method to reduce the risk is installing guards to separate the rotating parts from the employees.

90. C: There are three fundamental types of air-supply equipment used in the workplace for ensuring good-quality, breathable air during periods where respiratory protection is required. These are supplied-air respirators (such as hose masks, air-line modules, hoods, and air-supplied suits); self-contained respirators (i.e., portable SCBA tanks); and air-purifying respirators (such as full-face or half-face respirators).

91. D: The exposure is calculated using the flow rate and the total mass of silica detected. The total volume of air is determined using the flow rate and sampling time:

$$4 \text{ hours} \times \frac{60 \text{ min}}{1 \text{ hour}} \times \frac{1.7 \text{ L}}{1 \text{ min}} = 408 \text{ liters collected}$$

Liters must be converted, as the OSHA PEL is expressed in cubic meters (m^3).

$$408 \text{ L} \times \frac{1 \text{ m}^3}{1000 \text{ L}} = 0.408 \text{ m}^3$$

Since the exposure does not change over the 8-hour time frame, the 4-hour data can be used to determine the occupational exposure:

$$\frac{0.022 \text{ mg}}{0.408 \text{ m}^3} = 0.054 \frac{\text{mg}}{\text{m}^3} = 54 \text{ µg/m}^3$$

92. B: Although everyone should take ownership and accountability for job site safety, it is ultimately the employer's responsibility to ensure that only well-maintained and operable hand and power tools are utilized at a job site.

93. A: Whenever an OSHA inspection occurs, the following items are deemed mandatory: (1) the employer must always be advised by the compliance officer as to the reason for the inspection; (2) the employer must accompany the compliance officer on the inspection; (3) the compliance officer must show official identification upon inspection; and (4) the employer must be assured by the compliance officer that any trade secrets observed during the inspection will remain confidential.

94. A: The four crucial elements for implementing a successful PPE program are: (1) selection, (2) management, (3) use, and (4) maintenance. PPE should be selected for the hazards that are present and should be appropriate for those hazards. The inventory should be managed so that equipment is available when needed and properly sized for employees. PPE should be used when needed and used properly. Reusable PPE must be properly maintained in order to maintain its integrity and protective properties.

95. A: If the job site does not have a landline that provides location information, or a communication system that provides the latitude and longitude to an emergency services operator, the employer must post such information in a common area on the work site. The employer should convey the location of this information to all employees who can contact medical assistance in the event of a serious injury.

96. C: Welding, cutting, brazing, and other spark-producing operations present many hazards, which include burns, air contaminants (such as metal fumes), potential eye damage from UV light, and explosions in the presence of flammable materials.

97. B: The MIL-STD-8820 risk assessment has 5 probability levels: frequent (A), probable (B), occasional (C), remote (D), and improbable (E). Occasional is the middle level (C) and represents a likelihood that can be described as "likely to occur sometime in the life of an item."

98. A: As of January 2022, a willful violation can result in a fine of up to $145,027. OSHA defines a willful violation as one where the employer knowingly failed to comply with applicable standards or acted with indifference to employee safety. Ignorance is not an acceptable defense when facing such penalties.

99. C: A warning line system is defined in 29 CFR 1926.500 as a rooftop barrier that is erected to warn workers that they are approaching an unprotected side or edge. The warning line designates the area past which additional fall protection measures, such as lifelines or safety harnesses, are required.

100. A: Proper investigation of a work site accident will involve the use of rope or security tape to secure the scene, a camera to record the scene, and employee interviews to gather additional information about the incident.

101. B: The end result of a potential electrocution incident is generally determined by how much current passes through the body, the duration of time that a person is exposed to the current, and the actual path the current takes.

102. D: Electric shock can potentially result in a variety of injuries, including cardiac arrest due to electrical effects on the heart, muscle, and nerves. However, the most common injuries are burns from where the current enters and exits the body when the worker becomes part of the circuit.

103. A: Per OSHA 29 CFR 1926.96, foot protection for construction employees shall meet the requirements and specifications stated within the ANSI Standard for Men's Safety-Toe Footwear, Z41.1-1967.

104. B: Several safeguards can be employed to protect workers from accidents involving machines or tools. These include implementing a regular inspection and maintenance schedule; requiring workers have no loose clothing, jewelry, or hair while using machines and tools; and ensuring the proper functioning of machine and tool guards. Lockout/tagout procedures are implemented during maintenance and repair of machines, not while the machines are being used.

105. B: A facility emergency action plan would be activated in response to a potential local emergency resulting from a tornado or flood. The plan outlines evacuation procedures, medical treatment, and contact information.

106. A: Effective safety training can be instructor-led training, self-paced training (including online), or structured on-the-job training.

107. B: Employers are required to notify employees of changing conditions, including the introduction of new hazards into the workplace. This communication can occur by means normally used to convey information to employees, such as with bulletin boards, through designated representatives, or through electronic means, as long as all affected employees are notified.

108. D: If an employee is demonstrating either a lack of understanding or limited retention, quick action should be taken to retrain the employee in safe practices—for their safety as well as the safety of others. For you as a safety professional, there is a requirement to provide effective training for employees. In this example, the employee may have been inattentive during the training or may not have understood the language used very well. Thus, if the situation can be resolved on the job under the supervision of an experienced worker, the apprentice will be provided the skills to work safely.

109. A: A Code of Ethics, such as the one adopted by BCSP, is a document that outlines the expectations for member behavior, best practices as defined by the organization, and the context within which members should make decisions. The Code establishes the rules, principles, and values of BCSP with the requirement that those who achieve and want to maintain certification shall adhere to them in all their dealings as representatives of the organization. Failure to do so can lead to discipline, up to and including decertification.

110. D: Three types of shoring systems mentioned in 29 CFR 1926 Subpart P include aluminum hydraulic shoring, timber shoring, and pneumatic shoring. Dead shoring is used to support existing buildings.

111. A: As part of the Incident Command System, the Incident Commander is responsible for all aspects of an emergency response. The individual establishes a command center, determines an action plan, monitors response activities, and determines when the response will be terminated.

112. B: After a proposed standard or change has gone through an advisory committee and the language has been finalized, the standard is published in the Federal Register. In addition to the text of the new or revised standard, the publication contains an effective date after which the standard must be followed and can be enforced. The Federal Register is published daily by the National Archives and Records Administration under the Government Publishing Office.

113. C: OSHA has provided an interpretation that, as required under 29 CFR 1926.50, medical facilities should be "reasonably accessible," meaning a three- or four-minute drive time. If such a facility is not reachable within that time frame, the employer must have a person at the job site who is certified in first aid. The certification must be from the U.S. Bureau of Mines, the American Red Cross, or an equivalent.

114. A: A hazard and operability study (HAZOP) is a regimented and methodical evaluation of a planned or existing undertaking or process that attempts to identify and assess potential issues that may ultimately cause a risk to workers, equipment, and/or operational efficiencies.

115. C: The Environmental Protection Agency (EPA) establishes regulations for the proper management of waste, including hazardous chemical waste. While OSHA designates the requirements for the safety of the workers handling waste or cleaning up spills, the actual regulations for how to safely treat or dispose of the waste fall under the EPA.

116. A: A risk score is the multiplicative product of the frequency and the severity. Thus, jobs that have a high frequency value (meaning they either occur most often or are performed by the highest number of individuals) and have a high severity ranking (resulting in serious injury, fatality, or a large financial loss) have a higher resultant score. These are the jobs that should be prioritized for implementing control measures.

117. B: Brightly colored clothing increases the visibility of pedestrians who may work or walk near vehicle areas. High-visibility items are used to reduce the risk of workers being struck by delivery vehicles, heavy machinery, or materials being brought onto the site.

118. A: Injury trends can be identified in a variety of ways. Workers' compensation claim files will be a record of all injuries where an employee sought medical treatment for a work-related injury. Injury reports may capture not only cases where employees sought medical treatment but also minor injuries that only required first aid. First aid cases can be used as indicators of situations where a slight change could have resulted in a more serious result. But employees may not report all injuries, especially chronic injuries to joints and muscles, for fear of being put off work. Thus, conversations with employees may also identify hazardous conditions on the site.

119. B: The violations pictured are a blocked exit and a blocked fire extinguisher.

120. D: Employers must designate and train employees to assist in safe and orderly evacuation of other employees. They must review the emergency action plan with each employee covered by the plan. OSHA does not stipulate how training and review is achieved. Choice D is the best choice to meet OSHA requirements.

121. D: The hierarchy of controls is arranged so that the most effective method for hazard mitigation is selected before less protective measures. The first consideration under this principle is elimination — if there is no hazard, there is no risk. Replacing an oil-based paint with a water-based product can eliminate the risk of VOC exposure.

122. D: Whenever a company does not have a competent person or proper equipment to abide by OSHA standards, it is recommended to hire an expert to assist with the job. Planking ladders are not allowed for working at heights, nor are ropes allowed as safety lines. Taller ladders may be appropriate but introduce other risks. The safest option is to work from scaffolding installed by a competent person. If included in the job bid, this option will not add any unexpected costs.

123. B: Based on the program review and injury data, it appears as if this subcontractor is knowledgeable in safety regulations, prioritizes safety, and works in a safe manner. Thus, a brief review of the prime contractor's safety expectations, contact personnel for the prime contractor, and site-specific hazards would be appropriate in the form of a "tailgate" meeting before work begins.

124. B: Corrosive materials are defined as substances that will cause destruction of tissue wherever they contact the tissue. Thus, whenever employees are working with these substances, a means of drenching the skin and eyes with copious amounts of water must be available to rinse the material from an exposed worker as quickly as possible to reduce the damage.

125. C: Shelter-in-place procedures should include guidance on how to shut down ventilation systems and elevators, how to temporarily close off all exits and entrances, where building occupants should gather, and how to inform said occupants that an emergency has occurred. It is pivotal to note that during a shelter-in-place event, the primary intent is not to evacuate a work area or building (although at a later point it may ultimately be deemed by emergency response personnel that an evacuation may be necessary).

126. D: The emergency action plan (EAP), as required by 29 CFR 1926.35, must at least include procedures for evacuation, a method for accounting for employees after an evacuation, and systems for how employees shall report emergencies.

127. B: Per OSHA, the DART (Days Away, Restricted, and Transfer) rate is calculated using the following equation:

$$DART = \frac{(Days\ Away\ +\ Days\ Transferred\ +\ Days\ Restricted) \times 200{,}000}{Total\ employee\ hours\ worked\ in\ the\ year}$$

The sum of recordable injuries that resulted in days away from work, days transferred to another job due to the injury, or restricted work based on accommodations resulting from the injury is multiplied by 200,000. This product is then divided by the total number of employee hours worked during the year. In this example, there were 5 recordable injuries, which results in only 4 days away from the regular job. Therefore, the DART is:

$$4 \times \frac{200{,}000}{127{,}500} = 6.3$$

128. B: Behavior-based safety (BBS) is the concept that work sites are imperfect and employees must continually make choices between working safely or taking unsafe shortcuts. Under this model, an employee will make an unsafe choice based on the person (their experience, abilities, and capabilities), environment (available hazard controls, workplace, and company culture), and behavior (how the employee acts and responds while working).

129. D: Under the Hazard Communication Standard (29 CFR 1910.1200, as referenced in 29 CFR 1926.59), employers are required to maintain Safety Data Sheets (SDSs) for all chemicals used in their work. However, if an employee is exposed to a chemical and receives medical treatment, the employer must keep information about that specific chemical for thirty (30) years after the date of exposure, as some effects from chemical exposure are latent — they do not appear until time has passed. Employers may choose to keep the SDS, but, per the standard, must at least keep a record of the chemical to identify where and when it was used.

130. C: A bloodborne pathogen is any microorganism that is transmitted by way of blood or other body fluid that can cause diseases in humans. Common examples of such pathogens include, but are not limited to, HIV and hepatitis B. Universal precautions are the recommended method for protecting individuals providing aid to an injured worker from exposure to bloodborne pathogens.

131. B: OSHA fall clearance is calculated by adding the free fall distance, deceleration distance, D-ring shift, D-ring height, and safety factor:

$$\text{Fall clearance} = \text{free fall distance} + \text{deceleration distance} + \text{D-ring shift} + \text{D-ring height} + \text{safety factor}$$

$$\text{Fall clearance} = 6\text{ ft} + 3.5\text{ ft} + 1\text{ ft} + 4\text{ ft} + 2\text{ ft}$$

$$\text{Fall clearance} = 16.5\text{ ft}$$

132. C: Per the OSHA standard, one of the required elements of an EAP is "procedures to be followed by employees who remain to operate critical plant operations before they evacuate." For some industrial workplaces, hazardous operations must be shut down before evacuation to ensure the fire does not lead to major catastrophes (e.g., explosions or major chemical releases).

133. A: To calculate a fall distance, take the sum of the employee's height, length of the lanyard, and free-fall distance (6 feet + 6 feet + 3 feet = 15 feet). In general, a safety factor should also be added (i.e., 3 feet). Based on this calculation, the arrangement is not recommended as the employee will likely hit the ground before his fall is arrested.

134. D: Working above an elevation of six feet requires an employer to implement appropriate fall protection methods. Ventilation is required when there is a potential for toxic fumes or vapors, such as overspray from painting operations. The employer must train all employees on the dangers of working with chemicals, including paint. The theater is a large, open space that does not meet the definition of a confined space.

135. D: The concept of "universal precautions" is based on treating any bodily fluid as a potentially infectious material. Under this premise, barriers like disposable gloves are used to prevent the transmission of bloodborne diseases, also termed pathogens.

136. A: 29 CFR 1926.961 outlines the requirements for re-energizing lines and equipment. These include ensuring that all crews working on lines or equipment release their clearances, ensuring that all employees are physically clear of lines and equipment, and ensuring that all protective tags are removed from the point of disconnection. The standard requires that all grounds are removed prior to re-energizing.

137. D: OSHA regulation 29 CFR 1926.1053(a)(19)(iii) stipulates that a fixed ladder using a cage or well with multiple ladder sections must have landing platforms at maximum intervals of 50 feet.

138. A: Liability insurance is a means to protect a business from injuries or damage that occurs due to their acts or omissions (what they do or what they do not do). Typically, a contract will require that all contractors and subcontractors carry their own insurance to prevent the prime contractor from having to pay for something a subcontractor did. For example, if a pedestrian trips over a water line being used by Subcontractor B and breaks an arm, the prime contractor will not want to be responsible for the medical costs. Thus, all parties, including the property owner, should carry some type of liability insurance.

139. C: There are several measures that may be employed in the workplace for controlling exposures to chemical hazards. These include PPE, personal hygiene, and reducing worker exposure time.

140. A: Recommended weight limit (RWL) is the weight that healthy workers could lift for up to eight hours without increasing the risk of musculoskeletal injuries. It is calculated as the product of load constant, horizontal multiplier, vertical multiplier, distance multiplier, asymmetric multiplier, frequency multiplier, and coupling multiplier.

141. B: A properly labeled chemical under the Globally Harmonized System (GHS) has specified required information. One of those pieces of information is a pictogram: one of nine standard images used to nonverbally convey hazards presented by the chemical. The pictogram is a visual system to convey information regardless of the language of the handler or user. A chemical may have one or more pictograms present on a compliant label.

142. A: The Respirable Silica Standard (29 CFR 1926.1153) describes specific exposure control measures for various types of dust-generating equipment, task locations (indoors or outdoors), and established time frames (up to four hours per shift, or longer than four hours per shift). According to the table, for a walk-behind saw outfitted with a water delivery system being used outside, no additional controls are required.

143. A: The three fundamental categories of risk management (process) controls are physical, educational, and avoidance. Examples of physical controls are guarding, barriers, and signage; examples of educational controls are knowledge, training, and skills; and examples of avoidance controls are the initiation of programs or incentives by management in support of hazard-aversion reinforcement goals.

144. C: Workers can be injured if they are struck by falling materials that have been improperly stored and stacked. Falling materials can result in cuts, bruises, and even fractures.

145. B: Training employees on the OSHA Hazard Communication Standard is required whenever a new hazardous substance is brought into the work environment, or in conjunction with an employee's initial assignment to a specific work area. The Hazard Communication Standard, 29 CFR 1910.1200, is incorporated in the construction standard by reference via 29 CFR 1926.59.

146. D: The site-specific safety plan should be updated whenever there is a significant change in site conditions. Such changes include modifications in traffic patterns, progression in the phases of the project (e.g., pre-project planning, post-excavation, post-foundation, post-framing), and weather events that impact risk.

147. D: A good training objective identifies the audience, the desired behavior, the specific circumstances where the behavior is expected (condition), and the criteria to measure success (degree).

148. A: The ISO 45001:2018 occupational health and safety management system standard sets the basis for an effective management system to address the health and safety of workers. The purpose of such a management system is to take a systemic approach to how an organization manages its health and safety protocols, with the overall goal being to proactively identify and respond to hazards before an accident or incident occurs.

149. B: Powder-actuated tools use a system similar to a gun to drive a fastener. When the trigger is pulled, the powder-containing cartridge is struck, resulting in a small explosion that pushes the fastener into the material. This explosion is extremely loud, which poses a noise hazard. And, in the presence of flammable materials, the explosion can act as an ignition source. If used with frangible materials, the force of the fastener can eject fragments toward the user's face.

150. A: Per OSHA 29 CFR 1904.33, the time requirement for retention of documents (e.g., OSHA Form 301) is designated to be 5 years following the end of the calendar year that the records cover.

151. C: The multilinear events sequencing (MES) method usually employs a time line to illustrate the sequence of events or occurrences that contributed to or was the direct cause of an event.

152. D: During a crisis or emergency, media responses should be provided by a single, pre-designated person. This will prevent the release of conflicting information, allow all information to be properly vetted prior to release, and allow the company to control the information to avoid speculation and false information being released. The spokesperson should only provide messages that have been approved by the Incident Commander and parent organization.

153. A: An exit route has three components: exit access, exit, and exit discharge. An exit access is the portion of the exit route that leads to the exit (e.g., corridor leading to stair tower). An exit is a separated area that is a protected way to travel to the exit discharge (e.g., a stair tower with 2-hour fire-rated construction). An exit discharge is part of an exit route that leads to an outside area of safety (e.g., door at the bottom of a stair tower that leads outside).

154. A: Inspection checklists are a valuable tool for a safety professional. Such checklists should be developed using the relevant OSHA 1926 standards as well as the manufacturer's instructions. These instructions will provide equipment- and model-specific elements that should be checked routinely to protect workers from equipment malfunction and exposure to hazards.

155. B: An asphyxiant is defined as a material that displaces oxygen in air (physical asphyxiant) or interferes with the body's ability to transport oxygen in the blood (chemical asphyxiant). Physical asphyxiants can result in lowering the percent of oxygen available for breathing in the environment. Chemical asphyxiants starve the body of oxygen at the cellular level by inhibiting the blood's ability to transport oxygen to muscles and the brain.

156. C: By using the risk assessment matrix, it can be determined that the probability and severity of the task align to a high-risk ranking. The recommendation is for the level of risk to be reduced for this task.

157. D: The discipline of probabilistic risk assessment defines risk as the product of frequency and severity. When numerical values are applied to the variables, risk for any activity will have a value that can be compared to other activities. Using this method allows activities to be ranked so that an organization can allocate resources to the highest-risk activities.

158. D: In addition to changing the content of the training program based on the specific job and location, safety professionals should incorporate seasonal environmental risk factors into the program. During the summer months, workers should be reminded about hydration and the potential for heat illness. During wet weather, there is an increased risk of slipping and electrical hazards with the standing water. As fall turns into winter, cold stress should be addressed to keep it in the front of the workers' minds.

159. C: Typical means of reducing construction ergonomic risk factors include using such equipment as adjustable scaffolding to prevent reaching or overhead work, reduced-vibration power tools, and motorized screeding equipment to reduce the force required by workers.

160. A: According to the American Society of Mechanical Engineers (ASME) Standard B30.10, during a rigging inspection, if the throat opening of a hook (which attaches to the rigging) has increased in excess of 5 percent of the original throat opening (or ¼ inch), it should be replaced.

161. D: BCSP will issue certifications under the condition that the certified individual adheres to the published Code of Ethics. Any certified safety professional must (among other delineated obligations) work to protect property and the environment, act with integrity by avoiding conflicts of interest, and avoid misrepresenting their professional qualifications. Failure to do so could result in a forfeiture of certification.

162. B: When developing a maintenance checklist, manufacturer specifications and product literature should be the primary references from which checklist items are developed. Manufacturer recommendations will also include the frequency with which various items should be checked.

163. B: Using ladders or other items to increase a worker's reach while on any level of a scaffold is prohibited by 29 CFR 1926.452. Working on a ladder on a scaffold platform increases the chance of falling due to the instability of the ladder, and places the worker above the protective guardrail that is installed to prevent falls. Such behaviors must be addressed immediately to eliminate the risk of injury.

164. C: Per OSHA 29 CFR 1926.404, all 120-volt single-phase 15- and 20-ampere receptacle outlets on construction sites, which are not a part of the permanent wiring of a building or structure and which are in use by employees, shall have approved ground-fault circuit interrupters for personnel protection.

165. C: During the demolition phase of any project that contains asbestos-containing materials (ACM) as identified by a competent person or a laboratory, the contractor must follow the regulations outlined in the CAA to prevent the fibers from becoming airborne. The contractor must notify the appropriate state agency prior to initiating the project. Measures that must be implemented to keep fibers from being released include wetting the material, sealing the material in leak-tight containers, and removing the ACM from the site as quickly as possible. AHERA applies only to publicly owned buildings, while RCRA will determine how the waste is managed once removed from the site.

166. D: Errors in information transfer that cause a misunderstanding can occur at any of the three parts of communication: sender, information, or receiver. If the sender does not understand how to use the communication system, does not have a thorough knowledge of the topic, is not clear on what information they are transferring, or does not understand the audience, the communication process is impacted. If the information being sent is not complete or is not clear enough to be understood, communication does not occur. Finally, if the receiver is unable to understand the information, does not receive the information, or is unable to access the information, the message is lost. Failure to properly communicate during a safety training results in an ineffective session that can result in injuries or accidents.

167. D: Mandatory participation in safety program activities and initiatives (e.g., voluntary protection programs) is not typically employed as an enforcement measure for ensuring compliance with safety program protocols. Such measures may, however, include disciplinary actions for blatant disregard of safety rules, incentives for proactively following safety rules (e.g., providing gift cards to the entire staff for achieving zero lost time over a calendar year), and training on safety requirements.

168. C: The VPP program was established to promote effective health and safety systems that rely on a partnership between the employer, OSHA, and the workforce. By setting performance-based criteria, OSHA will approve those applicants whose safety management systems are effective at preventing injuries and controlling hazards. The highest level of achievement is the "Star" rating, wherein an employer has demonstrated exemplary achievement in their safety management system.

169. D: Electrical bonding is defined as the connection of two or more objects with a conductor. If two objects that can retain a charge are not bonded, then a person or object that connects the two items is subject to a shock, or the potential difference can lead to a static discharge, which acts as an ignition source. Typical bonding connectors include copper bonding straps, solder lugs, pressure connectors or bolts, and star washers. A typical bonding conductor is a copper braid with pressure-type clamps.

170. B: The risk of falling from a portable ladder can be reduced by ensuring that a ladder's rungs are slip-resistant, ensuring that no overhead electrical conductors are nearby (especially when using a metal ladder), and anchoring a ladder to a permanent support structure for stability. Ladders should also be placed far enough from a wall that the arch of the foot (not just the toes) can fit comfortably on the rungs.

171. C: The Health Insurance Portability and Accountability Act, better known by the acronym *HIPAA*, is a federal law that regulates who can access an individual's health records and information.

172. B: At the beginning of the project, the employer must designate a competent person to identify all confined spaces, as well as those that would require a permit prior to entry. Under 29 CFR 1926.1203, permit-required confined spaces must be identified with a "danger" sign that prohibits unauthorized entry. Additionally, the prime contractor must be told of the existence and location of the permit spaces.

173. B: 29 CFR 1926.652 mandates that excavations five feet (1.5 meters) deep or greater require a system to protect employees against cave-ins unless the excavation is made entirely in or of stable rock.

174. C: The elements of a chemical release emergency response plan are outlined in 29 CFR 1926.65. Required elements include pre-emergency planning, roles, communication, prevention, places of refuge, site security, evacuation procedures, decontamination procedures, medical treatment, post-incident critique, protective equipment, and notification systems.

175. D: Audit findings can be grouped into the following categories: (1) nonconformances, (2) areas of concern, and (3) opportunities for improvement. Nonconformances are areas where the system is not meeting a mandatory criterion. Areas of concern would be findings where the system is approaching a nonconformance. Opportunities for improvement are areas where the system can be improved but a nonconformance was not identified, or where an optional criterion can be achieved.

176. D: As stated in 29 CFR 1926.1418, crane operators must be given the authority to refuse to move a load if there is any concern for safety. Once a qualified person has reviewed the situation and addressed the hazard, the operator can resume the lift.

177. A: Scrap, unused materials, cords, and tools left on floors and walkways can present a variety of hazards. Scrap and excessive stored materials can provide fuel for sparks and other heat sources, resulting in a fire. Items in walkways can present trip hazards or, in extreme cases where scrap covers a floor, uneven walking surfaces that can cause workers to trip and fall. Piles can provide pests with areas to hide and live, thereby increasing the risk that they will interact with workers.

178. B: Even though they lack a power source, hand tools still carry a risk of injury. The amount of force, the material the tool is manufactured from, and the shape of the handles and actuators all impact the user. Bent handles that allow the hand and wrist to be in neutral positions reduce the negative impact on nerves, vessels, and tissue as well as the risk of chronic injury.

179. D: Pipes and bars should not be stored in racks facing main aisles, as they pose a hazard to others when removing supplies. Bricks and masonry block stacks that exceed heights designated in 29 CFR 1926.250 shall be tapered to prevent falling. Similarly, round-stock shall be blocked, and bagged materials shall be stepped back.

180. A: "White finger" syndrome (also known as "vibration white finger" or VWF) is damage to the vascular system in the hands or fingers. As capillaries become damaged, blood no longer easily flows into the fingers, resulting in a white coloration.

181. C: OSHA defines a leading indicator as an objective measure to assess proactive actions taken to improve organizational safety performance. Lagging indicators (metrics for what has already happened) can be used to measure a program's effectiveness, whereas leading indicators work to prevent incidents.

182. B: Before any project is started, a risk assessment should be performed specific to the jobs, tasks, and work site location. Once the risks are identified, emergency equipment should be staged on site so that it is available in the event of an emergency. Regardless of other equipment present, every site should at least have fire suppression equipment (such as portable fire extinguishers) and first aid kits available.

183. D: An emergency evacuation plan should clearly denote primary and secondary evacuation routes, the location of alarm pull stations, and primary and secondary assembly areas.

184. B: Accidents are more likely to occur when workers have recently started a new job; thus, it is imperative that safety training takes place before workers commence a new job or task. Employees conducting unfamiliar work must be shown how to safely perform procedures, be instructed on proper accident response and reporting procedures, and be instilled with the fact that safety is a priority with the organization.

185. C: OSHA is a division of the Department of Labor. Thus, the OSHA-related web links under the Department of Labor's website have the most up-to-date standards available.

186. D: Per 29 CFR 1926.1001 and 1926.1002, rubber-tired self-propelled scrapers, rubber-tired front-end loaders, rubber-tired dozers, wheel-type agricultural and industrial tractors, crawler tractors, crawler-type loaders, and motor graders that are used in construction work shall all be equipped with rollover protective structures. Rollover protective structures are designed to protect the driver of heavy equipment in the event that the vehicle tips or rolls over, preventing an injury due to the driver being caught between the vehicle and the ground. They can be integral to the structure of the cab or can be a frame attached to the vehicle.

187. B: A job hazard analysis (JHA) is a proactive approach to identifying the hazards associated with a job before they can result in harm. A hazard can be an object, such as a rotating saw blade, or a condition, such as high toxic-gas levels, that can cause harm or illness to a worker. If these hazards are not effectively controlled, an injury or illness will result.

188. C: Changes in weather can quickly impact the hazards present at a job site. Water and runoff can fill a trench, presenting a risk of drowning or entrapment to workers in the trench. Water can also change the characteristics of the soil, increasing the chance of excavation failure. Additionally, walking surfaces become slippery, and there is an increased risk when using electrical equipment or power tools.

189. A: Constructing trusses on the structure presents fall hazards (to both workers and materials), the need to use power tools at elevated locations, and the need to support the trusses while under construction, regardless of the method used. Nail guns present the hazard of using powder tools, misfires, and nails missing the truss members, presenting a struck-by hazard (flying nail). The use of hand tools introduces repetitive motion injuries, awkward positions, and the risk of striking the non-tool-bearing hand. Preparing the trusses on the ground eliminates the hazard of heights and falling structural members but retains the tool-related hazards. Purchasing and installing pre-built trusses reduces the fall hazards (as employees work at a lower elevation), removes most of the tool hazards, and reduces the falling material and struck-by hazards.

190. C: The ISO 14001 environmental management system is built upon the plan-do-check-act model. This model encourages evaluating the safety management system with a continual improvement mindset. By incorporating safety into the management system, creating communication lines across the organization, and outlining roles and responsibilities for all levels in the organization, the company reduces its negative impact on the environment.

191. A: Cutting through a concrete slab will require mitigation strategies to reduce the workers' exposure to airborne silica. Since the water line will require a trench more than five feet deep, an excavation safety program will have to be developed to protect workers down near the pipe. Additional measures will have to be considered to prevent water from entering the trench to reduce the risk of drowning while in the trench. Finally, every project should include an emergency response plan that outlines methods for contacting emergency services, responding to medical emergencies, and accounting for employees in the event of an emergency.

192. C: The Globally Harmonized System (GHS) is also known as the Globally Harmonized System of Classification and Labelling of Chemicals. Developed by the United Nations, the system replaced the different national systems with a single way to convey the hazards of chemicals that are shipped globally. The GHS standard, codified by OSHA in 29 CFR 1910.1200, outlines labeling requirements for chemicals, the ways that safety information is conveyed to the users of chemicals, and training requirements for the system.

193. A: There are three types of respiratory equipment used in the workplace. These are self-contained breathing apparatuses (SCBAs), supplied-air respirators (such as air-line modules, hose masks, air-supplied suits, and hoods), and air-purifying respirators (quarter-, half-, or full-face).

194. A: The "Five Whys" method is a root-cause determination process in which the investigator asks "why" at various choice points to determine the underlying cause of an accident, incident, or near miss that has already occurred. At the end of the chain of questions is typically a root cause that can be addressed to prevent future occurrences. Both failure mode and effects analysis (FMEA) and hazard and operability analysis (HAZOP) are designed to predict and prevent failures. A job hazard analysis (JHA) is not a root-cause determination method.

195. D: In group sessions, facilitators must keep one individual from dominating the discussion and thereby limiting participation of the other members. Introverts may not be comfortable participating or asking clarifying questions, which will impact the effectiveness of the training for them. Also, the facilitator must try to avoid groupthink, wherein participants will, to avoid becoming outsiders, make an effort to agree with each other in spite of the facts or their own personal opinions and beliefs.

196. B: The hierarchy of controls is based on the principle of selecting the most effective control possible. The most effective controls are those that eliminate the hazard, followed by those that separate the worker from the hazard (engineering controls). In this scenario, the use of a wet saw removes or reduces the airborne silica in a worker's breathing zone. Next in terms of effectiveness are policies and procedures (administrative controls), such as worker rotation, to reduce each person's time of exposure or reduce the number of employees exposed to a hazard. The least effective method of hazard control is protective equipment, such as respirators, safety glasses, and gloves. These items do not remove or reduce the hazard and are dependent on proper selection. Controls from various categories can be used together to reduce the risk to the lowest level possible.

197. A: Per OSHA, the Recordable Injury Rate is calculated by summing up all recordable injuries, multiplying by 200,000, and dividing by the total number of employee hours worked during the year. In this example, there were 7 recordable injuries, therefore the RIR is:

$7 \times 200{,}000/180{,}000 = 7.8$

198. D: On a site, the weather and the phase of the project can introduce hazards that may not normally exist. Extension cords used for power tools that are in good condition can have increased risk of shock if they are lying in puddles of water or if the worker unplugs the cord with wet hands. Concrete surfaces that are wet present an increased slip hazard, especially while carrying materials that may obscure vision or change the worker's center of gravity, increasing the chances of dropping the load. Workers who have clothing inappropriate for the conditions (for example, lacking a rain jacket) can have an increased risk of cold stress due to accelerated heat loss from wet skin.

199. C: Safety programs that reduce the frequency and severity of accidents will likewise reduce workers' compensation claims. As occupational injury rates decrease, the premiums associated with workers' compensation will likewise decrease over time.

200. C: Adults typically retain information in three ways: visually, audibly, and kinesthetically. An effective training program will incorporate all three modalities to increase the level of retention for the students. A slide will present information visually while the facilitator explains the information, appealing to the auditory mode. The kinesthetic mode is used by having students either complete a worksheet or complete a task based on the new information.

CHST Practice Test #3

1. Deterministic and stochastic describe the two categories of health effects related to which of the following?

a. Injuries due to falls
b. Arsenic toxicity
c. Ionizing radiation
d. Pinch point injuries

2. While conducting a job site inspection, the safety professional identifies temporary power panels at the worksite that will be used by multiple contractors. In reviewing 29 CFR 1926.417, the professional determines they must be able to "...have tags attached at all points where such equipment or circuits can be energized." Which of the following would NOT be a source to assist the professional in identifying the energy control points?

a. National Electrical Code
b. Certified electrician
c. Panel manufacturer instructions
d. The contractor who installed the panel

3. To prevent bricklaying materials from falling onto workers below, what is the minimum distance equipment may be staged from an unprotected overhead edge?

a. 4 feet
b. 3 feet
c. 2 feet
d. 1 foot

4. NIMS is a standardized system for which of the following?

a. Installation of masonry structures
b. Steel erection management
c. Large-scale chemical spill response
d. Mechanical systems installation

5. At a housing development, the general contractor's safety professional is planning the weekly safety briefing for the electricians who will be installing the light fixtures and fans before the units are turned over to the customers. Which of the following items should be included in the briefing?

a. Improper ladder use
b. Heat stress
c. Protection for the flooring
d. Proper use of flexible cords

6. Tinted lenses are required during arc welding to protect the eyes from ______ radiation.

a. UV
b. IR
c. Gamma
d. X-ray

7. A crew will be evaluating a slab for voids using radiographic instruments. Each worker must wear a _______ to measure their exposure to radiation.

a. multi-gas meter
b. Geiger counter
c. galvanometer
d. dosimeter

8. A logic diagram that depicts potential failures in each individual segment or component of a system is:

a. A fault-tree analysis
b. An event-tree analysis
c. A system reliability and test evaluation
d. A probabilistic risk assessment (PRA)

9. Which OSHA form is used to describe the details of a specific reportable injury?

a. 300
b. 300A
c. 301
d. 1904

10. A demolition company is working on a contract downtown. They construct a canopy over the sidewalk to protect pedestrians in the area. During an unexpected thunderstorm, the canopy supports fail, dropping debris on a vehicle and causing significant damage. Which insurance policy would be used to cover the loss?

a. Auto
b. Liability
c. Property
d. Builder's risk

11. During the remodel of a historic building, the contractor is generating lead paint waste. The contractor is properly storing and labeling the waste according to _________ requirements.

a. OSHA
b. EPA
c. DOT
d. NHSA

12. A large, multi-story parking structure project has just completed the excavation phase. The ramp for the haulers has been removed and ladders have been staged around the excavation for the workers to access the site. There are several steelworkers in the excavation installing rebar in preparation for the concrete pour in two days. They are using chop saws and cutting torches to trim and form the pieces. A summer thunderstorm passes overhead, lasting for about one hour. What additional hazard has been introduced into the worksite?

a. Drowning
b. Entrapment
c. Shock
d. Hazardous atmosphere

13. Which of the following would NOT be considered a benefit of commencing an accident investigation as soon as possible?

a. The sooner the causes are found, the sooner appropriate culpability can be assigned.
b. Future accident costs can ultimately be reduced via the prompt attainment of accident root causes.
c. A rapidly commenced investigation sends a message of robust corporate engagement.
d. Rapid investigations help ensure that the incident is still fresh in the minds of those from whom information is being gathered.

14. An ammeter measures __________.

a. voltage
b. noise
c. electrical current
d. radiation

15. Who is usually responsible for taking roll calls during a shelter in place or a building evacuation?

a. Emergency response team manager
b. Emergency coordinator
c. Office warden
d. Office manager

16. The safety professional for the prime contractor sends out short, daily training sessions on a specific topic using a text delivery method. This is an example of __________.

a. CBT
b. microlearning
c. remote learning
d. group instruction

17. In a failure mode effects analysis, the risk of a failure mode is identified with a risk priority number of 343. Which statement is most likely true?

a. A risk priority number of 343 is relatively low when using factor rating scales of 1-10.
b. The risk priority number is incorrect and should represent an even number.
c. The severity of the effect was rated as a 7. The likelihood of occurrence was rated as a 7. The detection ability of the failure mode was rated as a 7.
d. This number would not be able to be compared to other risk priority numbers to evaluate priorities for mitigation.

18. Which of the following is NOT a major hazard related to crane operation?

a. Electrical
b. Pinch points
c. Overloading
d. Falling objects

19. Wildfires are common around your workplace; therefore, this type of emergency is listed on the emergency action plan. How should the company use the emergency action plan to prepare for the next wildfire season?

a. Test the heating, ventilation, and air conditioning (HVAC) system.
b. Post signs to limit door and window usage.
c. Request unhealthy employees work from home.
d. Increase outside air intake to 100%.

20. What class of fire extinguisher should the contractor have for a remodel project of a residential house where the trim work will be stained and lacquered on-site using oil-based products?

a. A
b. B
c. C
d. D

21. A masonry subcontractor is staging mortar in preparation for installing a cinderblock retaining wall. The bags are piled too high and fall on an employee, resulting in a compound fracture of her leg. She is transported to the hospital via ambulance, is admitted, and remains in the hospital to surgically stabilize the leg. When should the subcontractor notify OSHA of the injury?

a. In March of the following year using the OSHA 300A form
b. After the employee is discharged and her return-to-work date is known
c. Within 8 hours of the incident
d. Within 24 hours of the incident

22. An air compressor emits an average noise level of 95 dBA. The preferred method of managing the noise exposure of the employees working inside a building is which of the following?

a. Earmuffs
b. Placing the compressor outside
c. Ear plugs
d. Rotating workers inside the building every two hours

23. Identify which of the following roles is NOT part of a chemical spill response team.

a. Supervisor
b. Cleanup competent person
c. Site safety officer
d. HAZMAT technician

24. The XYZ Construction Company has 127 full-time workers. Each employee works a full year, or 2,000 hours, at the company. During this time, six recordable incidents occur. What is the incident rate of the company?

a. 3.38
b. 4.72
c. 7.62
d. 2,116.7

25. A site superintendent is driving across a job site during a lunch break and sees the property owner and general contractor discussing payment for a vendor. Neither are wearing hard hats, safety glasses, or high-visibility vests. This is a failure of what behavior modification technique?

a. Praise
b. Shaping
c. Modeling
d. Self-monitoring

26. In an excavation shoring system, __________ are the horizontal members that distribute the compressive forces.

a. wales
b. uprights
c. cross-braces
d. berms

27. Who is responsible for maintaining the inspection logs for a tube and coupler scaffold?

a. A competent person
b. The safety professional
c. The superintendent
d. The waysman

28. Under which of the following circumstances is the job superintendent NOT required to brief all employees on the job site?

a. When a crane lift requires a temporary change to an assembly area
b. When utility surveyors come on site to mark gas lines
c. At the beginning of the job
d. When a new superintendent is assigned to the project

29. Which of the following is NOT considered a tangible benefit of commencing an accident investigation as soon as possible?

a. The sooner the causes are found, the sooner the lessons can be shared within the organization to prevent recurrences.
b. A rapidly commenced investigation usually sends a message of robust corporate response and engagement.
c. Liability costs can be reduced by rapid fact-finding and root cause determination.
d. Witness accounts are more reliable soon after the incident, and accurate recollection can fade as time passes.

30. Which of the following is generally true with regard to worksite safety inspections?

a. Unbiased inspections are best conducted by personnel who only have a generalist's knowledge of the subject area being inspected.
b. The most useful findings come from scheduled inspections.
c. Inspections, by design, should always be generically oriented in nature.
d. Inspectors are independent entities who do not have a direct affiliation with the organization or entity undergoing inspection.

31. A painting contractor has doubled its workforce over the past three months to accommodate the large increase in contracts secured by the new management team. The safety professional is charged with conducting hazard communication training for the growing workforce. While developing the training program, which of the following does NOT need to be considered?

a. Different familiarity levels with the Hazard Communication Standard
b. English proficiency
c. New paint products on the market
d. Different roller head types for exterior surfaces

32. An employee is working on an elevated surface that is 10 feet from the lower level. The employee is using a fall arrest system with a 6-foot tear-away shock-absorbing lanyard. Which choice best describes this fall protection system?

a. The system will not be effective.
b. The system will be effective.
c. The system is an example of fall prevention.
d. The system is an example of fall restraint.

33. Which of the following leadership personalities typically renders decisions on an autonomous basis and closely monitors employees?

a. Permissive autocrat
b. Directive democrat
c. Permissive democrat
d. Directive autocrat

34. How many attendants are required to be stationed outside of a permit-required confined space during entry?

a. One
b. Two
c. One per entrant
d. One per every two entrants

35. A new subdivision is approximately one hour from the nearest hospital. Based on that distance, OSHA requires that the general contractor have a designated individual who is certified in __________ on the job site at all times in the event of a medical emergency.

a. AED
b. CPR
c. First aid
d. FAA communications

36. The Department of Labor's website contains links to the most current versions of ______________.

a. ANSI standards
b. NFPA resources
c. OSHA standards
d. NIOSH recommendations

37. Which of the following is NOT a part of emergency response plan preparation?

a. Planning
b. Publishing
c. Practicing
d. Evaluating

38. Performing a(n) ________________ is an effective way of determining whether a specific training regimen is a means of solving a particular organizational problem.

a. training audit
b. fishbone analysis
c. event-tree evaluation
d. needs assessment

39. A work schedule calls for installing a masonry wall on the second floor of a building and the mortar is being delivered today. Which of the following is NOT a recommended action to reduce the risk of ergonomic injury?

a. Order smaller bags of mortar.
b. Implement a pre-shift stretching exercise.
c. Rent a lift to place the bags on the upper level.
d. Hire day laborers.

40. An excavation company is removing spoils for construction of a multi-story building using haulers. One of the trucks is improperly loaded and falls off the ramp. The driver is not wearing a seatbelt and is ejected from the cab, resulting in his death. How long does the employer have to contact OSHA regarding the accident?

a. After the police have conducted their investigation
b. Immediately after the accident is reported
c. Once the victim has been transported to the hospital
d. No more than eight hours after the accident

41. The parent company of a prospective general contractor is ISO 45001 certified. In their bid package for a major subdivision job, they provide their last audit report. Which of the following findings listed in the report is the least concerning?

a. Major nonconformance
b. Minor nonconformance
c. Deviation
d. Opportunity for improvement

42. The supervisor for a roofing crew requires that employees install a visual limit line to reduce the risk of falls. Using the theory of the hierarchy of controls, what type of protective measure are they relying on?

a. Elimination
b. Engineering
c. Administrative
d. Personal protective equipment

43. When an architect's rendering of a multi-story building includes a pulley system to support a suspended scaffold for window washing, he or she is implementing the concept of ________.

a. Heinrich's safety pyramid
b. safety by design
c. safety engineering
d. administrative design

44. EDTA chelation therapy is the recommended treatment for which of the following acute toxicities?

a. Beryllium
b. Nickel
c. Chromium VI
d. Lead

45. Which of the following is an appropriate response from an organization when an unsafe condition is found during a voluntary internal audit?

a. Initiate corrective action measures.
b. File a report with management.
c. Delay action until the following budget year.
d. Conduct a follow-up audit the next year to determine if the condition still exists.

46. ________ shoring systems are the most versatile, as they are custom fit into the trench.

a. Aluminum
b. Timber
c. Shield
d. Coffin-box

47. Spare forklift batteries and stored paint thinner must be ________ at a construction site.

a. labeled
b. prohibited
c. segregated
d. hidden

48. Which of the following is generally NOT true regarding worksite safety inspections?

a. Inspectors typically do not have a direct affiliation with the organization or entity undergoing inspection.
b. Inspections are usually conducted by personnel who have robust knowledge, training, and/or experience within the subject area that the inspection is concerned with.
c. Inspections can either be scheduled or unscheduled.
d. Inspections, by design, should only be detail-oriented in nature.

49. Which of the following injuries that occurred on a Friday during repaving of a roadway would need to be recorded on the OSHA 300 log?

a. Dirt in an eye that needed to be flushed out
b. Back sprain that lasted for 4 days
c. Sore wrist that healed over the weekend
d. Deep cut requiring multiple bandages, but employee returned to work Monday

50. A corporate safety professional produces a standardized site safety plan for a nationwide steel contractor. This SSP can be used at any job site if ________.

a. the projects are all similar
b. the state does not have its own health and safety administration
c. the name of the prime contractor is added
d. it is updated with site-specific information

51. A safety professional conducts an in-person training on lockout/tagout. The session finishes with a quiz that more than fifty percent of the class does not pass. The preferred resolution is to:

a. Make the test multiple choice.
b. Reduce the difficulty level of the test.
c. Reassess the training program delivery.
d. Choose another assessment method.

52. Which of the following is NOT considered an employee's personally identifiable information that must be safeguarded by an employer?

a. Email address
b. Marital status
c. Driver's license number
d. Biometric information

53. Which of the following accident investigation techniques is the least reliable?

a. Photographs
b. Sketches
c. Witness interviews
d. Written notes

54. ____________ is a human behavior theory used to explain what motivates workers to seek a safe work environment.

a. Hierarchy of controls
b. Heinrich's safety pyramid
c. Equity theory
d. Maslow's hierarchy of needs

55. The manufacturer of a rotary saw incorporates a "kill switch" into the tool. This is considered a(n) ________ control installed in the design phase.

a. protective equipment
b. administrative
c. foolproof
d. engineering

56. The foreman chooses to get an early start on an irrigation project in late November. Employees begin shoveling the trench just before sunrise. The weather is 45 degrees and foggy with a slight breeze. The foreman should continually monitor workers for ________.

a. blisters
b. hypothermia
c. proper body mechanics
d. fatigue

57. Employers with more than 10 employees are required to post an annual summary of all recordable injuries using the OSHA 300A form or state equivalent. When must this form be posted?

a. Dec 31
b. Jan 1
c. Feb 1
d. July 1

58. _______ connect muscles to bones, while ________ connect bones to bones.

a. Tendons/ligaments
b. Ligaments/tendons
c. Bursas/ligaments
d. Sheaths/tendons

59. Which of the following is NOT typically employed when identifying potential risks in the workplace?

a. Accident history evaluations
b. Quantitative risk assessments
c. Employee feedback
d. Cost-benefit analyses

60. A general contractor (GC) begins an excavation to install water and sewer lines eight feet below grade for a new subdivision. The competent person elects to use benching to prevent trench cave in. The project changes owners and is halted for a month. Once the project is restarted, what action should the GC take regarding the utility installation?

a. Select a new location for the trench.
b. Begin the pipe installation.
c. Have a competent person inspect the benching.
d. Hire a subcontractor to finish the utility installation.

61. An employer has more than 10 employees that report on-site. Which of the following documents must be in writing?

a. Emergency action plan
b. Continuity of operations plan
c. Active shooter plan
d. Bomb threat response plan

62. An employee gets epoxy in his eye and is rushed to the hospital. The supervisor grabs the SDS binder on her way to accompany the injured employee. Under Section 3, there is an entry noted as "trade secret." Who can request the exact composition of the epoxy from the manufacturer?

a. The injured employee
b. The supervisor
c. The procurement department
d. The treating physician

63. When is a written fire prevention plan required by OSHA?

a. It is required for all employers.
b. It is required for all employers with more than 5 employees.
c. It is required for all employers with more than 10 employees.
d. It is required by specific OSHA standards based on materials/processes used for employers with more than 10 employees.

64. The first phase of a subdivision is under construction. During the safety inspection, the superintendent sees that the roofers have daisy-chained extension cords through some of the units and onto the roofs. What is the primary hazard with this configuration?

a. Falls
b. Shock
c. Fire
d. Falling objects

65. A safety professional develops a noise survey that involves taking 10 measurements on the factory floor. Immediately following the eighth measurement, the noise meter battery dies. It takes four hours to recharge the battery and the factory is winding down for the day. Which of the following actions would compromise the professional's ethics?

a. Take 2 measurements the following day and generate 2 reports
b. Average the measurements to estimate the other 2 values
c. Make a note in the report and use only 8 values
d. Start the survey over again the following day

66. A safety professional is acting as the Subject Matter Expert (SME) for an external safety management system audit. During a records review, the SME notices an inconsistency in a training log that was missed by the auditor. On what principle is the SME obligated to notify the auditor?

a. Bias
b. Integrity
c. Protect health and safety
d. Professionalism

67. Which of the following individuals would ideally be the most qualified to measure a variety of radiological hazards in the workplace and resultantly provide recommendations to mitigate such hazards?

a. Certified health physicist (CHP)
b. Certified industrial hygienist (CIH)
c. Certified Safety Professional (CSP)
d. NSPE-registered nuclear engineer

68. At a pre-job briefing, the general contractor requests that all subcontractors provide the names and phone numbers of their designated first aid person. The framing contractor does not have a medically trained individual on site. The general contractor must:

a. Select a different framing contractor.
b. Get the project started immediately.
c. Post a contact number for emergency services.
d. Schedule the framing work for the weekend.

69. Who should be involved in emergency and disaster response planning?

a. Labor employees
b. Labor employees, supervisors, management, and response planning committee
c. Labor employees and management
d. Only the safety department

70. A job involves converting an abandoned warehouse downtown into a multi-use office/retail/residential facility. During the initial walk-through with the new property owner, the restroom is found to contain numerous used syringes. What should be used to remove the items?

a. Leather gloves and a garbage bag
b. Leather gloves and a bucket
c. A trash picker and a bucket
d. A trash picker and a garbage bag

71. Which of the following is considered the most effective method for documenting the scene of a safety incident at a construction site?

a. Photography
b. Video
c. Sketches
d. Witness interviews

72. A fencing company has multiple crews working every day. The owner of the company has provided the required first aid kits for every job supervisor. What is the required frequency for inspection of the first aid kits to make sure they are replenished?

a. Daily
b. Weekly
c. Monthly
d. Quarterly

73. A remodel of a car dealership involves applying epoxy to a 4,000-square-foot showroom floor. The product is delivered in multiple containers three days before it will be applied. Which of the following would NOT be a method to control any potential accidental releases?

a. Establishing a "no pedestrian" zone near the product
b. Placing the containers on spill skids
c. Constructing a temporary berm around the product staging area
d. Staging multiple spill kits near the staged product

74. Who is responsible for inspecting rigging?

a. Supervisor
b. Safety designee
c. Qualified rigger
d. Crane operator

75. What is the PEL for asbestos as an eight-hour time-weighted average?

a. 0.1 f/cc
b. 0.5 f/cc
c. 1.0 f/cc
d. 5.0 f/cc

76. Ironworkers are grinding burrs off steel beams. The activity is generating a noise level of 100 dBA. What is the maximum time that the employees can be exposed to this noise level without any ear plugs in a single day?

a. 0 hours
b. 1 hour
c. 2 hours
d. 3 hours

77. An employee texts his supervisor that he will not be able to work for the next few days and will be using sick leave. He offers to send a note from his doctor to validate the time. The supervisor texts back asking what type of illness the employee is suffering from. This is a violation of _________.

a. Employee's Right to Privacy Act
b. HIPAA
c. ADA
d. FLSA

78. _________ can be worn to reduce the risk of being struck by industrial vehicles.

a. Steel-toed shoes
b. Hard hats
c. Radio headsets
d. High-visibility clothing

79. What is the calculated risk of an accident event that has a frequency of 0.0001 per year and an associated consequence of 2.5 fatalities?

a. 2.5 fatalities per year
b. 0.0001 fatalities per year
c. 0.00025 fatalities per year
d. 0.000040 fatalities per year

80. Which of the following is used to determine a worker's exposure to lead?

a. Job activities
b. Work history
c. Blood test
d. Breathing test

81. Any crisis management plan should include a(n) __________ to manage the public relations aspect of the incident.

a. IC
b. PIO
c. PAI
d. CEO

82. A crane is hoisting a replacement HVAC unit onto the roof of a new restaurant. Once the unit is in the air, a landscaping crew moves in to replace a sprinkler head and finish spreading their last load of bark while a painting contractor is working in the parking lot nearby. The HVAC unit clips the parapet and falls to the ground, resulting in a leg amputation for one of the landscapers. Under OSHA's multi-employer worksite policy, who will likely NOT receive a violation for the incident?

a. The crane contractor
b. The landscape contractor
c. The painting company
d. The prime contractor

83. A safety professional is properly certified and has been appointed as the first aid responder at a construction site. A call comes in that an electrician has slipped while pulling wire, and the wire has gone all the way through the palm of his hand. Upon arriving at the scene, the first thing the safety professional should do is ______________.

a. Put on gloves
b. Take photographs
c. Identify witnesses
d. Apply a tourniquet

84. A welder who is complaining of a fever, difficulty breathing, muscle pain, and chills may be experiencing ___________.

a. asbestosis
b. hypoxia
c. emphysema
d. metal fume fever

85. Which of the following is a detailed and systematic process to evaluate potential risks associated with the design and operation of a multifaceted, complex technology?

a. Probabilistic risk assessment
b. Failure modes and effects analysis
c. Hazards assessment
d. Documented safety analysis

86. Which of the following activities does NOT involve a stress that can result in a work-related musculoskeletal disorder?

a. Kneeling while screeding concrete
b. Carrying a bag filled with bark
c. Using a putty knife for ceiling drywall seams
d. Bending at the waist to clear an irrigation trench

87. Within the Incident Command System (ICS), which member of the command staff exercises emergency authority to stop and prevent unsafe acts?

a. Public information officer
b. Safety officer
c. Resource officer
d. Liaison officer

88. What is an endeavor to uncover and reduce hazards that may be at hand for product or system users as well as to those who are accountable for their maintenance conventionally known as?

a. Preliminary hazard analysis
b. Occupational health hazard assessment
c. Documented safety analysis
d. Technical safety requirement evaluation

89. A general contractor has a large workforce managing several projects simultaneously across the Southwest. The CEO is working with the safety manager to develop a monthly email safety campaign for job site supervisors, since email is the company's main communication tool. The topic for the month of May is heat stress. The manager develops a bulleted list of heat stress indicators, provides one-page heat stress posters in multiple languages, and requests managers review the email and send any questions prior to delivering the training. However, in June, there is a spike of heat-related illnesses reported to the office. What is the likely reason for the failure of this initiative in communicating heat stress mitigation factors to employees?

a. Employees didn't understand the concept of heat stress.
b. Supervisors didn't read the email.
c. Supervisors don't understand the mitigation strategies.
d. Employees are not fluent in English.

90. Which of the following equipment is NOT required to be routinely inspected?

a. Ladders
b. Respirators
c. Safety glasses
d. Fall protection harnesses

91. The following is a compendium of noise-level measurements within an assessed work area over a two-week time span. From the provided data, which of the following conclusions should be drawn?

Measurement Day	Number of Hours Measured	Measurement Range	Measurement Average
1 of 10	6.75	74-81 dBA	77 dBA
2 of 10	8	67-79 dBA	71 dBA
3 of 10	3	86-88 dBA	87 dBA
4 of 10	8	71-92 dBA	79 dBA
5 of 10	7	81-85 dBA	84 dBA
6 of 10	3.5	77-80 dBA	78 dBA
7 of 10	8	58-72 dBA	67 dBA
8 of 10	6.5	66-70 dBA	69 dBA
9 of 10	2.25	89-94 dBA	93 dBA
10 of 10	7.5	75-83 dBA	78 dBA

a. The work area should be designated as a hazardous noise area as per OSHA 1910.95, with mandatory hearing protection PPE always worn.
b. The work area should be designated as a hazardous noise area as per OSHA 1910.95, with hearing protection PPE recommended.
c. The work area meets the minimum TWA threshold levels of a part-time hazardous noise area as per OSHA 1910.95; thus, hearing protection PPE is mandatory for at least 4 hrs per workday.
d. The work area does not meet the minimum TWA threshold levels of a hazardous noise area as per OSHA 1910.95, and thus, hearing protection PPE is not mandatory; however, it should be recommended.

92. __________ is the process of conveying information to a person or group.

a. Training
b. Communication
c. Email
d. Directing

93. A painting company is trying to move from residential to commercial projects. They accept a job to repaint the exterior of a grain silo. The painting contractor opts to subcontract the scaffold erection and will also be using scissor lifts. What training topic would the safety professional NOT need to address with the crew before beginning the job?

a. Confined spaces
b. Scaffolds
c. Aerial lifts
d. Hazard communication

94. Guardrail systems shall be designed to withstand application of _____ pounds by not deflecting to less than 39 inches from the working level.

a. 50
b. 100
c. 200
d. 250

95. Which of the following is NOT a health effect resulting from overexposure to radiation from an industrial radiograph?

a. Nausea
b. Vomiting
c. Hair loss
d. Tooth loss

96. The benefits of an equipment inspection record include all the following EXCEPT:

a. Standardizing the items evaluated
b. Meeting OSHA standards
c. Providing a permanent record of inspection
d. Outlining the approved condition for equipment

97. A production line has had an increase in defective parts over the past month, and managers of the line are meeting to analyze the failure. They will want to analyze the effect (defective parts) and formulate possible sources of variation. Which technique should be used to examine cause and effect in this scenario?

a. Fault tree analysis
b. Failure mode and effect analysis
c. Fishbone analysis
d. Root cause analysis

98. A radial saw is on-site to cut numerous pieces of trim. The subcontractor is asked to cut a larger post, but it does not fit on the saw because of the guard. To make the cut, the worker should:

a. Remove the guard to accommodate the large wood.
b. Secure pieces of plywood to extend the guard to hold the wood
c. Use wire to hold the guard out of the way to safely make the cut
d. Find another saw

99. Identify the most effective method of preventing vibration injuries from hand tools.

a. Train workers on vibration hazards.
b. Implement a policy work rotation.
c. Create a tool maintenance program.
d. Require guards remain on tools at all times.

100. Which section of the Safety Data Sheet lists the recommended engineering controls and personal protective equipment required to reduce the risk of exposure?

a. Section 8
b. Section 9
c. Section 11
d. Section 14

101. What atmospheric hazard must be tested FIRST to determine whether it is safe to enter a permit-required confined space?

a. Flammability
b. Oxygen level
c. Carbon dioxide
d. IDLH condition

102. Properly designated and identified __________ can reduce the risk of struck-by injuries at a construction site.

a. material laydown areas
b. evacuation areas
c. utility trenches
d. walking paths

103. Which of the following typically does NOT require use of a fall harness?

a. Manlift
b. Boom lift
c. Suspension scaffold
d. Scissor lift

104. Per the GHS, the lowest level of hazard is identified by the signal word _________.

a. Notice
b. Caution
c. Danger
d. Warning

105. The Board of Certified Safety Professionals' Code of Ethics expects that any member who achieves certification shall "...act with responsibility and integrity..." In this context, integrity can be defined as _________.

a. always telling the truth no matter who the customer is
b. being honest and doing the right thing regardless of the circumstances
c. not showing preference to one group over another
d. refraining from misrepresenting one's abilities

106. At a construction site, which of the following is NOT required to be stored in a closed, fire-resistant container until removed from the job site?

a. Used rags soaked with oil-based stains
b. Excess latex paint
c. Unused paint thinner
d. Paint thinner used to clean brushes

107. According to OSHA, which of the following is NOT a root cause analysis technique when investigating an accident?

a. Risk assessment
b. Checklist
c. Causal factor determination
d. Sequence diagrams

108. OSHA 300A logs must be kept by the employer for ______.

a. three months
b. one year
c. three years
d. five years

109. The process of installing a breaker lock on a service panel to prevent electricity from flowing through a circuit during maintenance is called _______.

a. blanking
b. blocking
c. lockout/tagout (LOTO)
d. hazard elimination

110. A training firm is contracted to perform hazard communication training. The facilitator begins the presentation with a slideshow and videos. The next segment starts with the instructor asking participants to share situations where improperly labeled containers resulted in incidents or near misses. The second segment is using which aspect of adult learning?

a. Self-direction
b. Internal motivation
c. Association
d. Immediate application

111. The finishing carpenters are complaining of back pain while installing trim in an apartment building. The supervisor observes that they are placing the miter saw on the floor to avoid damaging the tile on the countertops. Which of the following would be the best solution to reduce the ergonomic risk to their backs?

a. Rotate the crews every two hours to a different building.
b. Distribute back belts.
c. Provide a portable worktable for the saw.
d. Assign more carpenters to the job.

112. Which of the following should NOT be used to reduce elevated body temperature?

a. Shade
b. Cooling garments
c. Energy drinks
d. Rest break

113. While conducting an inspection of fire extinguishers at a job site, the safety professional discovers the inspection tag was not initialed for the previous month. She thinks she completed all inspections, and simply forgot to sign this tag. What action should the safety professional take?

a. Initial both this month and the previous month.
b. Start a new tag.
c. Leave the previous month blank and initial this month.
d. Make a note in the job file.

114. ___________ are protective garments used to shield the body from slag.

a. Level A suits
b. Welding leathers
c. Arc flash gloves
d. Tyvek suits

115. A landscaping crew arrives to work in a new subdivision that is opening the following week. Upon arrival, the supervisor sees that there is scaffold present on several units where a roofing contractor is replacing rain gutters. What additional PPE item should the crew chief mention during the tailgate meeting?

a. Ear plugs
b. Filtering facepieces
c. Hard hats
d. Cut gloves

116. An electrician's apprentice punctures his hand with a wire, and the injury is bleeding. While rendering aid, another electrician punctures her hand with the same wire. Both employees are treated at the local clinic and released, each missing two days of work due to their injuries. The safety professional correctly determines that these are recordable occupational injuries. When completing the OSHA 300 log, the safety professional is required to enter all the following EXCEPT:

a. The apprentice's name
b. The electrician's name
c. The employees' job titles
d. The days both spent away from work

117. Which choice best represents the next step for implementing training after assessing needs?

a. Acquire materials
b. Design
c. Pre-test
d. Deliver training

118. Under the behavior-based safety model, all the following are considered a potential root cause for an accident EXCEPT:

a. Tools
b. People
c. Work environment
d. Behaviors

119. Identify the most effective means of controlling exposure to silica dust during concrete cutting operations.

a. Rotating workers operating the saw
b. Using a wet saw
c. Requiring operators to wear respirators
d. Moving support workers upwind of the dust plume

120. Which of the following is the best control to implement when using a radiograph to locate conduit within a foundation?

a. Wearing a dosimeter
b. Limiting the amount of time spent using the device
c. Purchasing a properly shielded device
d. Replacing with ultrasonic testing

121. A safety professional is scheduled to take the CHST certification exam in three months. In the interim, he applies for a job and states that he is a CHST. What is this a violation of?

a. OSHA regulations
b. BCSP bylaws
c. Fair labor standards
d. BCSP Code of Ethics

122. __________ is a strategy that reduces the risk of shock and electrocution when different subcontractors are working on the same electrical system.

a. Labeling circuits
b. Lockout/tagout
c. Arc flash safety
d. Hierarchy of controls

123. A safety professional decides to evaluate a job site for adherence to the electrical safety training program that was completed last week. The professional can either conduct a(n) ________ assessment or a ________ audit.

a. informal/formal
b. program/training
c. in-person/remote
d. employee/management

124. The graphical representation on a GHS-compliant label is called a ________.

a. placard
b. pictograph
c. pictogram
d. hazard statement

125. All the following are classes of GHS precautionary statements EXCEPT:

a. Incompatibility
b. Prevention
c. Storage
d. Disposal

126. Which of the following is the most effective way to manage the risk to a construction company's employees during installation of trusses in a housing development?

a. Provide employees with fall protection harnesses to use while constructing the trusses
b. Rent boom lifts for the employees to work out of while constructing the trusses
c. Hire a subcontractor to install the trusses
d. Purchase prefabricated trusses

127. The job superintendent hangs a poster on her trailer with a sketch of the job site and two areas marked "Assembly Point." This is required by what program?

a. Emergency action plan
b. Medical emergency plan
c. Spill response plan
d. Hazardous materials business plan

128. A remodel job involves removing lead paint from iron work using a grinder. The employer must provide all the following EXCEPT:

a. Leather gloves
b. Safety glasses
c. Hand-washing facilities
d. APR

129. Which of the following injuries is NOT required to be reported to OSHA by telephone or using the online portal?

a. A worker using a hammer without safety glasses to remove tile loses use of an eye.
b. A framer using a table saw removes the guard and subsequently cuts off a finger.
c. A stucco contractor pierces her finger with chicken wire, resulting in stitches.
d. A roofer falls off a ladder, suffering an aneurysm, and is in the hospital for two days.

130. Which of the following activities is prohibited while removing asbestos-containing materials?

a. Spraying the material with water
b. Shoveling up dried debris
c. Using compressed air inside of containment
d. Using disc cutters with local ventilation

131. What is the preferred method of addressing harmful dusts, fumes, vapors, and mists to prevent worker exposure?

a. Air-purifying respirator
b. Self-contained breathing apparatus
c. Work rotation
d. Local exhaust ventilation

132. What end-product tool is conventionally derived from a failure modes and effects analysis?

a. An action list
b. A reliability plan
c. A fault tree
d. A frequency/severity concept summary

133. You are reviewing a supervisor's accident investigation form. You notice the only corrective action listed is "retraining." For this to be true, what must also be true?

a. Causes must be exclusively related to lack of skills or knowledge
b. Cause must be complex
c. Root causes must be related to physical deficiencies
d. Root causes must be related to management deficiencies

134. Construction standards are in which group of standards in Title 29 of the Code of Federal Regulations?

a. 1910
b. 1915
c. 1926
d. 1928

135. Asbestos fibers are identified in construction materials by ________.

a. visual inspection by a competent person
b. reviewing material markings and labels
c. sample collection and laboratory analysis
d. reactivity to specialized reagents

136. When hazards that can be addressed with personal protective equipment (PPE) are identified at a job site, who is responsible for providing the PPE?

a. The employer
b. The employee
c. The general contractor
d. The trade union representative

137. What is typically regarded as a very effective method or model for conducting a root-cause analysis?

a. The single variance method
b. The prevention-occurrence-prevention model
c. The five whys method
d. The triple event tree model

138. During construction of a multi-story industrial building, the general contractor directs the roofing contractor to stage their materials on the second floor. In order to comply with OSHA regulations, the general contractor must post which of the following?

a. Maximum number of pallets allowed
b. The exact storage location
c. Type of materials allowed to be stored
d. Maximum allowable weight

139. A ________ is the fundamental, system-related factor that resulted in an accident.

a. foundational reason
b. contributing factor
c. circumstance
d. root cause

140. Unlike an emergency action plan, an emergency response plan for hazardous chemical spills must include ________.

a. a written program
b. training requirements
c. designated roles
d. a medical surveillance program

141. Which of the following is NOT a mitigation strategy to reduce the risk of vibration injuries to the arms and hands while using a jackhammer over a period of two days?

a. Wearing padded gloves
b. Switching operators every two hours
c. Changing bits every four hours
d. Using a backhoe

142. What part of the body does a body harness typically distribute a significant amount of fall-arrest force to?

a. Bottoms of the feet
b. Thighs
c. Lower back
d. Forearms

143. A hazardous materials contractor arrives on-site to remove crates of household chemicals that have been abandoned at a construction site. The general contractor contacted the property owner, the fire department, and the carpenters' union. Which of the following parties does NOT have the right to ask for the cleanup contractor's health and safety plan?

a. Property owner
b. Carpenters' union representatives
c. The fire captain on-site doing an inspection
d. An electrician on-site

144. Flammable gas meters detect the presence of flammable vapors and display the result as a ___________.

a. flammable level
b. percent of lower explosive limit
c. flash point
d. permissible exposure limit

145. An emergency action plan outlines ___________ strategies for responding to emergencies, while a site-specific safety plan describes ___________ actions.

a. general/specific
b. management/employee
c. contractor/subcontractor
d. first responder/contractor

146. Which of the following is true regarding lanyards used within a fall-protection system?

a. Lanyards connect lifelines to pitons.
b. Lanyards connect anchoring points to safety harnesses.
c. Lanyards connect safety belts to fall arrestors.
d. Lanyards connect lifelines to safety belts.

147. Large construction projects have multiple contractors working at any given time, often working on different levels. This results in ________, a leading cause of falling object injuries in the construction industry.

a. overhead hazards
b. improperly stored materials
c. unstable structures
d. impeded walking surfaces

148. Which of the following gas detector cells is NOT typically included in a four-gas multi-meter?

a. Carbon dioxide
b. Carbon monoxide
c. Oxygen
d. Flammable

149. The safety professional for a paint company is reviewing injury reports and notices several reports of dermatitis from a project where employees were applying stain by hand. The root cause evaluation determines that latex gloves do not protect the skin from oil-based stains. Where should the safety professional seek information about a better choice of chemical protective gloves?

a. Safety catalog
b. OSHA standards
c. Manufacturer website
d. NIOSH publications

150. Non-ionizing laser systems, such as those in surveying equipment, shall not be used in snow, rain, or fog because of the potential for ________.

a. dispersion
b. uneven beams
c. overheating
d. lack of accuracy

151. A _______ is a document used to verify safety provisions are in place and being followed prior to any spark-producing activity.

a. fire prevention plan
b. welding permit
c. job hazard analysis
d. hot work permit

152. A crew is demolishing an outdoor patio, and one worker is wearing sunglasses instead of company-issued clear safety glasses. What marking should the supervisor look for on the sunglasses to determine if they are designed to protect eyes from flying debris?

a. Z87
b. Impact-resistant
c. 1926.28
d. Safety glasses

153. ______ can be quantified as the product of frequency, severity, and probability, and is used to allocate resources to reduce potential for injuries at a job site.

a. Risk
b. Hazard
c. Loss
d. Danger

154. The need for PPE and the type of PPE required at a job site are determined by which of the following?

a. Prime contractor's requirements
b. Job hazard analysis
c. Competent person's determination
d. Signed contract

155. A contractor is hired to lay a concrete pad for an oil derrick in a remote location. Most employees have their personal cell phones in their vehicles. To comply with the OSHA requirements for medical response, what information must the superintendent post for first responders?

a. Latitude and longitude of the job site
b. The name of the nearest town
c. The station number of the nearest fire station
d. The phone number of the company that hired the contractor

156. An HVAC contractor is scheduled to install new chillers on top of a three-story building the following week using a crane. To proactively identify potential hazards, all of the following would be important considerations EXCEPT:

a. Other contractor activities
b. Employee schedules
c. Weather forecast
d. Occupancy status of the building

157. A demolition subcontractor has set up containment around pipes that have been insulated with asbestos. The subcontractor's staff are wearing respirators and protective suits and are packaging the removed insulation in properly marked bags. Which piece of legislation is the contractor following?

a. Clean Air Act
b. Asbestos Hazard Emergency Response Act
c. Occupational Safety and Health Act
d. Resource Conservation and Recovery Act

158. A simple model of communication involves sender, message, and __________.

a. interpretation
b. evaluation
c. receptor
d. receiver

159. Which of the following training safety methods ensures that participants receive the same information?

a. Computer-based
b. Instructor-led
c. On-the-job
d. Policy review

160. A(n) _______ supplements the contractor's emergency action plan by describing the job-specific hazards at a particular site.

a. emergency response plan
b. spill prevention control and containment plan
c. site-safety plan
d. hazardous materials business plan

161. ___________ are the rules of conduct established by the leadership of an organization that reflect the values of the company.

a. Ethics
b. Morals
c. Values
d. Mission and vision

162. Which of the following activities requires use of a respirator outfitted with a HEPA filter?

a. Applying oil-based paint
b. Cutting a foundation indoors with a walk-behind wet saw
c. Steel plate welding
d. Dry sweeping wood dust

163. An AED is used to __________ during a medical emergency.

a. provide oxygen
b. treat anaphylactic shock
c. stop major bleeding
d. correct irregular heartbeats

164. Waste bins for which material listed below are required to have lids?

a. Metal scrap
b. Lumber with protruding nails
c. Garbage
d. Drywall scrap

165. A worker at a construction site is hit by a truck delivering trusses and is hospitalized for two days. The safety professional identifies the root cause as the lack of a policy requiring high-visibility clothing at the site. She writes the policy, trains all workers on-site, and begins site inspections to determine if workers are following the new procedure. What continuous improvement theory is she using to change the behavior of the workers?

a. PDCA
b. CAPA
c. Heinrich's safety pyramid
d. Hierarchy of controls

166. A full-face APR provides a higher assigned protection factor than a quarter- or half-face configuration. What is the main contributor to the higher level of respiratory protection?

a. The visor
b. Tighter straps
c. More straps
d. Facial variability

167. A company recorded 29 OSHA-recordable injuries. The company uses an index rate that does NOT match OSHA's index rate. The company uses a rate of 100 employees at 50 hours per week for 50 weeks of the year. What is the recordable injury rate for 1,454,000 hours?

a. 4.99
b. 3.99
c. 2.99
d. 1.99

168. A concrete company has been hired to pour the foundations for a housing development. It is the largest job the company has secured to date. The safety professional is concerned that the size and duration of the job will expose employees to hazardous levels of noise for several days in succession. To determine whether a hearing protection program needs to be implemented, the safety professional should ________.

a. review 29 CFR 1926.52
b. talk with the foreman
c. write a hearing conservation program
d. hire an industrial hygienist

169. Elevated body temperature is referred to as __________.

a. hyperthermia
b. hypothermia
c. hypoxia
d. hyperkalemia

170. The upper limit of allowable oxygen within a confined space is ________ %.

a. 19.5
b. 20.9
c. 23.5
d. 50

171. A __________ is a written program intended to protect people and the environment in the event of a fire, explosion, or unplanned release of hazardous waste.

a. Contingency Plan
b. Fire Prevention Plan
c. Emergency Action Plan
d. Emergency Response Plan

172. A masonry contractor has implemented a pre-work stretching program to reduce the number of shoulder injuries to the crew. How can the contractor evaluate whether the program has been effective?

a. Ask the workers if they feel better.
b. Review their workers' compensation data.
c. Observe the employees working.
d. Conduct an audit of the job site.

173. The prime contractor on a hotel construction project notices an increase in eye injuries from the drywall contractor. The superintendent asks the safety team to evaluate whether mandatory personal protective devices, particularly safety glasses, are being worn by the approximately 100 individuals finishing the lower floors. _______ is the most effective method of evaluating compliance with the prime contractor's PPE.

a. Sending out a survey to all subcontractors regarding PPE usage
b. Conducting on-site observations
c. Asking all subcontractors for their PPE invoices
d. Polling subcontractor employees as they come to work in the morning

174. A general contractor's safety professional is conducting a tailgate training on emergency procedures for the job site. During the session, she comments that everyone on-site should be aware of the local building material company's delivery drivers, as they always drive too fast and are "an accident waiting to happen." In this situation, the safety professional is conveying a(n) ___________.

a. Data-based fact
b. Stereotype
c. Bias
d. Observation

175. The prime contractor's safety professional is conducting an inspection for a shopping mall renovation project. During the inspection, an employee of the framing contractor is seen cutting metal studs near a stack of cardboard boxes containing fasteners. Who should the safety professional contact to control the risk of fire?

a. The prime contractor
b. The framing contractor's safety representative
c. The employee cutting the metal
d. The site superintendent

176. A process uses a highly volatile substance that can be toxic at high concentrations. Which of the following outlines the best approach to reduce risk?

a. Replace the chemical with a non-toxic substance.
b. Isolate the chemical with an enclosure.
c. Use job rotation to limit time of exposure.
d. Provide exposed employees with respiratory protection.

177. During a training session on fall protection harnesses for roofers, the safety professional notices that some attendees are on their phones, one has fallen asleep, and two are talking in the back of the room and not facing the instructor. Which of the following reasons is most likely NOT contributing to the lack of attentiveness?

a. It is too early in the morning.
b. The attendees are not proficient in English.
c. The material is too rudimentary.
d. The material is not applicable to their job.

178. An electrician on a 3-foot ladder running wire overhead in a ceiling all day is at risk for which of the following ergonomic stressors?

a. Forceful motion
b. Vibration
c. Excessive mass
d. Awkward posture

179. All the following are benefits of instructor-facilitated safety training EXCEPT:

a. Content can be adjusted to the experience level of the trainees.
b. In-person training incorporates all modes of adult learning.
c. Attendees can share their experiences with each other.
d. Sessions are usually convenient for all attendees.

180. During the framing of a house, a small fire ignites and is quickly extinguished. The safety professional investigates and finds that the fire appears to have started where three extension cords are plugged into a feeder extension cord. The extension cords are being used for a table saw, a miter saw, and a circular saw. The feeder cord is connected to the temporary power source located outside of the house. The overheated connection set a pile of sawdust on fire. Which of the following is the root cause of the fire?

a. Too many extension cords were plugged into a single feeder.
b. Not enough power supplies were on-site.
c. The feeder cord was too short.
d. There was no extension cord inspection program.

181. A general contractor's safety professional finds a "daisy chain" of extension cords being used by a finishing carpenter and their crew. The safety professional wants to consult the appropriate reference for proper extension cord use. Which organization's website should be searched?

a. OSHA
b. NIOSH
c. ANSI
d. NFPA

182. A tower crane collapses during construction of a multi-story building, resulting in multiple fatalities. After the initial incident has been stabilized, the general contractor should immediately deploy the provisions of their __________ to minimize the impact of the incident on the employees, the public, and the company.

a. fire prevention plan
b. emergency action plan
c. medical response plan
d. crisis management plan

183. A new road construction project has multiple subcontractors all reporting to the prime contractor. At this job site, who has the ultimate responsibility for the health and safety for this project?

a. The certified construction health and safety technician
b. The prime contractor
c. The pipefitters union representative
d. The subcontractors share responsibility

184. The most effective method to reduce the risk of fire at a job site is __________.

a. having fire extinguishers on-site
b. establishing fire watches
c. posting the contact number for the fire department
d. conducting routine inspections

185. What is the best air-purifying cartridge for oil-based spray paint?

a. Organic vapor
b. Organic vapor/HEPA combination
c. HEPA
d. Filtering facepiece

186. The primary route of exposure to asbestos fibers during demolition activities is ______.

a. skin
b. eyes
c. respiratory system
d. hands

187. A safety professional is conducting a job site walk when she notices a group of framers, some of whom are not wearing their safety glasses. The safety professional approaches those wearing safety glasses and thanks them for adhering to the site requirements for PPE. What behavior modification technique is she employing?

a. Modeling
b. Praise and approval
c. Shaping
d. Self-monitoring

188. In preparing a bid for a shopping center build, a contractor should include ________ insurance, which protects losses caused by fire, water, or other types of damage before the final building is turned over to the owner.

a. Workers' compensation
b. Liability
c. Property
d. Builder's risk

189. A systems safety team is selecting a process hazard analysis technique to use during process design. The team describes the process using the following phrases: fairly complex; characterized by multiple sources of energy that are a threat to potential targets; tolerance of system parameters is critical for safety (e.g., temperature, pressure, flow); human interaction has important implications for safety (e.g., protection from hazard sources, monitoring of parameters); the system's operations and functions are not intended to change over time. The team cannot easily identify a list of components nor failure rates for components, whether components are in series or in parallel, or likely ways components will fail. The goal of the project is to protect people from hazards and prevent process deviations and their consequences. Based on the team's description, which is the best choice to complete a full process hazard analysis during the design phase?

a. Subsystem hazard analysis and what-if analysis
b. Fault tree analysis and failure mode effects analysis
c. Energy trace barrier analysis and hazard operability study
d. Job hazard analysis and preliminary hazard analysis

190. Chemical labels compliant with the Globally Harmonized System (GHS) are required to include __________.

a. instructions for proper use
b. a precautionary statement
c. a hazardous chemical concentration
d. markings indicating container volume

191. Which of the following activities would be appropriate for use of a cartridge-style nail gun?

a. Joining two pieces of 2x4
b. Attaching plastic sheeting to stud lumber
c. Securing a footing to a concrete slab
d. Attaching a mounting bracket to a tiled wall

192. A steel contractor is preparing to cut a metal beam as part of a remodel project. There are cardboard boxes containing flexible conduit and wooden pallets nearby. What is the minimum distance those materials must be moved away from the beam while it is being cut?

a. 15 ft
b. 25 ft
c. 35 ft
d. 50 ft

193. A disaster recovery plan should designate a(n) ________________ who ultimately decides what recovery actions need to be taken and assigns responsibilities for achieving such actions.

a. recovery coordinator
b. emergency response team task leader
c. disaster preparedness liaison
d. contingency manager

194. As utilities are installed in a new building, an increase in __________ is critical to maintaining the safety of all subcontractors working on the project.

a. Communication
b. Inspections
c. Safeguarding
d. PPE

195. What is the recommended maximum weight a person should lift unassisted to avoid injury?

a. 61 pounds
b. 51 pounds
c. 41 pounds
d. 31 pounds

196. What employee information is an employer obligated to treat as confidential?

a. Medical information, including disabilities or medical conditions
b. Personal identifiable information
c. Financial information, such as salary or withholdings
d. All of the above

197. Which of the following aspects of an accident is NOT captured by digital photography?

a. The location of items relative to one another
b. The point of view of the victim
c. Overall weather conditions
d. The physical condition of equipment

198. Assuming that all biological fluids are contaminated with a human pathogen and using proper exposure controls is referred to as __________.

a. best practices
b. bloodborne pathogen protection
c. engineering controls
d. universal precautions

199. U.S. citizen employees who live and work in the United States have a legal right to _________ certain records under their name.

a. inspect
b. replace
c. purge
d. transfer

200. How frequently should rigging equipment such as ropes, wires, slings, and hooks be inspected?

a. Before every shift
b. Weekly
c. Monthly
d. Quarterly

Answer Key and Explanations for Test #3

1. C: Industrial radiographs use ionizing radiation in the form of x-rays or gamma rays to detect defects below the surface. Ionizing radiation exposure can result in acute and chronic health effects and are separated into two categories: deterministic and stochastic. Thus, the operator should wear a dosimeter, which measures the amount of radiation the operator is exposed to over time.

2. A: Whenever a safety professional comes across a piece of equipment they are not familiar with, they should consult relevant professionals for guidance. These professionals can include competent persons, certified persons, manufacturers, and equipment installers.

3. A: Per OSHA 1926.502, no materials related to overhead brickwork other than mortar and masonry can be located within 4 feet of the unprotected edge.

4. C: NIMS is an acronym for the National Incident Management System, which outlines the roles and responsibilities for multi-agency emergency responses, such as for a large-scale chemical spill. The system standardizes emergency response roles and responsibilities, relying on the incident command structure, to increase harmonization between responding agencies.

5. A: The safety professional needs to understand how the work is done to identify and proactively address potential at-risk behaviors. Electricians installing fixtures in finished structures have a low risk of heat stress and will likely not require power tools to install light fixtures and ceiling fans. Therefore, the primary hazard is improper use of ladders, including not following the "three points of contact" rule or not using ladders of the proper height, both of which expose workers to risk of falling.

6. A: Arc welding produces intense ultraviolet (UV) radiation. This radiation can result in a corneal burn, similar to a sunburn, referred to as "welder's flash" or "arc burn." Although most cases heal in a couple of days, some may result in an infection.

7. D: A dosimeter is a wearable device that is used to detect the amount of ionizing radiation a worker is exposed to over a period of time.

8. A: A fault-tree analysis is a logic diagram that depicts potential failures in each individual segment or component of a system. Fault-tree analysis can be used to proactively identify potential safety issues in a system or as a root cause analysis tool after an incident has occurred. Event trees, system reliability evaluations, and probabilistic risk assessments (PRAs), as well as failure modes and effects analyses (FMEAs), are also useful tools that are regularly employed in the realm of design-safety engineering. However, these tools do not match the description in the question.

9. A: OSHA provides three separate forms for recording injuries that meet the recordkeeping requirements of 29 CFR 1904.7. Recordable injuries are summarized annually on Form 300A, they are logged on Form 301, and the details are recorded on Form 300.

10. B: Liability insurance is a means to protect a business from damage or injuries that occur due to their acts (what they do) or omissions (what they do not do). In this case, the damage is caused by the canopy, and the company that installed the unit is responsible for paying all related repair costs through their liability insurance.

11. B: The Environmental Protection Agency (EPA) determines the requirements for the management of hazardous waste, including identification, labeling, and disposal of the materials.

12. C: The safety professional must continue to assess a worksite for any changing conditions that could introduce hazards not previously controlled. In this scenario, the rain being trapped in an excavation containing power cords introduces a shock hazard that was not present when the location was dry.

13. A: The purpose of an accident investigation is to prevent future accidents and return the company to confident functioning. The sooner an accident's root cause(s) is/are discovered, the sooner these lessons can be shared with the subject organization for preventing recurrences. A prompt response to an accident also sends a message of robust corporate response and engagement, and witness accounts are usually more reliable earlier as opposed to later.

14. C: An ammeter measures the intensity of current passing through an electrical circuit.

15. C: The office warden is usually responsible for taking roll calls during a shelter in place or a building evacuation. These roll calls help to ensure that all employees are accounted for during an emergency situation.

16. B: Microlearning is a training method that breaks a topic into smaller lessons that are distributed to the workforce via text or email. This differs from computer-based learning in that the content is delivered to the learner instead of the learner logging on to a training site using a computer or mobile device.

17. C: Risk priority numbers in failure mode effects analysis are calculated by taking the product of the severity rating, the likelihood rating, and the detection rating: $7 \times 7 \times 7 = 343$.

18. B: Electrical hazards, overloading, and falling objects are the three major hazards associated with overhead crane lifts. According to OSHA, over 50% of all crane accidents are related to the boom contacting overhead power lines. In addition, over 80% of all crane failures result from overloading. Finally, the highest risk to those in the vicinity of overhead operations is falling objects or dropped loads.

19. A: To prepare for this emergency, the facility should test the HVAC system to ensure it is operating correctly, filters are changed, and it is set to parameters conducive to receive increased concentrations of pollutants.

20. B: Fire extinguishers must be selected based on the type of fuel present. Since the lacquer and stain are both oil-based, an extinguisher designed for flammable liquids is required. Therefore, a Class B extinguisher would be the appropriate type. Another option would be a combination extinguisher that includes Class B fuels (e.g., A/B/C).

21. D: Employers are required to notify OSHA within 24 hours of any serious injury that results in hospitalization, amputation, or loss of an eye.

22. B: When managing an employee's exposure to a hazard, the preferred method of mitigation is eliminating the hazard. Personal protective equipment (such as earmuffs and ear plugs) reduces, but does not remove, the hazard. Rotating workers is an administrative control that only works if the employees adhere to the schedule. Moving the generator outside eliminates the hazard and places a barrier (engineering control) between the workers and the hazard.

23. B: Under the Hazardous Waste and Emergency Operations (HAZWOPER) construction standard (29 CFR 1926.65), there is no required role designated as "competent person."

24. B: Incident rate (in the below equation, IR) can be calculated using the following formula:

$$IR = \frac{\text{Number of OSHA recordable incidents} \times 200{,}000}{\text{Number of employee hours worked}}$$

We can calculate the incident rate for the company and time period in question using the numbers from the question, as follows:

$$IR = \frac{6 \times 200{,}000}{127 \times 2000} = \frac{1{,}200{,}000}{254{,}000} = 4.72$$

25. C: When a manager follows the rules and requirements of a job site, he or she is attempting to change the behavior of subordinates by modeling the expected behavior of every individual, regardless of title. Therefore, managers and safety professionals must consider the optics of not always following the health and safety requirements.

26. A: A wale is a horizontal member that distributes the force of the cross-braces to the uprights to hold back the soil in an excavation.

27. A: Under 29 CFR 1926.451, a scaffold must be inspected prior to each shift by a competent person, who should update the log after each inspection. The inspection is typically recorded on a tag attached to the scaffold.

28. B: All employees at a job site must be informed when the plan is initially created at the beginning of the job, whenever there is a change to a plan element such as moving an assembly location, or whenever an employee's responsibility under the plan changes, such as a new superintendent coming on-site.

29. C: Benefits of commencing an accident investigation as soon as possible include: the sooner an accident's root cause is discovered, the sooner these lessons can be shared within the organization for preventing recurrences; a rapidly commenced investigation usually sends a message of robust corporate response and engagement; and witness accounts are usually more reliable earlier as opposed to later. Any costs associated with litigation will not be reduced due to the pace of fact-finding.

30. D: Effective worksite safety inspections are usually conducted by inspectors who are independent entities that do not have a direct affiliation with the organization or entity being inspected and are likewise typically conducted by personnel who have expert-level knowledge, training, and/or experience within the subject area(s) being inspected. Furthermore, inspections can either be unscheduled or scheduled, but unscheduled inspections typically provide better insights than scheduled ones.

31. D: When developing a training program, the training professional needs to consider multiple facets to ensure the information is conveyed effectively. The education and experience level of the workforce, the primary language, and the hazards they will encounter should all be addressed in the program.

32. A: The system will not be effective because the fall distance using a 6-foot lanyard will be greater than 10 feet.

33. D: A directive-autocrat leadership personality does not typically provide employees the opportunity to contribute to decision-making processes and usually affords subordinates a minimal degree of latitude in carrying out their tasks (i.e., the micromanager). Other types of leadership personalities include permissive-autocrat (semi-liberal style), directive-democrat (semi-liberal style), and permissive-democrat (liberal, flexible, and open style).

34. A: Per 29 CFR 1926.1204, one attendant must be stationed outside the permit-required confined space throughout the duration of the entry operations. One attendant may be assigned to more than one space if he or she effectively monitors multiple spaces.

35. C: OSHA standard 29 CFR 1926.50 requires that a hospital or clinic be "reasonably accessible." In the event that a hospital or clinic is not "reasonably accessible," the employer must have an individual on-site who is certified in first aid. Certification must be from the US Bureau of Mines, Red Cross, or an equivalent, and must be documented.

36. C: The Occupational Safety and Health Act directed the Department of Labor to protect workers and ensure a safe workplace for all US employees. The responsibility for establishing the agency to develop and enforce safety regulations was given to the Department of Labor. Therefore, current OSHA standards are available as web links under the Department of Labor's website.

37. B: Effective emergency response plans are prepared by way of planning, practicing, and evaluating. Planning establishes roles and responsibilities in the event of an emergency while practicing allows for practical troubleshooting. Plans should be constantly evaluated for areas of improvement to increase efficiency and reduce losses during an emergency.

38. D: Performing a needs assessment is an effective way for determining whether a specific training regimen is a means of solving a particular organizational issue. A training needs assessment compares the current state of the safety program and the desired state to determine whether training will close any identified gaps.

39. D: Ergonomic injuries can impact muscles, tendons, and ligaments. Lifting more than the body can handle can result in muscle strains and ligament sprains. Reducing the mass of the load, preparing the body for the task, or eliminating the need to manually move the material can all reduce the risk of injury to the worker.

40. D: Under OSHA's reporting and recordkeeping rules, the employer must notify OSHA within eight hours of any fatality that occurs at the worksite.

41. D: An audit is when a company's safety management system is compared to standards set forth by a certifying body. A company that is ISO 45001 certified has demonstrated that they have an effective occupational safety program that has been vetted by a third party. During evaluation of their programs, findings are typically categorized as "major nonconformance," "minor nonconformance," or "opportunity for improvement." Items that have been identified as "opportunities" are not considered major and have been identified by the auditors as areas in which the company can improve, but do not have an immediate impact on employee safety.

42. C: Establishing a visual indicator as a method of hazard control is an example of an administrative control. It is a policy or procedure intended to reduce exposure to the risk but does not eliminate or otherwise control the risk itself.

43. B: The concept of "safety by design" is based on incorporating the hierarchy of controls during the design phase of a project. In this case, inclusion of pulleys in the initial construction will eliminate the need for locating appropriate support members and eliminate the cost of installing them after construction is complete.

44. D: Ethylenediaminetetraacetic acid (EDTA) chelation therapy is used to treat acute lead toxicity in the blood. Prolonged, repeated exposures to large amounts of lead-containing dust can result in seizures, coma, and even death.

45. A: OSHA encourages the use of internal audits as a proactive means of identifying unsafe conditions. If an unsafe condition is detected during an audit, the employer is required to take action to correct the hazard. Hazards that are identified during internal audits, that are not corrected, and that are subsequently identified during an OSHA inspection will be cited as "willful" since the employer had prior knowledge and did not act.

46. B: Timber shoring systems are custom built for each job and are, therefore, the most versatile. These types of systems are best suited for areas where work will occur for an extended period.

47. C: Lead-acid batteries contain sulfuric acid, which is an oxidizing material. Oxidizing acids are incompatible with solvents due to increased fire risk and must be segregated from one another.

48. D: Worksite safety inspections are usually conducted by inspectors who do not have a direct affiliation with the organization or entity being inspected and by personnel who have robust knowledge, training, and/or experience within the subject area that the inspection is concerned with. Moreover, inspections can be either scheduled or unscheduled and can be either detail-oriented or general in nature.

49. B: OSHA requires that certain injuries be recorded on an OSHA 300 log and reported annually to employees. "Recordable" injuries are those that result in fatalities, loss of consciousness, lost workdays, restricted or modified work, transfer to another job, hospitalization, amputation, loss of an eye, fractures including teeth, chronic occupational illness, and occupational exposure to transmissible diseases. Therefore, if an employee is injured on the last working day of his or her shift and can return the following workday without requiring any job modification, that injury does not meet the threshold to be reported.

50. D: A site-specific safety plan must reflect the hazards and risks that are specific to the job site it is being used for. Local impacts (such as rivers or nearby freeways), weather, topography, and on-site conditions must all be assessed for their risk to employee safety.

51. C: If several trainees demonstrate a lack of understanding of the material or are unable to execute the desired tasks post-training, the delivery methods must be re-evaluated to determine whether the material is being clearly conveyed.

52. B: An employee's personally identifiable information (PII) includes full name, email address, phone number, home address, Social Security number, driver's license number, biometric information, and medical records. The employee entrusts the employer with protecting any information that could result in an employee losing their expectation of privacy or allowing unauthorized access to their finances.

53. C: Although witness statements are critical in an incident investigation, the investigator must consider that they tend to be the least reliable source of information. Recollections are subject to memory lapses due to the stress of the accident, gap-filling (inserting details that did not exist), and lack of recall since the witness may not have been in an appropriate state of mind to commit details to memory. Verbal statements are also subject to positive and negative biases. Witness statements should always be collected, as they can provide critical information, but that information must be evaluated with caution.

54. D: Maslow's hierarchy of needs is a model used to explain human behavior. In this model, a person will seek to fulfill the needs at the base of the pyramid (physiological and safety) before progressing upwards to other needs, such as belonging and esteem. If the lower-level needs are not met, a worker will seek to change his or her environment or circumstances.

55. D: When a tool or equipment manufacturer considers safety systems during the development of a product, this is referred to as "safety by design." A kill switch is an engineering control designed to cut power to the saw in the event the user loses control or releases their grip.

56. B: Cold stress, such as hypothermia, can be a risk even at temperatures above freezing. Wind and moisture accelerate the body's loss of heat. In addition, physical exertion can produce a layer of sweat, which can also result in heat loss and a drop in body temperature.

57. C: Employers must post an annual summary of the prior year's OSHA recordable incidents using an OSHA 300A form from February 1 through April 30.

58. A: Tendons are the fibrous structures that attach muscles to bones, allowing muscle contractions and extensions to move a joint. Ligaments connect bones to bones, creating stability for the skeleton.

59. D: Accident history evaluations, employee feedback, and quantitative risk assessments are regularly employed in industry for identifying potential risks in the workplace. Cost-benefit analyses are usually done to determine the economic viability of a hazard control or mitigation strategy, not to identify potential risks.

60. C: OSHA standard 29 CFR 1926.651 requires that a competent person conduct inspections of excavations to identify conditions that may impact the protective measures installed. Therefore, the competent person must verify that the soil conditions have not changed before work can resume in the excavation.

61. A: According to the OSHA standard, an emergency action plan must be in writing, kept in the workplace, and available to employees for review. However, an employer with 10 or fewer employees may communicate the plan orally to employees.

62. D: Under the Hazard Communication Standard (29 CFR 1910.1200), if it is necessary to know the composition of a hazardous material component listed as a "trade secret" to render treatment to an exposed employee, the treating physician or nurse can contact the importer or producer to request that information.

63. D: Per 1910.39, an employer must have a fire prevention plan when an OSHA standard requires one (e.g., ethylene oxide standard or methylenedianiline standard). The plan must be in writing if the employer has more than 10 employees.

64. C: It is prohibited to plug an extension cord into another extension cord, a practice referred to as "daisy-chaining." The primary hazard is that the cords can overheat and become a source of ignition if fuel, such as paper, flammable liquids, or sawdust, is present, resulting in a fire.

65. B: When collecting data that will be used to determine whether hazards exist or controls must be implemented, the sampler must ethically collect their data. Using estimates as a matter of convenience due to an error is not the proper way to complete the survey.

66. B: Integrity is defined as the quality of being truthful and possessing high moral principles. Therefore, when an SME notices an error that was missed by another party, they must not hide the mistake to ensure that all gaps in the safety management system are addressed.

67. A: A certified health physicist (CHP) is generally considered as a go-to subject matter expert regarding radiological hazards and doses in the workplace. A certified industrial hygienist (CIH), alternatively, would likely be the most qualified individual to assess a variety of hazards in the workplace and resultantly provide recommendations on how to effectively mitigate such hazards. A certified safety professional's (CSP's) breadth of expertise and training is typically commensurate with that of a CIH. A nuclear (professional) engineer's main area of focus is usually centered around innovating processes and systems that ultimately improve the quality, productivity, and safety of nuclear power plants.

68. C: The emergency action plan requires denoting those employees who are medically trained to assist injured employees. However, if medically trained employees are not available, the employer must post the contact number for emergency services or the number for the nearest medical facility.

69. B: At a minimum, impacted employees, their supervisors, and management should be involved in emergency and disaster response planning. As a best practice, employers should create a response planning committee to facilitate planning activities.

70. C: Whenever someone is faced with items that could potentially be contaminated with blood, engineering controls such as remote pickup devices, including a trash picker or broom and dustpan, should be used to remove the items. Sharp items, such as needles, should be placed in rigid containers to eliminate the risk of the needle poking an employee, thus preventing exposure to bloodborne pathogens.

71. B: A video camera allows the safety professional investigating an incident to record the context and environment of the scene while also providing an opportunity to capture verbal notes. In contrast, photography may lack depth and scale, while video allows the investigator to retrace steps, provide an accurate point of view, and incorporate the time element when determining what happened.

72. B: Employer-provided first aid kits must be stored in weather-proof containers and must be inspected at least weekly to replenish used items.

73. A: Any contractor who has hazardous materials or hazardous waste on-site must also have controls in place to contain and control any potential spill of the material. This includes placing the material on spill skids to catch any leaks or spills, constructing a berm to prevent any spill or release from reaching a storm drain, or placing spill kits on the area to address any liquid that spills from the containers.

74. C: Under 29 CFR 1926.753, a qualified rigger shall inspect the rigging for proper labeling, damage, and working load.

75. A: The primary route of exposure for asbestos fibers is inhalation with a permissible exposure limit of 0.1 fiber per cubic centimeter of air over a an eight-hour time-weighted average. Thus, an air-purifying respirator equipped with high-efficiency particulate air (HEPA) filters is necessary when working with or near friable ACM.

76. C: Although the time-weighted average (TWA) for noise exposure is 90 dBA, employees can be exposed to higher levels for shorter durations. According to Table D-2 in 29 CFR 1926.52, an employee can be exposed to 100 dBA for 2 hours without requiring any controls, as long as they are not exposed to any other sources of excessive noise for the rest of the day.

77. C: The Americans with Disabilities Act (ADA) prohibits an employer from asking an employee about the nature of his or her disability, which can include health conditions (both permanent and temporary) such as illnesses. The Health Insurance Portability and Accountability Act (HIPAA) prohibits sharing of such information.

78. D: Industrial vehicles have different configurations of driver compartments and may be much larger than passenger vehicles, making it more difficult for the driver to see pedestrians during operation. Thus, high-visibility clothing increases the chances of a worker being seen by the driver.

79. C: Risk is equal to the product of frequency and consequence (also referred to as severity). The overall calculated annual risk of an accident event which has a frequency of 0.0001 per year and an associated consequence of 2.5 fatalities is derived by multiplying these two values. Hence, (0.0001/year) × (2.5 fatalities) = 0.00025 fatalities/year.

80. C: Under the OSHA construction lead standard, 29 CFR 1926.62, employees exposed to airborne lead above the action level for more than 30 days in a 12-month period must enter a program for monitoring of blood lead levels.

81. B: The National Incident Management System (NIMS) outlines standardized roles and responsibilities of the Incident Command System (ICS) that can be used as an organizational structure to manage responses to an emergency. This command structure includes a public information officer (PIO), who coordinates all media releases with the incident commander and conducts all media briefings.

82. C: According to OSHA's multi-employer worksite policy, violations that are related to incidents that occur at construction sites with several subcontractors will depend on the circumstances of the accident. In this case, the crane company could be the "creating employer," as they did not adequately enforce the exclusion zone underneath a live lift, and their activities directly resulted in the injury. The landscape company may be considered an "exposing employer," in that they did not adequately inform their employees to avoid areas under live loads. Finally, the prime contractor could be considered the "controlling employer," as they have overall health and safety responsibility for the job.

83. A: The primary concern in the immediate aftermath of an injury is to stabilize the injured person and render first aid. Whenever blood is present, first responders should provide a barrier between themselves and the potentially infectious blood, as there is no time to ask the individual if he or she has a contagious disease. Therefore, "universal precautions" dictate that the first responder should don gloves and assist the injured person.

84. D: When metals are superheated, they may form inhalable fumes consisting of vapors and small particles. Certain metals, such as chromium VI, can cause "metal fume fever," and can cause lung cancer if exposure is not prevented.

85. A: A probabilistic risk assessment (PRA) is a detailed and systematic process to evaluate potential risks associated with the design and operation of a multifaceted, complex technology. A chemical plant or a new transportation system, for example, would both undergo PRAs during its conceptual design stage.

86. B: Improper ergonomics can result in musculoskeletal disorders in the construction industry. Any position where the body is not in a neutral, unstressed position, such as kneeling, reaching, or bending, can result in muscle, tendon, or ligament injuries over time.

87. B: The safety officer in the ICS has many roles, including, but not limited to, having authority to stop and prevent unsafe acts, preparing and presenting safety briefings, identifying hazardous situations and actions to mitigate risks and hazards, and ensuring the safety of assigned incident personnel.

88. B: An endeavor to uncover and reduce hazards that may be at hand for product or system users, as well as to those who are accountable for their maintenance, is conventionally known as an Occupational Health Hazard Assessment. Such assessments also usually attempt to evaluate ways to reduce risks or impacts resulting from such hazards if an associated undesirable event was to subsequently unfold.

89. B: Effective communication strategies consider the message, the audience, and the mode of delivery. Even if the message is clear, is available to the employees, and is understood by management, communication can fail if the message is not delivered effectively or received clearly. Site managers for large companies that rely on email communications will have many messages with priority information to sort, review, and answer. In this instance, the message can easily get lost in an inbox. Priority messages should be delivered in a way that sets them apart from normal communication modes, such as by way of a conference call or digital meeting space.

90. C: Ladders (29 CFR 1926.1053), respirators (29 CFR 1910.134), and fall protection harnesses (29 CFR 1926.501) must be inspected on a routine basis. Records should be maintained to demonstrate that the required inspections have been completed at the appropriate intervals.

91. D: Per the work-area noise-level measurements provided (assessed over a two-week time span), a CSP, ASP, or IH should conclude that the area does not meet the minimum TWA threshold levels of a hazardous noise area as per OSHA 1910.95 (i.e., an 8-hr TWA greater than 85 dBA); thus, hearing protection PPE is not mandatory. However, it should be nevertheless recommended due to average levels being close to this threshold, with occasional exceedances at hand.

Measurement Day	Number of Hours Measured	Measurement Range	Measurement Average
1 of 10	6.75	74-81 dBA	77 dBA
2 of 10	8	67-79 dBA	71 dBA
3 of 10	3	86-88 dBA	87 dBA
4 of 10	8	71-92 dBA	79 dBA
5 of 10	7	81-85 dBA	84 dBA
6 of 10	3.5	77-80 dBA	78 dBA
7 of 10	8	58-72 dBA	67 dBA
8 of 10	6.5	66-70 dBA	69 dBA
9 of 10	2.25	89-94 dBA	93 dBA
10 of 10	7.5	75-83 dBA	78 dBA

92. B: Communication can be defined as the process of a group or individual conveying necessary information to another person or group. This process involves one party sharing or sending information that is received by another party.

93. A: The safety professional should tailor the training program to the specific hazards related to the site at which the work will occur. For the silo, painters should be made aware of the scaffold inspection requirements, how to manage aerial lifts, and the hazards associated with the chemicals they will be using.

94. C: Per 29 CFR 1926.502, a guardrail system must be able to withstand a vertical load of 200 pounds and not deflect to a height less than 39 inches above the walking/working surface.

95. D: Industrial radiographs use either gamma rays or x-rays to inspect opaque objects (such as concrete foundations) for voids or defects. Both gamma and x-rays are ionizing radiation that can cause nausea, vomiting, and hair loss as a result of chronic exposure.

96. B: While an inspection may be required by various OSHA construction standards, there is typically no requirement to maintain those inspection records. However, if a written record is not kept, there is no reviewable proof that the inspection ever occurred. In addition, a record or checklist can standardize the elements that are examined, especially if different employees will conduct the inspection. Finally, the inspection record can outline the "pass/no pass" criteria for each element.

97. C: A fishbone analysis (as shown below) identifies various potential factors or causes of an overall event or effect. This is a common analysis method for product design and defect prevention.

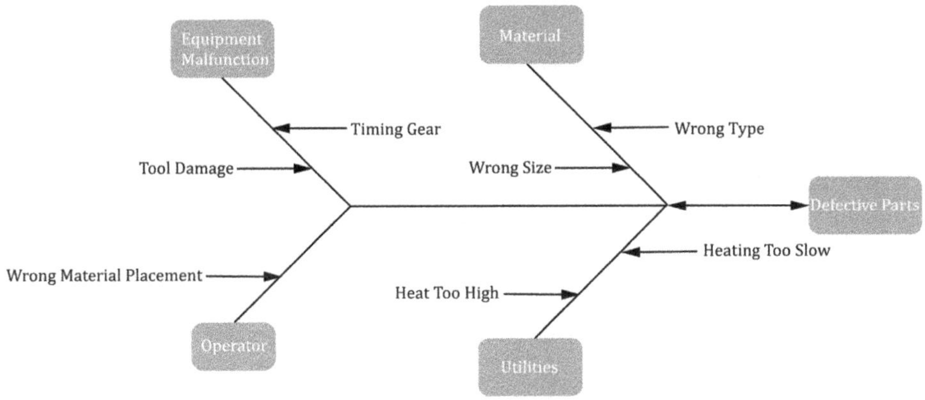

98. D: When present, a tool guard must always be used and may never be removed or modified. Disabling or otherwise impeding use of the guard is also prohibited. If a guard or other safety device would be compromised to accomplish a specific task, a more suitable tool must be used.

99. C: Hand tools that are poorly maintained can result in vibration-related chronic injuries. Although training and policies can be effective, they both require the individual worker to adhere to the policies and retain the lesson. However, a proper maintenance program will reduce the risk of vibration injuries.

100. A: A material's Safety Data Sheet (SDS) contains health and safety information in specific categories and sections. Section 8 lists the recommended engineering controls and personal protective equipment required to reduce the risk of exposure.

101. B: OSHA standard 29 CFR 1926.1204(e)(3) requires that testing of the atmosphere in a permit-required confined space be done in the following order: oxygen, then combustible gases and vapors, and finally toxic gases and vapors. This is because gas testing equipment may not work properly without oxygen, thus a false reading might be produced if tests for combustible or toxic gases are done before testing for oxygen.

102. D: Vehicle travel in a construction site is difficult to control even if roads are paved, as various trades require equipment and materials at various locations. Therefore, properly designated and identified walking paths reduce the risk of workers getting hit by vehicles on the job site.

103. D: Scissor lifts that are equipped with a compliant guardrail system do not require that an employee wear a fall protection harness.

104. D: The Globally Harmonized System of Classification and Labelling of Chemicals (GHS) uses two signal words: Danger and Warning. Danger indicates a higher risk level than Warning.

105. B: One way to describe a person who has integrity is that he or she does not waver from his or her morals or principles from one situation to another. Integrity is the quality of always being honest and doing the right thing, regardless of the circumstances.

106. B: Oily rags, flammable liquids, and used solvents must be kept in fire-rated containers to reduce the risk of fire until they are removed from the job site. Unwanted materials must be properly managed as hazardous waste.

107. A: A risk assessment is a tool used to proactively identify hazards in the workplace and to determine their potential impact upon worker safety.

108. D: Under 29 CFR 1904.33, the employer must retain OSHA 300, 300A, and 301 logs for five years after the calendar year the records address.

109. C: Lockout/tagout (LOTO) is a process used to control hazardous energy that, if released, can result in injury during maintenance, service, or repair activities. The different energy types that must be controlled include chemical, electrical, mechanical, and gravity.

110. C: Unlike child learners, adult learners have experiences to draw from when learning new information. By associating new information with past events, adults are better able to retain presented material.

111. C: Bending at the waist puts strain on the ligaments of the spine, which can impact the muscles of the lower back. Raising the work surface from the floor to a standing height can eliminate the need for bending over to make each cut and lower the risk of back injuries.

112. C: Energy drinks typically contain caffeine, which is a diuretic. Diuretics increase production of urine, which can result in dehydration. When someone is experiencing elevated body temperature, sweat is the body's primary cooling mechanism. Sweating can be inhibited by dehydration.

113. C: The Board of Certified Safety Professionals has adopted a Code of Ethics that requires those certified by the Board to "act with responsibility and integrity." Therefore, if evidence of an inspection is missing, the professional cannot make assumptions and must accept her mistake. There is a possibility that this extinguisher was removed from the location and may not have actually been inspected.

114. B: Welding leathers include coats, pants, chaps, sleeves, or hoods that are used to protect the welder from molten sparks, referred to as "slag," which can cause burns to the skin.

115. C: Supervisors need to assess each job site for non-routine hazards that could be introduced by other contractors. For example, if a crew is unaccustomed to being near overhead work, they may not be used to protecting themselves from falling objects. Therefore, the supervisor should address hazard mitigation strategies, such as wearing hard hats to reduce the risk of struck-by injuries when workers are near scaffold operations.

116. B: The OSHA 300 log records information for each recordable injury. However, employers are prohibited from recording the names of individuals who experienced injuries classified as "privacy cases," including sexual assault, injuries to intimate body parts or reproductive organs, bloodborne pathogen exposures, or cases where an individual suffered an injury from an object contaminated with someone else's blood.

117. B: The overall sequence of implementing training is to assess needs, design training, acquire materials, deliver training, then evaluate.

118. A: Behavior-based safety (BBS) assumes that work sites are imperfect, and employees must continually make choices regarding working safely. Under this model, an employee will make an unsafe choice based on the person (his or her experience, abilities, and capabilities), environment (available hazard controls, workplace, and company culture), and behavior (how the employee acts and responds while working).

119. B: Engineering controls are methods of eliminating a hazard or placing a barrier between the worker and the hazard. A wet saw uses a stream of water to significantly reduce the amount of dust generated by the cutting. Personal protective equipment, such as respirators, are not considered effective barriers.

120. D: Whenever possible, risks at the job site should be eliminated. Replacing a device that uses ionizing radiation with an alternative and equally effective method eliminates the risk of exposure for the employees.

121. D: The Board of Certified Safety Professionals (BCSP) Code of Ethics prohibits any member from misrepresenting his or her professional qualifications. Applicants are not allowed to use BCSP certifications until their test results have been certified. Those found violating the Code of Ethics can have their certifications stripped.

122. B: Lockout/tagout (LOTO) is the electrical safety concept wherein every individual who is working on a system applies their own lock to an isolation device (such as a knife switch or breaker) to prevent the circuit from being activated without their knowledge. This is critical when multiple groups are working on the same system so that when one group completes their work, they do not energize the system, thus protecting others from risk of shock or electrocution.

123. A: Assessments are informal methods of determining adherence to programs, policies, and procedures. Non-adherences are typically corrected on the spot with no formal follow-up procedure, unless a gross violation is observed. Audits are more formal processes that typically rely on checklists and record reviews.

124. C: The graphic on a label that is compliant with the Globally Harmonized System of Classification and Labelling of Chemicals (GHS) is called a pictogram. Every hazardous material must have at least one of the eight approved graphics (OSHA does not enforce the environmental hazard pictogram).

125. A: OSHA recognizes four categories of precautionary statements under the Globally Harmonized System of Classification and Labelling of Chemicals (GHS): prevention (such as exposure minimization), response (including spill response and first aid), storage, and disposal.

126. C: By hiring a subcontractor to install the trusses, the contractor is transferring the risk of injury from his company to another while eliminating the risk to their employees.

127. A: Under 29 CFR 1926.35, the mandatory elements of the emergency action plan (EAP) are outlined. Every EAP must include a means to account for employees after an evacuation, such as designating assembly points where head counts can be conducted to make sure everyone is safely out.

128. A: Grinding activities can create airborne debris, necessitating use of safety glasses. In addition, lead is poisonous to humans, and fine lead dust must not be inhaled. Finally, the OSHA lead standard requires that hand-washing facilities be made available to prevent hand-to-mouth exposure (ingestion) of any lead-containing dust.

129. C: OSHA requires reporting of serious workplace injuries, defined as loss of an eye, amputation (meaning loss of bone), and any in-patient hospitalization of one or more employees.

130. B: OSHA standard 29 CFR 1926.1101 outlines activities that are prohibited while working with known or suspected asbestos-containing materials, regardless of the measured levels. These activities include disc cutting without point-of-cut ventilators, shoveling dry debris, using compressed air outside of containment, and dry sweeping.

131. D: Inhalation of dusts, fumes, vapors, and mists can result in negative health consequences for the worker, up to and including death. The most effective control method is to remove the hazard from the breathing area. The best method of removal is local exhaust ventilation. Other methods of hazard control have failure potential that could result in an exposure.

132. A: Normally, an action list is the sought-after end-product tool that is rendered by conducting a failure modes and effects analysis. Actions receive a risk rank to allow for prioritization. If these actions are executed well, risk in the workplace is often substantially reduced.

133. A: Training only provides skills and knowledge. It is only an appropriate corrective action when investigation reveals that employees lack training or knowledge.

134. C: Title 29 of the Code of Federal Regulations (29 CFR) outlines the scope and requirements of OSHA. The subparts apply to specified industries and are enforced by OSHA. Part 1926 applies to any "construction, alteration, and/or repair, including painting and decorating" activity an individual is employed to conduct.

135. C: Materials that could potentially contain asbestos must be sent to a certified laboratory to confirm the presence of the fibers. Fibers are first identified using special microscopic techniques, then a report is issued regarding the presence or absence of asbestos.

136. A: It is the duty of the employer to provide appropriate personal protective equipment (PPE) for hazards that are identified at their worksite.

137. C: The five whys method is typically regarded as a very effective method or model for conducting a root-cause analysis. This approach essentially advocates the use of consecutive, logically connected *why* questions that are used to ultimately reach the root cause or justification behind an occurrence. There is no requirement that the process be limited to five iterations—more or fewer may be used to reach a conclusion.

138. D: OSHA standard 29 CFR 1926.250 requires that the load limit of floors within buildings be conspicuously posted in terms of pounds per square foot.

139. D: A root cause is a failure in the safety management system that created the condition for an incident to occur. If the root cause is identified and corrected, the likelihood of a similar incident recurring is minimized.

140. D: In addition to outlining a training program and the command structure for a chemical spill response, a written emergency response plan must outline a medical surveillance program. The medical portion is designed to evaluate those responders who will wear protective equipment and to monitor for chronic exposure to highly toxic materials, such as lead.

141. C: Padded gloves can be used to reduce the impact of a vibrating tool on nerves and blood vessels. Work rotation is a method to reduce the total amount of vibration exposure to an individual worker. Replacement of a hand tool with a machine eliminates this vibration hazard.

142. B: A body harness consists of straps attached to other components of a personal fall system. In case of a fall incident, the subject straps mainly distribute the fall-arrest forces over the chest, shoulders, waist, thighs, and pelvis. A properly designed (and donned) harness should yield minimal impacts to the lower back, arms, and feet.

143. A: Under the Hazardous Waste Operations construction standard (29 CFR 1926.65), the contractor conducting the cleanup operations must make their written health and safety plan available to contractors involved in the cleanup, employees, designated representatives (such as union reps), OSHA personnel, and any other agency with authority over the site.

144. B: Flammable gas meters have the capability to display the measurement as "% LEL," or percent of the lower explosive limit. The lower explosive limit (LEL) is the minimum concentration of vapors in air at which a spark will result in combustion. The meter will not identify the gas, but it will indicate the risk of fire.

145. A: An emergency action plan (EAP) delineates how a business will respond to typical situations in very general terms. A site-specific safety plan (SSSP) identifies specific actions based on job site location, topography, local services, and project-specific conditions that will impact employee safety.

146. B: A lanyard's primary role within a fall-protection system is to connect a safety harness to an anchoring point. Lanyards absorb fall energy, so they reduce the impact load on a person when a fall is ultimately arrested.

147. A: Large projects often involve contractors working overhead, such as framers, roofers, glaziers, masons, and painters. Equipment and materials being used at an elevation always carry a risk of falling and striking those below.

148. A: Portable gas detection systems, often referred to as "multi-meters," can have a variety of gas detector cells installed. However, the most common configuration of four-gas multi-meters includes cells that can detect carbon monoxide, oxygen, flammable gases, and sulfur dioxide.

149. C: Chemical protective glove manufacturers produce permeation charts for the different products they offer. These charts are available on the internet and describe how a glove material, such as nitrile or butyl rubber, performs against different chemicals. The safety professional identifies the chemical or chemicals their staff will be working with, locates these chemicals on the chart, and selects the glove that provides the best overall protection to their workers.

150. A: The primary hazard associated with non-ionizing radiation, such as the kind used for industrial lasers, is damage to the eyes. When rain, snow, or fog are present, the air contains several reflecting and dispersive surfaces. This means that the radiation is spread over a wider area than the intended beam and may expose more workers to the hazard.

151. D: Under NFPA 51B, "Standard for Fire Prevention during Welding, Cutting, and Other Hot Work," a hot work permit is a document used to identify the provisions required to protect a temporary hot work location from sparks that could initiate a fire. A hot work permit is also used to verify that mitigation measures have been implemented prior to initiation of work.

152. A: ANSI Z87.1 outlines the testing specifications for impact-resistant eyewear. Eyewear that is classified as impact-rated will be marked with either "Z87" or "Z87+" on the frame.

153. A: The safety professional can conduct a risk assessment when determining resource allocation for mitigation of hazards. Risk is calculated by multiplying the frequency of the hazard (or the number of individuals exposed) by the severity of the potential injury and the likelihood of an injury occurring. The results can be used to identify higher-risk jobs that require more immediate attention.

154. B: A job hazard analysis (JHA), also referred to as a job safety analysis (JSA), is used to identify the hazards associated with any job or task at a construction site. Once the hazards are identified, the types of controls needed to reduce risk, including appropriate personal protective equipment (PPE), are also identified.

155. A: Most 911 call centers pull location data from the cell towers receiving the signal, and not from phones. Therefore, the exact location information of a caller may not be available to the operator. To address this, 29 CFR 1926.50 requires that the employer post either the latitude and longitude of the work location or other effective location identification information (e.g., State Route 37, mile marker 52).

156. B: Safety professionals should be well versed in recognizing hazards and identifying potential hazards based on future activities at the site. For example, the presence of other contractors in the lift area can introduce risk of injury if a load falls. Also, weather impacts on the activity, such as rain or wind, should always be taken into consideration. Finally, the safety of occupants in the building would need to be addressed by either adjusting the lift schedule outside of normal business hours or establishing an internal exclusion zone in the event a chiller falls through the ceiling.

157. A: When asbestos-containing materials (ACM) are disturbed, the hazardous fibers have the potential to break into smaller, inhalable pieces that can cause cancer. Therefore, asbestos abatement is regulated by the Environmental Protection Agency under the Clean Air Act (CAA). The Asbestos Hazard Emergency Response Act (AHERA) applies only to schools and requires ACM be identified and removed or otherwise properly mitigated.

158. D: The simplest model for communication involves three parts: the sender of the information, the message containing the information, and the receiver of the information. Failure with any one of these parts will negatively impact the desired effect of the communication.

159. A: One of the main advantages of computer-based training is that all participants receive the same information regardless of where or when they complete the course.

160. C: The employer's emergency action plan (EAP) outlines response measures for generalized emergencies, such as floods, fires, and medical emergencies. The site-specific safety plan (SSSP) expands on the EAP by including job-specific hazards, such as overhead lifting by cranes, and emergency elements, including designating the exact locations of the evacuation assembly areas, which are required by the EAP.

161. A: Ethics can be defined as the expectations of behaviors and conduct that leadership has for its employees. The code of ethics establishes the organization's definitions of honesty, loyalty, and fairness.

162. B: Exposure to respirable silica from cutting concrete can result in lung damage. OSHA's silica standard (29 CFR 1926.1153) lists specific tasks that require use of a high-efficiency particulate air (HEPA) filter, such as any concrete cutting indoors regardless of engineering controls, to reduce the risk of silicosis.

163. D: An automated external defibrillator (AED) is a device that applies an electric shock to the heart to correct an irregular heartbeat or restart a heart that has stopped beating. The unit automatically evaluates the victim's heart rhythm to determine whether a shock should be applied.

164. C: Under 29 CFR 1926.25, garbage bins must be outfitted with lids to reduce the risk of rodents or other vermin.

165. A: The Plan, Do, Check, Act (PDCA) theory of continual improvement is a tool that is used to determine if a solution is effective at correcting an identified problem. The safety professional identifies the need (Plan), implements a change such as a new policy (Do), evaluates if the desired changes have occurred (Check), and determines if additional work is needed (Act).

166. D: Full-face air-purifying respirators provide a higher level of respiratory protection than quarter- or half-face styles. This is directly related to the face-seal interface, as there is less variability on the forehead and chin than on the nose and mouth, resulting in a better seal. This is reflected in an assigned performance factor of 10 for half-face respirators compared to 50 for full-face units.

167. A: We first have to calculate the index. Then, we can calculate the incident rate. The incident rate will tell us how many people got hurt per 100 employees.

$$\text{Incident rate} = (\#\text{ incidents} \times \text{index})/\text{hours worked}$$

$$\text{index} = 100 \times 50 \times 50 = 250{,}000 \text{ hours}$$

Note: An OSHA incident rate uses 100 employees working 2,000 hours a year. A company does not have to use the same index but then would not be able to benchmark against companies that use the OSHA index.

$$Incident\ rate = \frac{29 \times 250{,}000}{1454000} = 4.99$$

168. D: An industrial hygienist is a safety professional who identifies, evaluates, and controls occupational hazards. They are trained in using instruments and tools to quantify hazards in the workplace. An industrial hygienist would evaluate the concrete operation, measure the sources of noise, and produce a report. This report would give the company's safety professional direction on whether a hearing conservation program is needed and would likely recommend how to control the noise exposure of their employees.

169. A: Hyperthermia is a failure of the temperature-regulating system of the body, resulting in an elevated body temperature. If the temperature is not reduced, workers may experience heat stress.

170. C: Normal oxygen in the atmosphere is 20.9%. An oxygen-rich environment is considered any level above 23.5%, and employees may not enter under those conditions. Entry into an oxygen-enriched environment can present a fire hazard, as objects that may not normally burn under normal conditions may burn in an enriched-oxygen environment. Elevated oxygen levels may also increase risk of cell damage.

171. A: For sites that store hazardous waste, the Environmental Protection Agency requires the organization to have a written contingency plan. This plan outlines the resources available for use in the event of an accidental release to reduce the impact on the environment.

172. B: A method of evaluating the effectiveness of a control designed to reduce injuries is to review data, such as workers' compensation claims, from before and after the control was implemented. If there is a decrease in similar reported injuries, then the corrective action was effective.

173. B: The most effective way to evaluate adherence to safety policies and programs is by observation. Polls, surveys, and equipment consumption rates do not provide an accurate picture of the true safety culture at a job site. The validity of the responses is tied directly to the honesty of the responders, who may not want to appear as if they are not following the established policy. Watching people working is the only way to determine whether they are following safe practices and if a policy is being followed or ignored.

174. C: A negative bias can be defined as a personal preference against a group. Unlike a stereotype, which is an overgeneralization shared by a group, bias involves pre-judgement. Such a bias could interfere with the safety professional's ability to determine the root cause of an incident involving the delivery company.

175. C: Although others in the chain of command should be notified of the failure to adhere to the fire prevention standards for welding and cutting (29 CFR 1926.352), the safety professional should immediately stop the employee from working and have him or her address the fire hazard.

176. A: A hazard should be first eliminated, or a non-hazardous alternative should be used as a substitute. The next preferred method of risk reduction is use of an engineering control or enclosure. Then, administrative controls and personal protective equipment should be considered.

177. A: The fact that the training is early in the morning most likely does not contribute to the lack of attentiveness. When planning a transfer of information, the trainer must evaluate whether the information is being communicated clearly. Failure of communication can be tied to language proficiency, misaligned level of difficulty (either too advanced or too rudimentary), and the attendees' perceived ability to apply the information to their job. Time of day is unlikely to be a factor that contributes to a lack of attentiveness.

178. D: Awkward posture is when the body is in an unnatural or unsustainable position. Unnatural positions include arms above the shoulders, bending at the waist, and twisting. When the body is placed in these positions, particularly for extended periods of time, it can impact the muscles, tendons, and ligaments. This results in aching, pain, weakness, and swelling.

179. D: One of the detriments of instructor-led training is that the session is at a fixed location at a specified time, which may not be convenient for all attendees. In addition, all required attendees are removed from their normal duties to complete the training.

180. B: The need to connect so many extension cords can be traced to the lack of available outlets for the multiple workers on-site. If additional temporary power supplies had been installed on-site, the cords would not have needed to be connected, and the cords would not have overheated and started the fire.

181. D: The National Fire Protection Association publishes the most current version of the National Electrical Code on their website. This code addresses proper use of extension cords, referred to as "flexible cords."

182. D: While an emergency action plan should address the immediate actions to be taken during an emergency to reduce the unforeseen event's impact on employees and those in the immediate area, a crisis management plan extends beyond the initial response. After the situation has been stabilized, an organization's crisis management plan should be implemented. This plan identifies the individuals who will manage the situation, outlines public relations and media strategies, describes the step-by-step process to restore the site to normal operations, and identifies external partners who can provide assistance if needed.

183. B: At any multi-employer work location, the prime contractor is responsible for all health and safety aspects of all activities under the contract. As the "employer," they may designate a safety professional to monitor and enforce the programs or allocate responsibilities to subcontractors, but they are ultimately responsible for the safety of everyone working on the project.

184. D: A fire requires fuel, an ignition source, and oxygen. Of those three elements, referred to as the "fire triangle," oxygen cannot be removed from the equation. However, fuel and ignition sources can be addressed. By establishing routine inspections of the job site, the safety professional can identify fuel—such as wood, paper wrappings, and oily rags—that needs to be removed to lower the fire risk. In addition, job walks can be used to identify and control potential sources of ignition, such as frayed cords and hot work. If the fuel and ignition sources are adequately identified and controlled, the risk of fire is greatly reduced.

185. B: Oil-based paints produce an aerosol of the solvent medium (organic vapor) and paint particles. Thus, a combination organic vapor and high-efficiency particulate (OV/HEPA) cartridge would provide the best protection.

186. C: Asbestos fibers are friable, meaning they break into smaller and smaller fibers. The fibers are small enough to be inhaled when made airborne during removal or demolition. Once inhaled, the fibers can enter the lungs, where they irritate the lungs and can cause cancer.

187. B: In this example, the safety professional is using the "praise and approval" approach to influence change in the behavior of those not wearing PPE, increasing the chances they will adhere to the rules at the job site in the future.

188. D: Builder's risk insurance is a policy that covers the project against losses incurred during construction. The policy protects against fire, flood, and certain types of damage that would impact the timeline for turning over the building to the owner upon completion.

189. C: Energy trace barrier analysis is a technique for identifying hazards specifically associated with energy sources. It evaluates unwanted flow of hazardous energy to personnel or equipment and helps identify necessary barriers or control measures to prevent hazardous energy from reaching targets. A hazard operability study is a structured process that looks for hazards resulting from potential deviations in system parameters. It helps identify hazard controls and human interaction elements to prevent deviations from causing adverse effects. Subsystem hazard analysis and what-if analysis are more suited to preliminary assessments and are not comprehensive methods. Fault tree analysis and failure mode and effects analysis would not be appropriate given the team cannot easily identify components, their interactions, or failure modes. Job hazard analysis and preliminary hazard analysis is not a good choice given the intent of the question and the information provided by the team.

190. B: A fully compliant GHS label must include, among other information, a hazard statement and a precautionary statement.

191. C: Powder-actuated tools use a controlled explosion to drive the fastener and should only be used with hard, non-brittle materials such as poured concrete to avoid the fastener penetrating or shattering the material.

192. C: NFPA 51B outlines the fire prevention measures required for any spark-producing activity, including welding and metal cutting. The minimum distance that combustible materials must be moved from the source of hot work is 35 feet.

193. A: A disaster recovery plan should designate a recovery coordinator who ultimately decides what recovery actions need to be taken and assigns responsibilities for achieving such actions.

194. A: As utilities are installed and activated at a job site, regular briefings or other forms of communication should be used to relay the status to all impacted crews. As electrical, gas, and water lines are energized, the risks to all workers increase, and depending on their activities, additional control measures may need to be implemented.

195. B: The maximum weight a person should lift unassisted is 51 pounds, as determined by the NIOSH lifting equation. Anything over that weight should be lifted with either a partner or a mechanical device.

196. D: An employer is obligated to protect all confidential employee information that they collect or maintain. This includes medical information, personal identifiable information (e.g., Social Security number), and financial information used for payroll purposes. Employers may be held liable if they do not properly protect the information in their possession.

197. C: Digital photography can capture multiple critical elements during an accident investigation, such as relative locations of items, victim point of view, and equipment condition, such as frayed wires. However, digital images must be supplemented with notes or other methods to record such things as temperature and mist that may have contributed to the incident.

198. D: The approach of assuming that any bodily fluid, including blood and vomit, contains an infectious human pathogen, and using appropriate precautions, such as gloves and mouth barriers, is referred to as "universal precautions." The intent is not to offend an injured person, but to protect those providing aid to that person, or those responsible for cleaning areas exposed to bodily fluids.

199. A: U.S. citizen employees who live and work in the United States have a legal right to inspect certain records under their name.

200. A: Rigging equipment should be inspected before use on each shift as required by 29 CFR 1926.251.

How to Overcome Test Anxiety

Just the thought of taking a test is enough to make most people a little nervous. A test is an important event that can have a long-term impact on your future, so it's important to take it seriously and it's natural to feel anxious about performing well. But just because anxiety is normal, that doesn't mean that it's helpful in test taking, or that you should simply accept it as part of your life. Anxiety can have a variety of effects. These effects can be mild, like making you feel slightly nervous, or severe, like blocking your ability to focus or remember even a simple detail.

If you experience test anxiety—whether severe or mild—it's important to know how to beat it. To discover this, first you need to understand what causes test anxiety.

Causes of Test Anxiety

While we often think of anxiety as an uncontrollable emotional state, it can actually be caused by simple, practical things. One of the most common causes of test anxiety is that a person does not feel adequately prepared for their test. This feeling can be the result of many different issues such as poor study habits or lack of organization, but the most common culprit is time management. Starting to study too late, failing to organize your study time to cover all of the material, or being distracted while you study will mean that you're not well prepared for the test. This may lead to cramming the night before, which will cause you to be physically and mentally exhausted for the test. Poor time management also contributes to feelings of stress, fear, and hopelessness as you realize you are not well prepared but don't know what to do about it.

Other times, test anxiety is not related to your preparation for the test but comes from unresolved fear. This may be a past failure on a test, or poor performance on tests in general. It may come from comparing yourself to others who seem to be performing better or from the stress of living up to expectations. Anxiety may be driven by fears of the future—how failure on this test would affect your educational and career goals. These fears are often completely irrational, but they can still negatively impact your test performance.

Elements of Test Anxiety

As mentioned earlier, test anxiety is considered to be an emotional state, but it has physical and mental components as well. Sometimes you may not even realize that you are suffering from test anxiety until you notice the physical symptoms. These can include trembling hands, rapid heartbeat, sweating, nausea, and tense muscles. Extreme anxiety may lead to fainting or vomiting. Obviously, any of these symptoms can have a negative impact on testing. It is important to recognize them as soon as they begin to occur so that you can address the problem before it damages your performance.

The mental components of test anxiety include trouble focusing and inability to remember learned information. During a test, your mind is on high alert, which can help you recall information and stay focused for an extended period of time. However, anxiety interferes with your mind's natural processes, causing you to blank out, even on the questions you know well. The strain of testing during anxiety makes it difficult to stay focused, especially on a test that may take several hours. Extreme anxiety can take a huge mental toll, making it difficult not only to recall test information but even to understand the test questions or pull your thoughts together.

Effects of Test Anxiety

Test anxiety is like a disease—if left untreated, it will get progressively worse. Anxiety leads to poor performance, and this reinforces the feelings of fear and failure, which in turn lead to poor performances on subsequent tests. It can grow from a mild nervousness to a crippling condition. If allowed to progress, test anxiety can have a big impact on your schooling, and consequently on your future.

Test anxiety can spread to other parts of your life. Anxiety on tests can become anxiety in any stressful situation, and blanking on a test can turn into panicking in a job situation. But fortunately, you don't have to let anxiety rule your testing and determine your grades. There are a number of relatively simple steps you can take to move past anxiety and function normally on a test and in the rest of life.

Physical Steps for Beating Test Anxiety

While test anxiety is a serious problem, the good news is that it can be overcome. It doesn't have to control your ability to think and remember information. While it may take time, you can begin taking steps today to beat anxiety.

Just as your first hint that you may be struggling with anxiety comes from the physical symptoms, the first step to treating it is also physical. Rest is crucial for having a clear, strong mind. If you are tired, it is much easier to give in to anxiety. But if you establish good sleep habits, your body and mind will be ready to perform optimally, without the strain of exhaustion. Additionally, sleeping well helps you to retain information better, so you're more likely to recall the answers when you see the test questions.

Getting good sleep means more than going to bed on time. It's important to allow your brain time to relax. Take study breaks from time to time so it doesn't get overworked, and don't study right before bed. Take time to rest your mind before trying to rest your body, or you may find it difficult to fall asleep.

Along with sleep, other aspects of physical health are important in preparing for a test. Good nutrition is vital for good brain function. Sugary foods and drinks may give a burst of energy but this burst is followed by a crash, both physically and emotionally. Instead, fuel your body with protein and vitamin-rich foods.

Also, drink plenty of water. Dehydration can lead to headaches and exhaustion, especially if your brain is already under stress from the rigors of the test. Particularly if your test is a long one, drink water during the breaks. And if possible, take an energy-boosting snack to eat between sections.

Along with sleep and diet, a third important part of physical health is exercise. Maintaining a steady workout schedule is helpful, but even taking 5-minute study breaks to walk can help get your blood pumping faster and clear your head. Exercise also releases endorphins, which contribute to a positive feeling and can help combat test anxiety.

When you nurture your physical health, you are also contributing to your mental health. If your body is healthy, your mind is much more likely to be healthy as well. So take time to rest, nourish your body with healthy food and water, and get moving as much as possible. Taking these physical steps will make you stronger and more able to take the mental steps necessary to overcome test anxiety.

Mental Steps for Beating Test Anxiety

Working on the mental side of test anxiety can be more challenging, but as with the physical side, there are clear steps you can take to overcome it. As mentioned earlier, test anxiety often stems from lack of preparation, so the obvious solution is to prepare for the test. Effective studying may be the most important weapon you have for beating test anxiety, but you can and should employ several other mental tools to combat fear.

First, boost your confidence by reminding yourself of past success—tests or projects that you aced. If you're putting as much effort into preparing for this test as you did for those, there's no reason you should expect to fail here. Work hard to prepare; then trust your preparation.

Second, surround yourself with encouraging people. It can be helpful to find a study group, but be sure that the people you're around will encourage a positive attitude. If you spend time with others who are anxious or cynical, this will only contribute to your own anxiety. Look for others who are motivated to study hard from a desire to succeed, not from a fear of failure.

Third, reward yourself. A test is physically and mentally tiring, even without anxiety, and it can be helpful to have something to look forward to. Plan an activity following the test, regardless of the outcome, such as going to a movie or getting ice cream.

When you are taking the test, if you find yourself beginning to feel anxious, remind yourself that you know the material. Visualize successfully completing the test. Then take a few deep, relaxing breaths and return to it. Work through the questions carefully but with confidence, knowing that you are capable of succeeding.

Developing a healthy mental approach to test taking will also aid in other areas of life. Test anxiety affects more than just the actual test—it can be damaging to your mental health and even contribute to depression. It's important to beat test anxiety before it becomes a problem for more than testing.

Study Strategy

Being prepared for the test is necessary to combat anxiety, but what does being prepared look like? You may study for hours on end and still not feel prepared. What you need is a strategy for test prep. The next few pages outline our recommended steps to help you plan out and conquer the challenge of preparation.

Step 1: Scope Out the Test

Learn everything you can about the format (multiple choice, essay, etc.) and what will be on the test. Gather any study materials, course outlines, or sample exams that may be available. Not only will this help you to prepare, but knowing what to expect can help to alleviate test anxiety.

Step 2: Map Out the Material

Look through the textbook or study guide and make note of how many chapters or sections it has. Then divide these over the time you have. For example, if a book has 15 chapters and you have five days to study, you need to cover three chapters each day. Even better, if you have the time, leave an extra day at the end for overall review after you have gone through the material in depth.

If time is limited, you may need to prioritize the material. Look through it and make note of which sections you think you already have a good grasp on, and which need review. While you are studying, skim quickly through the familiar sections and take more time on the challenging parts.

Write out your plan so you don't get lost as you go. Having a written plan also helps you feel more in control of the study, so anxiety is less likely to arise from feeling overwhelmed at the amount to cover.

Step 3: Gather Your Tools

Decide what study method works best for you. Do you prefer to highlight in the book as you study and then go back over the highlighted portions? Or do you type out notes of the important information? Or is it helpful to make flashcards that you can carry with you? Assemble the pens, index cards, highlighters, post-it notes, and any other materials you may need so you won't be distracted by getting up to find things while you study.

If you're having a hard time retaining the information or organizing your notes, experiment with different methods. For example, try color-coding by subject with colored pens, highlighters, or post-it notes. If you learn better by hearing, try recording yourself reading your notes so you can listen while in the car, working out, or simply sitting at your desk. Ask a friend to quiz you from your flashcards, or try teaching someone the material to solidify it in your mind.

Step 4: Create Your Environment

It's important to avoid distractions while you study. This includes both the obvious distractions like visitors and the subtle distractions like an uncomfortable chair (or a too-comfortable couch that makes you want to fall asleep). Set up the best study environment possible: good lighting and a comfortable work area. If background music helps you focus, you may want to turn it on, but otherwise keep the room quiet. If you are using a computer to take notes, be sure you don't have any other windows open, especially applications like social media, games, or anything else that could distract you. Silence your phone and turn off notifications. Be sure to keep water close by so you stay hydrated while you study (but avoid unhealthy drinks and snacks).

Also, take into account the best time of day to study. Are you freshest first thing in the morning? Try to set aside some time then to work through the material. Is your mind clearer in the afternoon or evening? Schedule your study session then. Another method is to study at the same time of day that you will take the test, so that your brain gets used to working on the material at that time and will be ready to focus at test time.

Step 5: Study!

Once you have done all the study preparation, it's time to settle into the actual studying. Sit down, take a few moments to settle your mind so you can focus, and begin to follow your study plan. Don't give in to distractions or let yourself procrastinate. This is your time to prepare so you'll be ready to fearlessly approach the test. Make the most of the time and stay focused.

Of course, you don't want to burn out. If you study too long you may find that you're not retaining the information very well. Take regular study breaks. For example, taking five minutes out of every hour to walk briskly, breathing deeply and swinging your arms, can help your mind stay fresh.

As you get to the end of each chapter or section, it's a good idea to do a quick review. Remind yourself of what you learned and work on any difficult parts. When you feel that you've mastered the material, move on to the next part. At the end of your study session, briefly skim through your notes again.

But while review is helpful, cramming last minute is NOT. If at all possible, work ahead so that you won't need to fit all your study into the last day. Cramming overloads your brain with more information than it can process and retain, and your tired mind may struggle to recall even previously learned information when it is overwhelmed with last-minute study. Also, the urgent nature of cramming and the stress placed on your brain contribute to anxiety. You'll be more likely to go to the test feeling unprepared and having trouble thinking clearly.

So don't cram, and don't stay up late before the test, even just to review your notes at a leisurely pace. Your brain needs rest more than it needs to go over the information again. In fact, plan to finish your studies by noon or early afternoon the day before the test. Give your brain the rest of the day to relax or focus on other things, and get a good night's sleep. Then you will be fresh for the test and better able to recall what you've studied.

STEP 6: TAKE A PRACTICE TEST

Many courses offer sample tests, either online or in the study materials. This is an excellent resource to check whether you have mastered the material, as well as to prepare for the test format and environment.

Check the test format ahead of time: the number of questions, the type (multiple choice, free response, etc.), and the time limit. Then create a plan for working through them. For example, if you have 30 minutes to take a 60-question test, your limit is 30 seconds per question. Spend less time on the questions you know well so that you can take more time on the difficult ones.

If you have time to take several practice tests, take the first one open book, with no time limit. Work through the questions at your own pace and make sure you fully understand them. Gradually work up to taking a test under test conditions: sit at a desk with all study materials put away and set a timer. Pace yourself to make sure you finish the test with time to spare and go back to check your answers if you have time.

After each test, check your answers. On the questions you missed, be sure you understand why you missed them. Did you misread the question (tests can use tricky wording)? Did you forget the information? Or was it something you hadn't learned? Go back and study any shaky areas that the practice tests reveal.

Taking these tests not only helps with your grade, but also aids in combating test anxiety. If you're already used to the test conditions, you're less likely to worry about it, and working through tests until you're scoring well gives you a confidence boost. Go through the practice tests until you feel comfortable, and then you can go into the test knowing that you're ready for it.

Test Tips

On test day, you should be confident, knowing that you've prepared well and are ready to answer the questions. But aside from preparation, there are several test day strategies you can employ to maximize your performance.

First, as stated before, get a good night's sleep the night before the test (and for several nights before that, if possible). Go into the test with a fresh, alert mind rather than staying up late to study.

Try not to change too much about your normal routine on the day of the test. It's important to eat a nutritious breakfast, but if you normally don't eat breakfast at all, consider eating just a protein bar. If you're a coffee drinker, go ahead and have your normal coffee. Just make sure you time it so that the caffeine doesn't wear off right in the middle of your test. Avoid sugary beverages, and drink enough water to stay hydrated but not so much that you need a restroom break 10 minutes into the test. If your test isn't first thing in the morning, consider going for a walk or doing a light workout before the test to get your blood flowing.

Allow yourself enough time to get ready, and leave for the test with plenty of time to spare so you won't have the anxiety of scrambling to arrive in time. Another reason to be early is to select a good seat. It's helpful to sit away from doors and windows, which can be distracting. Find a good seat, get out your supplies, and settle your mind before the test begins.

When the test begins, start by going over the instructions carefully, even if you already know what to expect. Make sure you avoid any careless mistakes by following the directions.

Then begin working through the questions, pacing yourself as you've practiced. If you're not sure on an answer, don't spend too much time on it, and don't let it shake your confidence. Either skip it and come back later, or eliminate as many wrong answers as possible and guess among the remaining ones. Don't dwell on these questions as you continue—put them out of your mind and focus on what lies ahead.

Be sure to read all of the answer choices, even if you're sure the first one is the right answer. Sometimes you'll find a better one if you keep reading. But don't second-guess yourself if you do immediately know the answer. Your gut instinct is usually right. Don't let test anxiety rob you of the information you know.

If you have time at the end of the test (and if the test format allows), go back and review your answers. Be cautious about changing any, since your first instinct tends to be correct, but make sure you didn't misread any of the questions or accidentally mark the wrong answer choice. Look over any you skipped and make an educated guess.

At the end, leave the test feeling confident. You've done your best, so don't waste time worrying about your performance or wishing you could change anything. Instead, celebrate the successful completion of this test. And finally, use this test to learn how to deal with anxiety even better next time.

Review Video: Test Anxiety
Visit mometrix.com/academy and enter code: 100340

Important Qualification

Not all anxiety is created equal. If your test anxiety is causing major issues in your life beyond the classroom or testing center, or if you are experiencing troubling physical symptoms related to your anxiety, it may be a sign of a serious physiological or psychological condition. If this sounds like your situation, we strongly encourage you to seek professional help.

Online Resources

Due to our efforts to try to keep this book to a manageable length, we've created a link that will give you access to all of your online resources:

mometrix.com/resources719/chst

It's Your Moment, Let's Celebrate It!

Share your story @mometrixtestpreparation

www.ingramcontent.com/pod-product-compliance
Lightning Source LLC
LaVergne TN
LVHW062146200926
848484LV00007B/1869